The Dealer

Comprising:

The Watcher
Pulse
Rage

Ed Adams

a firstelement production

Ed Adams

First published in Great Britain in 2022 by firstelement
Copyright © 2022 Ed Adams
Directed by thesixtwenty

10 9 8 7 6 5 4 3 2 1

A CIP catalogue record for this book is available from the British Library.

ISBN 13: 978-1-913818-28-9

eBook ISBN: 978-1-913818-29-6

Printed and bound in Great Britain by Ingram Spark

rashbre
an imprint of firstelement.co.uk
rashbre@mac.com

ed-adams.net

The dealer sits in her darkened room.
Some may not realise she is playing
nor that they are in the game.

Thanks

A big thank you for the tolerance and bemused support from all of those around me. To those who know when it is time to say, "Step away from the keyboard!" and to those who don't.

To Julie for understanding that only comes with really knowing.

To thesixtwenty.co.uk for direction.

To the NaNoWriMo gang for the continued inspiration and encouragement.

To Topsham, for being lovely.

To the edge-walkers. They know who they are.

And, of course, thanks to the extensive support via the random scribbles of rashbre via http://rashbre2.blogspot.com and its cast of amazing and varied readers whether human, twittery, smoky, artistic, cool kats, photographic, dramatic, musical, anagrammed, globalized or simply maxed-out.

Not forgetting the cast of characters involved in producing this; they all have virtual lives of their own.

And of course, to you, dear reader, for at least 'giving it a go'.

About Ed Adams Novels

Triangle Trilogy		About	Link
T1	The Triangle	Dirty money? Here's how to clean it. Money laundering	https://amzn.to/3c6zRMu
T2	The Square	Weapons of Mass Destruction – don't let them get on your nerves. A viral nerve agent being shipped by terrorists and WMDs	https://amzn.to/3sEiKYx
T3	The Circle	The desert is no place to get lost. In the Arizona deserts, with the Navajo; about missiles stolen from storage.	https://amzn.to/3qLavYZ
T4	The Ox Stunner	The Triangle Trilogy – thick enough to stun an ox Triangle, Circle, Square in one heavy book. all feature Jake, Bigsy, Clare, Chuck Manners	https://amzn.to/3sHxlgh

	Archangel Collection		
A1	Archangel	Sometimes I am necessary. Icelandic-born, Russian trained agent Christina Nott, learns her craft.	https://amzn.to/2Y9nB5K
A2	Raven	An eye that sees all between darkness and light. Big business gone bad and being a freemason won't absolve you.	https://amzn.to/2MiGVe6
A3	Card Game	Raven Part 2 – The power of Tarot whilst throwing oil on a troubled market	https://amzn.to/2Y8HLgs
A4	Magazine Clip	The above three in one heavy book.	https://amzn.to/3pbBJYn
A5	Play On, Christina Nott	Money, Mayhem, Manipulation. Christina Nott, on Tour for the FSB	https://amzn.to/2MbkuHI
A6	Corrupt	Parliamentary corruption. Trouble at the House	https://amzn.to/2M0HnOw
A7	Sleaze	Autos, Politics, Gstaad	https://amzn.to/3sE3UDt
A8	Ignoble	Corrupt and Sleaze omnibus – double album	https://amzn.to/3sp6EUL

	Big Science		
B1	Coin	Get rich quick with Cybercash – just don't tell GCHQ	https://amzn.to/3o82wmS
B2	An Unstable System	Creating the right kind of mind	https://amzn.to/2PRJciF
B3	The Watcher	We don't need no personal saviours here. From the Big Bang to the almighty Whimper	https://amzn.to/3kTFWjq
B4	Jump	Some kind of future.	https://amzn.to/3sCzK3h
B5	Pulse	Sci-Fi dystopian blood management with nano-bots. Want more? Just stay away from the edge	https://amzn.to/3qQlBvL
B6	Rage	A madman's war	https://amzn.to/3MEuKlL
B7	The Dealer	Dealt from a darkened room	Jump, Pulse and Rage

Blade's Edge Trilogy			
E1	Edge	World end climate collapse and sham discovered during magnetite mining from Jupiter's moon Ganymede.	https://amzn.to/2KDmYOW
E2	Edge Blue	Endgame, for Earth – unless?	https://amzn.to/2Kyq9au
E3	Edge Red	Museum Earth an artificially intelligent outcome – unless?	https://amzn.to/2KzJwjz
E4	Edge Condition	Edge Trilogy aka Edge of Forever	https://amzn.to/3c57Ghj

TABLE OF CONTENTS

PULSE

RAGE

The Watcher

Ed Adams

a firstelement production

Ed Adams

Champion Angel

Throw up your voice but not your mind
While them agents of change go monopolize
Their colours and their faces are just shades of the same
All lost in the game.

And we don't need no personal saviours here
Just a warm hearth and water. It's purely biological.
No posturing mannequin man or woman
Shall receive my hand.

Among all you angels is a champion angel
Among all you devils, there's a free soul
Up from the disenfranchised, the engine cries
Up from the circle there's a hole.

The child insubordinate disrupts the pecking order
So go marry young while you can
'Cause the weave of the rug and the cut of the throne
Testify before the ocean's open hand.

I promise you this promise we are not alone
But why is it I alone that promise this
Deny the forces that would hurry men
If you still can.

We come now to a fracture in the road
Here time has taken her toll
The endless freezing and the thawing of the heart
Would eventually divide us apart.

What's that you found in the pocket of your coat?
Looks like a small sentiment that she wrote
Don't be my personal saviour I would not be saved
I chose to walk alone.

We come now to a fracture in the road
Here time has taken her toll
The endless freezing and the thawing of the heart
Would eventually divide us apart.

Jocelyn Jager Adams / Benjamin Knox Miller / Jeffrey Carl Prystowsky

19

Dragon Lines Campfire

We meet at Point Reyes, a little to the north-west of San Francisco. A quiet scenic spot, where you'd make a fire on the beach and look out towards the Pacific as the sun sets.

It was a convention that we'd always meet at an intersection of ley lines, and this point was a doozy.

It was Limantour's idea to call us together. Just four of us. That's Drake, Tomales, Limantour and me. The ones we knew about with a stake in the future of Earth. I thought it amusing that Limantour had chosen a kind of Grand Central Station of ley lines for our meeting.

Tomales commented about the ley lines. She stood, balanced upon a tall rock, her hiking boots and three colour swirly shorts (brown, orange, and muted cream) offset with a long-sleeved red tee shirt over which she wore a short-sleeved pink tee. The colours played to her smooth brown skin well.

She begins, "Human ancestors knew about these lines for thousands of years. The native Indians of the United States used to call the ley lines spirit lines and their

Shamans used to use the electro-magnetic energy in these lines to help them contact the spirits. They even designed their medicine wheel on the spirit lines, as they knew these lines followed a straight round line. How did they know about these lines and the energies that they give off? The answer's simple: the sky gods told them."

Hold that thought about the sky gods.

We look up in the air as four silent brown pelicans flew along the edge of the shoreline.

Limantour continues the conversation. She wears a dark sailor costume, small top, red scarf in a vee and pleated skirt. Today, this mistress of chaos sports ginger bobbed hair.

"European druids called them mystical lines. In Wales, they used the same name as eastern countries; they called them dragon lines - the red dragon - Y Ddraig Goch - overturned the white dragon of the Saxons. And we know Eastern countries called them dragon lines as the sky gods flew in dragons along these lines. "

"Just like we do now?" asks Tomales, "In our dream-like state?"

Limantour replies, "Yes, the aboriginal people of Australia even called these lines 'dream lines'. They claim that knowledge passed on to them from the sky gods. That'll be us, I guess."

I shivered. I was aware of a recurrent dream in my downtime. It featured a gigantic serpent with gaping jaws which was always just a few steps behind me.

Drake in his khaki 'I'm going camping' look adds, "And

the lines link well. Throughout history, all megalithic structures were built on top of ley lines."

A few sparks from the campfire popped and spiralled up into the air. Drake knew how to make a good fire.

He continued, "From the pyramids of Giza to Stonehenge, Notre Dame, Solomon's Temple, Parthenon, Oracle of Delphi, Rennes Le Chateau, Ziggurat, the Vatican, DC Capitol, Mecca, Agia Sophia, Aztec Pyramids, Bermuda Triangle, Coral Castle, even Nikola Tesla's lab."

"I think we know too much," I interrupted, thinking about ley lines and trade routes.

Drake laughed and continued to fiddle with the fire. We all knew he could just wave to start it, but he proved that he still knew the primitive ways.

There was a crackling, and I saw the air de-mosaic. As the ripple subsided, I recognised Abbott appear on a rock. I'd known Abbott for a long time, but he diverted toward the counterforces in nature. Think of the four riders of the Apocalypse. He would be a supporter of all of them. Dressed in a black tee shirt, black shirts, and black sneakers, I hoped this was not a sign. Abbott from the dark.

He spoke, as if already in the conversation, "I think of all of the obelisks. Humans obsess over pointy things. Many of the sections where two or more ley lines intersect are marked with obelisks, such as Washington DC monument, Vatican Courtyard, and Cleopatra's Needle in Central Park."

"Even the one in London, England," I added.

"Although Cleopatra's Needles have all moved, I guess two out of three remained on a ley after they were re-sited," said Tomales. She was unwrapping a small backpack. I guessed she'd brought food along.

Four large gulls overhead this time. Swooping, noticing the food potential but staying aloof.

Tomales ignored the birds and continued, "Those electromagnetic lines of the Earth are its veins and receive its energies from the sun that connects and affects every living thing on this planet. Humans are surrounded by electrons, and through these auras and by the activation of the seven energy chakras the humans still have potential to connect with their higher self. Remember the golden halos depicted on all spiritual figures throughout history?" She made a triangle shape with her hands.

Limantour flashed a smile, "Fascinating though this is, let's get back to the main purpose. Thank you all for coming.

"I'm going to propose another Intervention."

The Intervention

Now, as we Watchers all knew, Interventions were strictly off-limits. Lepton, who had inadvertently sped up the introduction of light, had to pay by being secluded for the rest of eternity, although he did figure out a way around that difficult situation.

There had only been a handful of Interventions in all the universe's 13.8 billion years, and I wasn't sure I was ready to take part in one of them.

Another intervention by Leonardo, had occurred back in the 15th Century. Parachutes, fighting vehicles, hydrodynamics, helicopters, revised astronomy. A series of holistic system theories, so ahead of their time that they were often overlooked. Leonardo wasn't a Watcher though; he was a Wakener. But he showed the defect of not having a universal compelling event.

Compelling event? Something that galvanises the discovery provided towards its implementation.

Limantour begins, "You've done the math? You can all see that we are approaching an Earth endgame?"

I looked at others - who were all nodding. This was a

truly compelling event. Do something or Earth goes horribly wrong. Overpopulation, energy and food shortages, climactic decline.

Tomales jumps down from the rock. "Yes, I estimate it is something like a few hundred years - probably less - before the entire planet gets destroyed in another climate catastrophe."

Drake adds, "Yes, and this time there's no passing asteroid to take the blame."

Limantour next, "Tomales and I have been looking at other options. Minimal interventions that could turn all of this around."

"Like what?" asks Abbott, "Even that global pandemic didn't give the humans a hint that things were bad."

"We need to turn it on its head," says Tomales, "Instead of a threat, give them something to grasp at, to change behaviour."

"It'll take a long time," replies Abbott. "Probably one evolutionary cycle. That is too long."

"We were thinking about a discovery," says Limantour, "One that speeds up their thinking,"

"Oh, yes?" asks Drake. "What kind of discovery?"

"Magnetite," answers Tomales, "It could solve so many of their problems. I talked to Lekton about it."

"Lekton?" asks Drake, "But I thought he wasn't welcome anymore?"

Tomales answers, "Yes, but he can report on the future. Remember, he is one of us that can go forward and backward through time."

"He's in a different, dark, room, for sure," answers Limantour, "And he told Tomales about magnetite."

"Huh?" I ask. "The nearest magnetite source is right out at Jupiter, and they haven't discovered it yet, let alone worked out how to make it work for them?"

I wonder how I even knew this. The bad dream hotline must have been sending me messages.

Tomales continues, "That's where asteroid (153814) 2001 WN5 comes in. It's more than a Navajo flying rock. It was once part of Ganymede, the largest moon of Jupiter. If we tilt Hubble Four just a degree or so, it will collide with the asteroid and send back data about the extent of Magnetite available. Magnetite which could be a lasting clean source of power for the Earth."

She kicked the fire, and more sparks rose into the air.

"Okay, but what about the law of unintended consequences?" asks Abbott.

"I hear you, but I think this is for the greater good. The most that would happen is a few timespace ripples from the impact. It is hardly going to start a chain-reaction," answers Limantour.

"We don't even need to touch the Earth for this to work," says Tomales, "We can hardly be accused of Intervention when it is so indirect." She looked around at us all. Limantour was drawing something with a stick in the sand. A circle with a diagonal line through it, over to the left-hand side.

"Here," she says, "The asteroid (153814) 2001 WN5 passes within 250,000 km of the Earth. During the close approach, it should peak at about apparent magnitude 6.7, and will be visible in binoculars."

"If we tip Hubble Four ever so slightly, we'll have a direct course and the Hubble's own detectors will soon enough to pick up on the mineral composition."

"But aren't we forgetting something?" asks Abbott, "Earth does not understand about magnetite and its ability to make clean, powerful energy from lightweight structures."

"Okay, that is part two," answers Limantour, "We'll need to send in a few shards containing knowledge for the humans."

"I thought this was a simple plan?" asked Abbott." Now we have an asteroid collision and a rain of knowledge-shards hitting Earth?"

"Yes, the knowledge-shards were a later embellishment of the plan by Lekton, but I'm afraid they are an essential part," explains Tomales.

"And I suppose you know where to find sufficiently advanced civilisations that already know about magnetite mining?" asked Drake.

Limantour answers, "Yes, we had to find an exoplanet, and there is one on the Norma Arm of the Milky Way. They have the knowledge, but it would come in a bundle with several other discoveries, which cannot be easily separated."

"And I guess it's somewhere between 45,000 and 60,000 light years away?" asks Drake.

"More or less, " says Tomales, "Although remember we can manipulate time, at least going forward. Using travel time of a light year per second it is a 2000 second round trip or just over thirty-three Earth minutes, plus time at the far end for negotiation and so on. Call it a day altogether.

We all looked at one another. By tomorrow, we could have the secret of how to stop Earth from crashing headlong into oblivion, based upon a rather simple sounding plan by created by Lekton and replayed by Limantour and Tomales.

"But," I start, "There's so many things that could go wrong!"

"Faint heart never won fair ambition, " answers Limantour. "Has Earth taught you nothing?"

The others laughed. I sense they have already bought into the plan.

I looked toward the sky. The birds have all gone. It was the embers of sunset. So pretty, it would be a shame to lose this. I realised Limantour and Tomales had chosen this time for its theatricality.

"So do we get some of those shorts too?" I ask.

Opening

But I should wind back.

This is a history of the Universe kind of deal. The Earth doesn't even feature for the first 8 billion years. But I was there from the start and so were some of my dimensionless associates.

The first few moments were intensely interesting, then the next few billion years were screamingly boring. And the next couple after that. It didn't really pick up until about 50,000 years ago, when the ice was finally retreating from the northern hemisphere of Earth for the last time.

I could edit my time exposure for most of it. Like 21st century streaming devices, I could fast forward over the dross and freeze frame and play when something interesting appeared.

And some of you know that I could go forward through time, but I've not yet discovered any way to back-track, so I've had to make the most of any specific exposure to events.

Don't get me started on the Morrison premise, that we can only travel backward in time because it is only the past which we have experienced. Or that of Amis, with a secondary consciousness apparently living within. Even something like Kurt Vonnegut's Slaughterhouse Five with Billy Pilgrim watching bombs being retrieved by American bombers un-bombing Dresden. We all know it is the film being run backwards. Maybe also 'from swerve of shore to bend of bay' in the opening paragraph of Finnegan's Wake, obliquely denoting the curving of time.

Yes, we Watchers have time to browse most things.

Of course, I'm invulnerable, but not in a freaky-deaky superhero kind of way. It's just a function of my existence as a Watcher. We are all made of stars and that has a cosmic enough ring to it. My star turned out to be a planet, and it is the one known as Earth. They say every Watcher is given something specific, but I think it was a more chaotic allocation in the first phase of the so-called Big Bang.

Yes, I've been around for 13.8 billion years through the coming of the elements and the galaxies and the planets.

Look up to a dark part of the sky and in human science you'll see the remnants of the Big Bang rushing towards us all. Deep time and wide time, we live in both surrounded by dark energy and the conical curves of space in the humans' metaverse.

That's where I get a bonus pass because I can see gravity as a part of hyperspace. Most human minds can't comprehend this hybrid, heterotic disturbance of fundamental objects and think of it as something from a nightmare.

But on some levels, the dream is real.

Scientists talk about monster moonshine and cosmic strings. They are onto something, but their own mathematics have difficulty expressing it. Thus, the monstrous moonshine can refer to the perceived craziness of the relationship between M and the theory of modular functions expressed through a hyperbolic plane of prime factors.

Excuse the mathematics. Some say there is still a bottle of bourbon riding on discovering the proof.

To me and those like me, it really doesn't matter. We've seen the start of the universe, lived through billennia of emptiness and now that it is all getting genuinely exciting, we can see toward another event horizon as the Earth burns up through over-population and over-consumption.

We Watchers are not too impressed that the show which never ends will one day soon run out of content.

The thing is the scientists got it wrong. Their theory of the start of the universe is about as probable as the story of a world on the back of four elephants perched on the shell of a giant turtle. The human mathematics looks plausible, but someone should properly examine the big numbers needed to make everything work.

Yes, human mathematics needs around a decillion degrees Celsius to explain the start of the universe. That's a 10 with 33 zeros after it.

10,000,000,000,000,000,000,000,000,000,000,000.

Of course, the drop off from this number would release huge amounts of energy as it plummets in less than a second to a quadrillion and then to a trillion, billion and finally the temperature of the sun, some 6,400 degrees Celsius. These convenient assumptions written in by earnest straight-faced scholars reverse engineer a lie.

But these big numbers are not new, and Amedeo Avogadro had a go at calculating the number of particles of molecules, atoms, or ions in samples of compounds. But he and Stanislao Cannizzaro used a dimensionless number - still called the Avogadro number - to express the number of particles in one mole.

Dimensionless numbers: now we are getting somewhere.

Such a look into mathematics becomes a distraction. Turn the handle, crank the sums and however implausible it seems, if the numbers say so, then what can the scientific community do except believe? They have followed their thought using the guiderails of an incomplete mathematical system.

It is just like when the world was supposed to be flat, or the scientists thought that the sun orbited the Earth.

Yes, the cosmic dealer smiles all the time within her darkened room. She doesn't let on that there are other darkened rooms next door. You just must think of them.

I repeat.

Other darkened rooms, with different frames of reference. Lepton may have stumbled upon Light, but only in one of the frames of reference.

At least one of these dark areas of deep shadow contains a key to explain what happens when supergravity exerts its presence.

In string theory, a brane is a physical object that generalizes the notion of a point particle to higher dimensions. Branes (derived from the descriptions of a 2-dimensional membrane) are dynamic objects which can propagate through spacetime according to the rules of quantum mechanics. They have mass and can have other attributes such as charge.

How cool is that? Think of parallel universes.

The wrinkle in human thinking is that although multiple branes exist, it is not possible to travel between them.

Limited human thinking again, which conveniently misses the point? You can't come into my dark room if you are already in your own different one. That film-maker Kubrick had a go at explaining it, based upon the writings of Arthur C. Clarke, but it flew right over most people's heads as they said, "I don't understand what that was about."

Imagine one of those novelty celluloid fishes and how they distort with a hand's heat to tell one's fortune. With the fish we get humans imposing rules, written on the explanatory packet. Jealousy, Indifference, Love, Fickleness, Passion, Death. For a 'brane we are not given the rules like with a celluloid fish, but instead we are given a D-Brane world volume which sets the outer limits of possibility, like a big gauge running from -100 to +100. That's the frame of reference.

So come into another darkened room away from the constraints of mathematics and there's a whole other way to view things.

Maybe the universe was a freak occurrence of supergravity, whose ripple was larger than expected and broke through spacetime pushing the rubble of an alternative universe with it? I should declare an interest. I know something of supergravity.

Or maybe gravity left the room entirely, taking away the tight binding of everything and throwing it across space like toys from a pram?

Even in the theological and spiritual version, where Earth is created, then Lucifer runs amok, Earth gets judged and then is ruined. This causes a re-boot and another six unknowable days to create what becomes an Earth 2.0?

See the frames of reference? Traditional science, new science, theology. Suspend disbelief as one moves between each of the rooms.

Now I could fast-forward through the first half-billion years of Earth, after the Big Bang, when Earth was still covered in water. To a time of tiny mammals, ancestors of humans, and how they took advantage of the disappearance of the great lizards, or dinosaurs.

Big Bang and Beyond

I said I could fast forward over everything, but I'll pause awhile.

I won't start with humans. I'll rewind back to the beginning of their scientific belief system.

I think the grounded shape-shifting Tomales, and the chaos known as Limantour agree with me about the universe and Earth formation, but I'm not so sure about the analytical and serious Drake and the increasingly dark Abbott.

To be honest, I think Abbott (even beyond his clothing) is showing signs of defection. He has recruited two whom he refers to as Cardinal and Bishop and I'm certain they are making a play to rule a newly formed Earth Council.

Those are not the moves of a Watcher.

Something about the Earthside scientific belief system: Before time, there were four known fundamental interactions or forces—first, the outrider known as gravitation, and later the two electromagnetic forces with their weak and strong interactions; then there is the expansion of space itself and the super-cooling of the still immensely hot universe because of cosmic inflation.

That's a big one. Cosmic Inflation. It is how human science explains the aftermath of the Big Bang. How all the universe gets set to a temperature at one time, to put it into equilibrium.

I don't think so. A space so vast that equilibrium is set faster than the speed of light? Someone needs to keep a watch over those human scientists, bending logic to suit their explanations.

I suppose we could view it that the two prevailing electromagnetic forces brought me and my kind into being. A quirk of physics and the small ripples flooding across an undetected universe. These ripples were in a different dark room from the physicists and mathematicians. Ripples that were the basis of the large-scale structures that formed much later.

Stages of the very early universe are understood to differing extents. The earliest parts are beyond the grasp of practical experiments in particle physics and can only be explored through other means.

Stay with me as I describe the human view of the start of the universe. The kind that floppy-haired earnest-looking physicists describe over waves of growling bass-line synthesiser music. I'll add the bongs where suppositions should be challenged.

We get the first trillionth of a second of cosmic time (Bong!), riding the Planck epoch and experiencing a super-heated starting point, with an estimated temperature of 10^{33} degrees Celsius. (Bong!) I couldn't be physical during this, (who could?) and I would need a super-fast power of observation to see events that would otherwise pass in a blink.

Then, Time unpacks itself (Bong?). A convenient magic.

Then the second stage, also within minute fractions of the first second (Bong!). The Grand Unification Epoch, when the force of gravity separated from the three other fundamental forces, and the earliest particles (and anti-particles) were created (Bong - we should not discount gravity's effect on the whole creation process). Gravity separation allowed the formation of the 'branes which permeate alternate realities.

See what I said about gravity as the outrider.

I could handle all of this in my consciousness. I was equipped from the very start to handle metaverses, the non-persistence of gravity and the curious dilation of time. The fundamental force ripples that flooded the early universe became my friend and I, and the others like me, could adapt our own perception to appreciate what was happening.

And now, we could all detect time and gravity in this new and very particular dark room. As Watchers, we had been imbued with a unique gift to ride out the event horizon. We would exist for as long as Time existed,

On to the third stage, also within minute fractions of the first second (Bong!) It was the Inflationary Epoch which was triggered by the separation of the strong nuclear force.

We'd now gained a universe as it underwent an extremely rapid exponential expansion, known as cosmic inflation. The size of the universe during this tiny fraction of a second (Bong!) increased to around the size of a small insanely hard apple. And particles remaining from the Grand Unification Epoch, comprising hot, dense quarks, become distributed thinly across the universe.

As a Bong! to all of this, spacetime ripples would make an altogether plausible alternative theory.

Instead, sticking with Earth-science explanations, as the strong nuclear forces separated, particle interactions created large numbers of exotic particles and Higgs bosons in the Electro-weak Epoch. The Higgs bosons slowed down these particles and attributed mass to them, allowing a universe that was made entirely out of cosmic radiation to support particles with mass. This is when things became tangible rather than simply riding phase distortions.

Even now, Higg's boson isn't fully understood and additional theories like gauge invariance and spontaneous symmetry breaks have been wrapped around it. It is to the extent that analogies are used to describe it. That gets a Bong! from me. Think of a shady politician's accounts with a few balancing items to make everything tally.

Still in that first moment and we get the Quark Epoch, when quarks, electrons and neutrinos formed in large numbers as the temperature of the Universe cooled off to below 10 quadrillion degrees Celsius (Bong), and the four fundamental forces assume their present forms.

Or three of them do, with gravity as an outrider. Quarks and anti-quarks annihilated each other with contact, but in a process known as baryogenesis, a surplus of quarks survived and ultimately combined to form the first chemical matter.

We were experiencing the birth of tangible matter, admittedly still from the confines of one darkened room.

It's all about perception.

The Hadron Epoch followed, still within this first second of the Big Bang (Bong!). During this epoch, the temperature of the infant primordial universe cooled to around one trillion degrees Celsius (Bong).

Okay, CERN's Project ALICE are now thought to have 'momentarily' cooked up 5.5 trillion degrees Centigrade crashing gold nuclei at near light speed. That's hotter than the centre of the sun, by 250,000 times.

This super-hot Hadron Epoch was enough to allow quarks to combine to form hadrons such as protons and neutrons. Electrons colliding with protons fused to form neutrons and give off massless neutrinos, which continue to travel freely in space to this day at around the speed of light.

Now, this is where I should introduce my associate Lepton. He, like me, is a Watcher, although he graduated on to other things.

Lepton was trying to free himself. He could see it was a very long one-way trip along Earth's timeline and wanted to be able to exercise greater freedom to roam.

So, he decided to intervene. That is a strict no-no for all of us Watchers. Don't interfere.

What followed next was the Lepton Epoch, which occurred from one second to three minutes after the Big

Bang in what was a still very elastic form of time. Lepton figured out how to sweep up most of the hadrons and anti-hadrons and balanced the books by annihilation of pairs of them.

At the end of the Hadron Epoch, leptons, such as electrons, and anti-leptons, such as positrons, dominated the mass of the Universe.

Lepton was pleased, thinking this would allow him to break free from the immutable timeline, but didn't realise that he had inadvertently created light. The newly dominant electrons and positrons collided creating new energy in the form of photons. Lepton had switched on a visible universe, without intending to.

Light - An unintended consequence.

Something else blew my mind. Since Lepton's discovery of Light, he also acquired another power. He could skate in either direction along the timeline.

"I never go back to the point before I gave my gift," he explained to me, I don't want to lose it and then not be able to provide it again."

But Lepton could be a scout for the rest of us about the future. He could slip forward but then slide backward with his report of what happened. He told me that there were a few others with similar powers. It is because of this that we could devise our plan around the beach fire.

We all knew that because of Lepton we had seen the wave of space surf crashing down and it made it more manageable to assimilate.

This is the point where creation scientists are convinced

that stars cannot form spontaneously. And despite claims to the contrary, star formation seems to be nothing more than a secular attempt to explain the universe without invoking God.

The Bible's Genesis runs: *'Let there be lights in the expanse of the heavens to separate the day from the night. And let the lights be for signs and for seasons, and for days and years, and let there be lights in the expanse of the heavens to give light upon the Earth. And it was so.... And there was evening and there was morning, the fourth day.'*

Now we have secular 21ˢᵗ Century science vs Moses (who is the traditional author of Genesis in the 5ᵗʰ Century BC).

No explanations in Genesis. More a case of blind faith.

Now that was a big bump in logic to traverse but let us move onward.

The Nucleosynthesis Epoch, which supposedly lasted from three minutes to twenty minutes after the Big Bang showed the temperature of the universe falling to around one billion degrees Celsius (Bong) at which point atomic nuclei could begin to form as protons and neutrons; and combine through nuclear fusion to form the nuclei of simple chemical elements such as hydrogen, helium and lithium.

Watch out for H to He.

After around twenty minutes, the temperature and density of the universe had fallen to the point where such nuclear fusion could not continue. But we are still into Miracle territory.

Unlike Lepton, I was still reeling from this immensely

fast activity and the intensity of the physical reactions. I had no idea that everything would soon slow down for billions of years.

I'd like to run an audit on those time suppositions to see if things could be shown to unfold in a more plausible way.

Take Population I, II, and III stars. Population I are the metal-rich stars (like the sun), then Population II are the metal-poor globular clusters like the centre of the Milky Way.

Now - here's the thing - Population III stars are allegedly supermassive, luminous and hot with no metals. Except that Population III stars are undetected. They are inferred - like they are from another dimension.

That's where we go back to Lepton. These Population III stars also get referred to as Dark Matter. That will increase in significance as we go along further. For now, let's just say it is where the dragons live.

It raises the question about whether the scientists are observing it all from the correct frame of reference / right room to devise their conclusions? Did Lepton only switch on one form of light? Maybe there are others that would allow us to see the other 85% of the matter in the universe?

After Lepton's Intervention which caused the Universe to become visible, the longer Photon Epoch was next, starting from three minutes and ending around 240,000 years after the Big Bang. During this long period of gradual cooling in temperature, the Universe filled with plasma - a hot, opaque soup of atomic nuclei and electrons.

Most of the leptons and anti-leptons had annihilated each other at the end of Lepton's Epoch, so that the energy of the Universe was dominated by photons, which continued to interact frequently with charged protons, electrons, and nuclei.

The Recombination and Recoupling Epoch followed, from 240,000 to 300,000 years after the Big Bang; during which the temperature of the universe fell to around 6,000 degrees Celsius, which is around the same as the surface of the Sun.

This is where we get to something more understandable, and I'll believe the human numbers from this point, after which the convenience of reverse engineered assumptions becomes less important.

The density of the universe continued to fall as ionised hydrogen and helium atoms captured electrons and neutralised their electric charge. With the electrons now bound to atoms, the Universe finally became transparent to light, making this the earliest epoch observable today.

And now, the dark room used by science believers was suddenly bestowed with visibility. There was a comfortable frame of reference to explain many things after this point.

It was similarly convenient for science that the earlier part from zero to 300,000 years was all in complete darkness. By the end of this epoch, the known matter in the Universe consisted of about 75% hydrogen and 25% helium, and minute traces of helium.

There, I said Hydrogen to Helium was coming.

Come into my world

Come in, come in
Come into my world I've got to show
Show, show you
Come into my bed
I've got to know
Know, know you

I have dreams of Orca whales and owls
But I wake up in fear
You will never be my...
You will never be my fool
Will never be my fool

Regina Spektor

Fireside

It is unusual for Watchers to get together. We have other ways to communicate, so the physical manifestation of several of us together, such as in Point Reyes, is an exception.

One problem for us is to operate at Earth's speed. We can get close to it, but because we are used to striding through millennia, the small matter of a minute or a second can be troublesome. It's why we are more usually seen alone.

Fortunately, humans are not that observant. If we appear in the right clothes, the tell-tale signs we are Watchers become less apparent. An occasional blur or glitch when our timing goes adrift, or that strange moment when a human thinks she has seen one of us but then looks again and there is no one there.

Put several of us together on the beach and it is a different matter. We can all sync with one another, so any temporal drift is minimised as we self-correct from each other's vibes.

Limantour knew this as she leaned towards us. She was tending a kamado barbecue.

"So, will you support Tomales and me for this

Intervention? When we try to turn this situation around?"

I was taken aback. Being asked to be part of an Intervention is a big deal. There are other beings, the Wakeners, who are better placed for this task. And even for them it doesn't always end well.

Darnell - I'll tell you about him later - suffice to say he was really a Watcher who stepped too far, but also the Wakeners Leonardo and Galileo. They took something that was dormant in the culture of Earth and surfaced it, to either a kind of apathy or a violent dispute.

Leonardo's fortunes were mixed. Alongside his brilliance, he linked to a secret society known as the Priory of Sion, which started in 1099 and includes illustrious Grand Masters such as Isaac Newton.

The Priory of Sion is devoted to restoring the Merovingian dynasty (Mary Magdalene's bloodline), which ruled the Franks from 457 to 751, on the thrones of France and the rest of Europe. They are said to have created the Knights Templar as its military arm and financial branch. The Catholic Church tried to kill off Mary Magdalene's sacred royal bloodline (sometimes referred to as the Holy Grail) and their supposed guardians, the Cathars and the Templars, for popes to hold the Episcopal throne through the apostolic succession without fear of it ever being usurped by an antipope from the hereditary succession of Mary Magdalene.

So, for now, I'll use Galileo as the example. Galileo's championing of Copernican heliocentrism (Earth rotating daily and revolving around the sun) met with opposition from within the Catholic Church and from some astronomers. The Roman Inquisition investigated the matter in 1615, and concluded that heliocentrism was foolish, absurd, and heretical since it contradicted Holy Scripture. We saw a great scientist trying to give gifts to the world but being stamped upon by another belief system. Might commands over what is Right.

While I was thinking, we Watchers were running a back-channel of communication. It is faster than talking and can be used for difficult decisions.

Fireside support for Tomales and Limantour was forming. We would intervene to save the Earth. I had a feeling it wouldn't end there.

Early Universe

I had hung around in the early universe. It's where I first met Darnell. After all, it was all new. We didn't expect it to last for 370,000 years, so we both used judicious linked fast-forwarding during our combined process of inspection.

To begin with, it was still all kinds of subatomic particles re-balancing. Matter and anti-matter, against a background of persistent cosmic neutrinos. Then we get to where the opaque plasma cleared because of cooling, plus Lepton's considerable addition of photonic light.

Darnell was hooked by what Lepton had achieved, even if it was accidental. Darnell decided he wanted to leave a similar mark, despite being a Watcher.

I already mentioned Hydrogen and Helium. They were the fuel needed to fire up the first stars, but the clouds of hydrogen only collapsed slowly to form stars and galaxies, so there were no alternative sources of light. Those decoupled photons filled the universe with a brilliant pale orange glow at first, gradually red shifting to non-visible wavelengths after about 3 million years. One could say it was the cosmic Dark Ages.

I could see that Darnell was going out of his mind. It was a combination of the boredom of having to deal with such long periods of nothing much happening followed by short bursts of frantic activity. Both Darnell and I used our ability our fast-forward through time, but it still took a seemingly long time to get to the next interesting occurrence.

Fast forward then, to when the earliest generations of stars and galaxies formed and large structures gradually emerge, drawn to the foam-like dark matter filaments which had already begun to weave together throughout the universe.

The dark matter filaments were the ones the star gods used for travel.

Those early stars were huge, around 100 to 300 times the size of the Sun, but they sputtered out quickly, exploding as highly energetic unstable supernovae after mere millions of years.

Through all of this, we Watchers sat on the outside, literally watching the Universe like a very long cosmic documentary.

We eventually got to a billion years from the start and the universe took on its recognisable shape. The Dark Ages of galaxy clusters, superclusters, dwarf galaxies and quasars led to the period of deionization and the transitioning of the entire universe into its recognisable form.

This is where we really needed the three arrows fast-forward button to jump over the next eleven billion years.

The thin disk of Earth's galaxy formed at about 5 billion years from zero and the Solar System formed at about 9.2

billion years, with the earliest traces of life on Earth emerging by about 10.3 billion years.

That is a long time to watch and to wait. I guess that is how we acquired the name - Watchers.

Except Darnell wanted more. He was first plotting and then looking for his chance.

The thinning of matter over time reduced the ability of gravity to decelerate the expansion of the universe. In contrast, dark energy is a constant factor tending to accelerate the expansion of the universe. The universe's expansion passed an inflection point about five or six billion years ago, when the universe entered the modern 'dark-energy-dominated era' where the universe's expansion is now accelerating rather than decelerating.

Right now, human scientists say they understand the universe well, but I see gaps in their knowledge. Metaverses like the ones that we Watchers can inhabit are still seen as the stuff of comic books and cartoons. And even worse, jumped-up billionaire businessmen are trying to commercialise their own shoddy form of metaverse.

One thing is certain. The current Stelliferous Era will end when stars are no longer being born, and any continued expansion of the universe will mean that the observable universe will become limited to local galaxies.

Darnell and I could see and hear scientists explaining the Big Bang and not even mention the other theories. Edwin Hubble may have spotted the redshift of an expanding universe, but there could be so many other explanations. Similarly, the lightest elements of hydrogen, helium and lithium were rising to the top, much like the way small pieces fall to the bottom in a cereal box- the path of least resistance.

As for that background temperature of 2.75 Celsius, it could be worked backwards into the proof calculations.

Add in clashing 'branes of space and you could have whole different theories available, but humans have insufficient mathematics to express them.

For example, the black hole bounce-back theory could create a whole different science system in a different universe - another different darkened room.

Or there is the vortex theory of superfluid space-time, with space and time flowing with zero friction.

Darnell had asked what this could all mean. He asked me several times, because he knew I had some kind of special interaction with gravity.

Darnell's thought was that humanity sought its comfortable answer and, in doing so, has limited its outer range of thinking. The scientific explanation used is still mumbo-jumbo to most people, but plausible enough for well-meaning scientists to buy into it.

The Great Deception

Moon's teeth marks

And the moon's teeth marks are on the sky
Like a tarp thrown all over this
And the broken umbrellas like dead birds
And the steam comes out of the grill like
The whole goddamned town is ready to blow.

And the bricks are all scarred with jailhouse tattoos
And everyone is behaving like dogs.
And the horses are coming down Violin Road
And Dutch is dead on his feet
And all the rooms they smell like diesel
And you take on the dreams of the ones who have slept
here.

And I'm lost in the window
And I hide on the stairway
And I hang in the curtain
And I sleep in your hat.
And no one brings anything
Small into a bar around here.

Tom Waits

Dark forces

We'd been at the fireside for well over an hour. This was a long and comforting time to spend in the presence of other Watchers. Remember, we Watchers are not supposed have prāṇa - that 'life force' thing - so I was slightly surprised that I felt anything at all.

Limantour spoke again, "We'd have to avoid any of the dark forces. Just because Lepton showed us Light doesn't mean that there are no problems waiting for us."

Limantour was referring to the darkness which dominated the universe. Only 20% of it was matter and the rest was still, well, undefined. Another literally darkened room.

Drake spoke up. "Can we be sure of Lepton's report? I mean, he might try anything to be back on the inside with all of us?"

Tomales spoke. "No, Lepton has gone beyond us all. Just like we are beyond all the humans. We can envisage alternate realities, but Lepton is already in one of them."

Abbott spoke forcefully, "I can discuss this with my other associates: Bishop and Cardinal. They will also be certain to have a source for the knowledge-shards."

Knowledge shards - the elements for new knowledge which could be positioned into humanity's awareness to speed up certain discoveries.

"But if the k-shards are a bundle, we won't know exactly what we are sending towards Earth. How can we say that the other components included are all benign?" I asked.

Limantour answered, "Yes. I wondered about that too. We need to be sure that the k-shards can't do anything drastic."

"It's cool," answered Abbott. He was clutching a small branch, which he dropped onto the fire, "The k-shard bundle we use will have the magnetite as its pinnacle discovery; everything else will be contained in a knowledge triangle underneath it. Think of the magnetite as the transcendent growth need, and the other items below it as the deficiency needs."

"Hah," said Drake, "Like Maslow?"

"Kinda," answered Abbott, "Think of the magnetite use as the outer boundary of what can be achieved in this bundle of k-shards."

"So, we are not introducing any unwanted payload?" asked Tomales.

"No," answered Abbott. I had an uneasy feeling about just how quickly he had replied. I think Tomales did as well.

"Okay, well, that's settled," said Limantour. "We have a plan for intervention. We'll deflect Hubble Four by one degree and simultaneously rain some new relevant knowledge toward Earth. The humans will piece it together. Magnetite to supply clean, almost free power and be used to build tiny propulsion units capable of interstellar transportation."

Tomales, "Yes, they should be able to see that magnetite enables whole new inexpensive power sources coupled with an end to pollution. An ability to reverse climate catastrophe."

Limantour added, "This could be a great Intervention."

Watchers

I'll have to explain myself as a Watcher. I'm not some kind of superhero with special strength and flying powers. Some might attempt to classify me as an angel, or as a super-watchful guardian. Those people are trying to define me and thus to take control of my label.

It doesn't work like that. I'm outside of those classifications, my personal metaverse is altogether more permeable. Whether Aramaic, Theodotian, Greek, from the Book of Daniel or the Books of Enoch, humankind wants it to become all about classification.

None of these religion-like attempts are even close and yet, menacingly, they extort control over the belief system using stories of fallen angels, mass defections and even a race of hybrids which mankind called Nephilim and who are referenced in many versions of Numbers 13:33.

Now I can say that Erwin Schrödinger got close to understanding the Watchers. He defended what scientists dismiss as metaphysics — that realm of knowledge that lies beyond the current scientific tools and modes of truth-extraction, which always reveals more about the limitations of the human tools than about the limits of nature's truths.

Schrödinger wrote:

'It is relatively easy to sweep away the whole of metaphysics, as Kant did. The slightest puff in its direction blows it away, and what was needed was not so much a powerful pair of lungs to provide the blast, as a powerful dose of courage to turn it against so timelessly venerable a house of cards.

'But you must not think that what has then been achieved is the actual elimination of metaphysics from the empirical content of human knowledge.

'In fact, if we cut out all metaphysics, it will be found to be vastly more difficult, indeed probably quite impossible, to give any intelligible account of even the most circumscribed area of specialisation within any specialised science you please.'

Even as Schrödinger made his reality-reconfiguring contributions to science and its search for fundamental truth, he never relinquished his passionate curiosity about philosophy and the ongoing questions of meaning that kernel every truth in the flesh of consciousness.

Schrödinger was as drawn to Spinoza and Schopenhauer as he was to the ancient Eastern traditions, and especially in their untrammelled common ground of believing that everything has a mind.

As a Watcher, I've progressively discovered others like me. I mentioned Darnell, who stood out as one who was attempting to change things. And that Lepton is not Lekton (to those that know); Lekton was an altogether different and darker being, and I only ran into him much later.

Darnell wanted to copy the process of Lepton, to accelerate a discovery and somehow escape the tedium he felt he experienced as a Watcher. Darnell also inhabits a different realm now and, like Lepton, can move both ways along the time axis.

As Watchers we are not supposed to change anything in case it trips off parallel realities but more especially because of unintended consequences.

You know the kind of thing. The building correction at Pisa that didn't work, leading to an unintended consequence that the Leaning Tower became a tourist attraction. Introducing 100 non-native starlings into Central Park, because they were like the birds that Shakespeare saw. Now there's 60 million of them across the USA. I could go on, but it is best to remember when we try to pick out anything by itself, we find it hitched to everything else in the universe.

Second Level thinking

I've already mentioned the opposing questions to a few scientific facts. The Big Bang's implausibly packed first second of time. The immense heat of the apple-sized crushed universe. So-called Population III stars.

Roll into second order thinking. Ask "And then what?"

Watchers learn that first-level thinking is simplistic and superficial, and just about everyone can do it. All the first-level thinker needs is an opinion about the future, as in 'The outlook for the company is favourable, meaning the stock will go up.' It is often espoused by pundits and the people propping up a corner of a bar-room.

First-order thinking is fast, easy and some could say lazy. It happens when we look for something that only solves the immediate problem without considering the consequences. For example, think of this as 'I'm hungry, so let's eat a chocolate bar.'

Second-order thinking is more deliberate. Watchers think in terms of interactions and time, understanding that despite intentions, the interventions often cause harm. Second order thinkers ask themselves the question

'And then what?' This means thinking about the consequences of repeatedly eating a chocolate bar when hungry and using that to inform your decision. If you do this, you're more likely to eat something healthy.

The road to out-thinking people can't come from first-order thinking. It must come from second-order thinking and beyond. Seeing things that other people can't see.

It raises a question about proof. A simple example would be the very ley lines where we meet around the fire.

Some might contest the validity of ley lines. It would be another way of toppling some scientific investigations. It doesn't fit into the right box, so the theories are mocked and dashed with subtle passive-aggressive asides.

"Ley lines? What've you been smoking, at Glastonbury / Coachella?" and so it runs.

It suits the darker forces to remain in stealth. To reject nascent discoveries. It could become a problem for any new knowledge shards installed by Watchers or Wakeners. Would Earth-science simply repel all attempts at knowledge that doesn't fit its self-absorbed framework, or would humankind, like with Leonardo, take hundreds of years to realise the truths in what was being described?

I realised that what Limantour and Tomales were proposing would require more than a gentle intervention to be successful. It would probably need major event to act as a catalyst.

Tomales continued, "We will use the Swiss Research Lab CERN with its hadron collider to produce the needed accelerant for our deeds. It, coincidentally, is also sited on a well-known ley line."

 Limantour added, "To emphasise its significance, the entrance to CERN already has a colossal statue of the Hindu Goddess of destruction 'Kali', right in front of its reception."

We all smiled. Limantour could be obsessive about symbolism.

Tomales continued, "Now we could also locate a few other equivalent labs. The Brookhaven Lab in New York also sits on ley lines and has a Hadron Collider. And similarly there's a lab in Alaska and another in Bodø, Norway. Even a lab north of Moscow, run by the Moscow University."

"Are you about to start a nuclear war?" asked Drake. "There is only so much you can do with a handful of particle accelerators."

"Fear in a handful of dust?" asked Tomales.

"Highly accelerated dust!" answered Drake.

Limantour smiled, "We need to bypass general science and get directly to the smart people who work with properties like strangeness and charm. That way, we can keep things moving along. A gift of new theoretical physics, unlocking secrets of gluons and providing ways to harness hypercharges."

Limantour continued, "This could show them how to use the intrinsic power of magnetite to create small yet impressively powerful motors. They don't use fuel because of an entirely safe electromagnetic interaction. In sensible hands, this is a massive game-changer."

An annoying thing about humans is they have self-interest at the front of their cerebral cortex. Even by simultaneously planting k-shards around the Earth, it would still take a miracle for someone to interpret their findings and make the link to magnetite. Otherwise, they could all be competing to listen to different tunes on the radio.

What happened to Darnell

Limantour continued to tend the barbecue, which looked like a big green egg. Wafts of delicious cooking were blowing through the wind. This was a good day to be operating at Earth's speed, but I could sense that Limantour and Tomales must really want something if they were going to this much trouble.

"Farallon?" said Limantour; she was looking at me.

This was the first time she'd called me by name in maybe a millennium. I waited for the next line. I could sense it would be a request.

"We'd like you to do something, Farallon," says Limantour. She was casually plating the barbecued food.

"It'd be like something that Darnell did once before," adds Tomales.

"But we all know what happened to Darnell," I say, "After all, he was on the same timeline as the rest of the Watchers."

"Yes," says Tomales, "But Darnell became enthused with

an idea about ways to accelerate the process of Earth's creation and, more importantly, how to introduce sentient beings. Now we are trying to save that same Earth from extinction."

I spoke, "Remember that the period from the beginning to the formation of Earth? All 8.3 billion years of it? That's 5.4 billion years in the past. Darnell didn't take to it well."

I thought to myself; even at a speedy fast-forward rate, it had been an immense amount of time to traverse. Take a year as 31.5 million seconds. Forwarding at the rate of a year every second, it still would still take around 264 years to get from the beginning of the Universe to the beginning of the Earth. And that's if you skip the bits where the Andromeda Galaxy forms, or the Milky Way, or even Alpha Centauri is first observable.

Fortunately, we could bypass such constraints to move through time even faster.

Drake spoke, "Let's face it, we all knew that Darnell wasn't prepared to sit out billennia and that is why he started to obsess over ways to accelerate evolution."

Abbott interjected, "We have to remember that the original Earth wasn't that pretty, nor sweet smelling either."

"No, it was disgusting," added Limantour, "Today, humans take for granted that they live among diverse communities of animals that feed on each other. Earth's ecosystems are structured by feeding relationships like killer whales eating seals, which eat squid, which feed on krill. These and other animals require oxygen to extract energy from their food. But that's not how life on Earth used to be."

Drake added, "Darnell knew Earth needed adjustments and wanted to boost the oxygen content and reduce the amount of methane."

Abbott continued, "The earliest life forms we know of were microbes - microscopic organisms that left signs of their presence in rocks about 3.7 billion years old. Humans can look backwards through time to see these signals of a type of carbon molecule that is produced by living things. Evidence of microbes was also preserved in the hard sticky mats known as stromatolites that they made and which date to 3.5 billion years ago."

Drake added, "Eventually, some microbes began a remarkable transformation. One that is in part down to Lepton's inadvertent release of photons."

Drake continued, "These microbes became Earth's first photo-synthesizers, making food using water and the Sun's energy, and releasing oxygen as a result. This catalysed a sudden, dramatic rise in oxygen, making the environment less hospitable for other microbes that could not tolerate oxygen."

Abbott interrupted, "It also stiffened Darnell's resolve to expedite more oxygen for Earth."

He continued, "Today, scientists can see the evidence for this in Earth's seafloor rocks. With oxygen around, iron gets oxidised and removed. Rocks dating before Darnell's event are striped with bands of iron. Rocks dating after the transformation do not have iron bands, showing that oxygen was now in the picture. We are still talking about 2.4 billion years ago."

Drake continued, "Darnell had plenty of time to think

about what he would do. He was convinced that adding more oxygen to the Earth, to counteract the effects of the methane, would bring forward the next evolutionary phase by perhaps a couple of billion years."

Now Limantour spoke up whilst handing me a plate of barbecued deliciousness, "Darnell also noticed that profound innovations were occurring. Microbes can process lots of chemicals, but they did not have the specialised cells needed for complex bodies."

I sensed they were all quietly ganging up on me. Pushing me toward a corner.

Tomales spoke, "Darnell noticed that something revolutionary was happening as microbes began living inside other microbes, functioning as organelles for them. Mitochondria, the organelles that process food into energy, developed from these mutually beneficial relationships. For the first time, DNA became packaged in nuclei. The new complex cells - eukaryotic cells - incorporated parts playing specialised roles that supported the whole cell."

Limantour continued, "Darnell was looking for an early organism that could rapidly multiply and additionally produce large quantities of oxygen."

Now Tomales and I knew I was being overwhelmed, "Even today, school children on Earth are taught about the different parts of the single cell organism known as Amoeba, but the significance of its component parts maybe isn't emphasised enough."

Tomales continued, "This type of cell surrounds and engulfs its food and has a nucleus to carry DNA and a vacuole to regulate water ingestion. Now, groups of these cells began living together because certain benefits could be obtained. Groups of cells might feed more efficiently or gain protection from simply being bigger."

I could see what they were doing. The equivalent of 'get him saying Yes'. They would produce a string of incontestable facts in a row. Ones that it was easy for me to agree with. Then they would slip in the trick fact.

Tomales was still talking, "Darnell began to look beyond single cells towards slightly more complex organisms, and was always on the lookout for something small, that multiplied quickly and created exceptional quantities of oxygen."

I was trying to concentrate on something else. I realised I could slow down their fast forward of the facts, so I wouldn't miss the treats of the last 800 million years as Earth's inhabitants changed from resource-devouring sponges into politicians.

Tomales was on roll, and added, "The combined clusters of specialised, cooperating cells eventually became the first animals, which DNA evidence suggests evolved around 800 million years ago. Sponges were among the earliest animals. While chemical compounds from sponges are preserved in rocks as old as 700 million years, molecular evidence points to sponges developing even earlier."

Tomales and Limantour were still operating as a tag team, and now Limantour took over, "Oxygen levels in the ocean were still low compared to today, but sponges are able to tolerate conditions of low oxygen."

Then Drake piped up, "The sponges were a disappointment to Darnell because, like other animals, they required oxygen to metabolize, but they need little because they were not very active. They feed while sitting still by extracting food particles from water that is pumped through their bodies by specific cells. The simple body plan of a sponge comprises layers of cells around water-filled cavities, supported by hard skeletal parts. The evolution of ever more complex and diverse body plans would eventually lead to distinct groups of animals."

"This is all wrong," Darnell had said. "Earth can't just be inhabited by idle sponges." He set about looking at the genetics of the sponges.

I thought, "Or can it?"

Drake added, "Darnell realised that assembly instructions for the body plan are in a sponge's genes. Some genes act like orchestra conductors, controlling the expression of many other genes at specific places and times to correctly assemble the components. While they were not played out immediately, there is evidence that parts of instructions for complex bodies were present even in the earliest animals."

I realised that Darnell's supposition was to find something to kick evolution over the edge.

Tomales continued, "Darnell knew it was a slow process. Another 220 million years passed, and we reached the Ediacaran Period, some 580 million years ago. It is easier to count backwards now, instead of time elapsed, saying how long-ago various events occurred. Now at 1 second per year, that's only around seven years, although I'm guessing Darnell did all of his work in about half of that time."

She added, "No longer did we find just sponges, but also varied seafloor creatures - with bodies shaped like fronds, ribbons, and even quilts which lived alongside sponges for 80 million years. By the end of the Ediacaran, oxygen levels rose, approaching levels sufficient to sustain oxygen-based life."

Limantour took over again, "The early sponges may actually have helped boost oxygen by eating bacteria, removing them from the decomposition process. Darnell's idea was to engineer a major environmental change by boosting the oxygen balance on Earth. It was through the introduction of several handfuls of trilobites. These were armoured creatures he identified which could burrow along the seabed, aerating the sediment. Being small scale, they multiplied rapidly, and it led to the Cambrian explosion of new life forms."

I recollected. Some 541-485 million years ago we didn't just get the burrowing trilobites. We also got hard body parts like shells and spines which allowed animals to engineer their environments more drastically, such as digging burrows. This was because the oxygen rich Earth could now provide fuel for these new and more active animals, with defined heads and tails for directional movement to chase prey.

We were witnessing Darnell's trilobites leading to the inadvertent introduction of insects too.

As Darnell's trilobites transitioned to crustaceans, the first land-bound insects emerged. Then, about 400 million years ago in the Devonian period one lineage of insects evolved flight, the first animals to do so.

The trilobites became extinct and were replaced with molluscs, arthropods, and annelids. Now we had diversity and the emergence of food chains (well food webs, at least).

Darnell accelerated the introduction of lifeforms to Earth through the much greater generation of oxygen.

But here's the thing. The insects didn't know when to stop. They proliferated, but also grew bigger. Many of these insects were the size of small mammals, and it was only their lack of body structure that stopped them from growing even larger. It is why so many humans today have an instinctive fear of crawly things. These early insects could bring down large mammals - instinctively attacking their neck and injecting toxins. It was the playing out of the law of unintended consequences, in this case Darnell's consequences.

Some might say it was fortunate that global climate conditions changed several times during the history of Earth, and along with it the diversity of insects. Winged insects underwent a major expansion in the Carboniferous (356 to 299 million years ago) while insects that go through different life stages with metamorphosis underwent another major expansion in the Permian (299 to 252 million years ago).

Darnell had rolled the evolutionary dice with a few handfuls of trilobites. He'd accelerated oxygen creation on Earth, brought through the development of insects and seen the rapid development of many now extinct Earth species.

He'd also spent a considerable time planning it, although I'm dubious whether he could have foreseen the evolution from trilobite to crustacean to the burgeoning classes of insect.

None of us are clear how Darnell, like Lepton, ceased being a Watcher and seemed to acquire some new powers after this Intervention. But he moved to the same plane as Lepton and could skitter along the planetary timeline.

The Great Dying

I suspected I was being asked to do something similar to Darnell, by Limantour, but without the millennia of thought ahead of the deed. If Darnell had introduced trilobites, and they had led to oxygen creation to fuel the rise in life, then it was a pretty cool Intervention.

Let's face it, Darnell's Intervention ended when most insects had developed, but many early groups became extinct during the mass extinction event at the Permian and Triassic boundary. It was the largest extinction event in the history of the Earth, often referred to as The Great Dying and occurred around 252 million years ago.

There were two concurrent events. An asteroid hit the Earth; there is even evidence in the form of fullerenes containing noble gases in the form of 3He and 36Ar atoms -- Helium and Argon isotopes that are more common in space than on Earth.

It was at the same time as huge volcanic eruptions in Siberia created immense choking dust in the atmosphere.

World geography was also changing. Plate tectonics pushed the continents together to form the super-continent Pangea and the super-ocean Panthalassa.

Weather patterns and ocean currents shifted, many coastlines and their shallow marine ecosystems vanished, sea levels dropped. Do all of that to the Earth and inevitably life forms take a tumble.

This Great Dying formed the boundary between the Permian and Triassic geologic periods, as well as between the Palaeozoic and Mesozoic eras, approximately 252 million years ago.

It was the Earth's most severe known extinction event, with the extinction of 57% of biological families, 83% of genera, 81% of marine species and 70% of terrestrial vertebrate species. It was the largest known mass extinction of insects.

It pulled the curtain around what Darnell had done, and left no evidence of the wilder excesses, except for a latent genetic instinctive fear in humans of insects and arachnids.

There were three distinct pulses of extinction. The causes of extinction were elevated temperatures and widespread oceanic anoxia and acidification due to the large amounts of carbon dioxide that were emitted by the massive eruption of Siberian volcanoes.

Then there was the burning of hydrocarbon deposits, including oil and coal, by these volcanoes and emissions of methane by methanogenic microorganisms contributing to the extinction. The planet was coated with dust, so there was no light, which limited oxygen production and created and a toxic surface. No wonder little could survive. But what did survive were some of the most robust life forms.

The survivors of these terrible events evolved in the

Triassic (252 to 201 million years ago) to what are essentially the modern insect orders that persist to this day. Most modern insect families appeared in the Jurassic (201 to 145 million years ago). No wonder humankind thinks the ant will survive everything.

In an important example of co-evolution, several highly successful insect groups — especially the Hymenoptera (wasps, bees and ants) and Lepidoptera (butterflies) as well as many types of Diptera (flies) and Coleoptera (beetles) — evolved in conjunction with flowering plants during the Cretaceous (145 to 66 million years ago).

And so it goes on; many modern insect genera developed during the Cenozoic that began about 66 million years ago; insects from this period frequently became preserved in amber, often in perfect condition. Such specimens are easily compared with modern species, and most of them are recognisable as part of the same species.

Out of Space

I'll take your brains to another dimension

I'm gon' send him to outer space
To find another race

I'll take your brains to another dimension
Pay close attention

Cedric Miller / Keith Thornton / Lee Perry / Maurice
Smith / Max Romeo / Trevor Randolph

Cherry wood chips and a handful of pecans

We'd all eaten by this time; Limantour had used cherry wood chips and a handful of pecans to slow cook the short ribs. She'd added butter and bourbon and glazed it all with garlic, ginger, soy, maple syrup and cider vinegar. Then served it with leek. It was mind-blowing, but Limantour knew good ideas were best consumed with the best food.

"So, we'll be asking for your help?"

I looked around the circle. They knew my special skills would help.

A) To deflect the Hubble Four by one degree.

B) To set off a gravity ripple which would bring humankind a shower of knowledge-shards.

"Okay, if I help you, how do I know I won't end up like Darnell?"

Tomales answered, "We don't know that, but think of it. The ability to pass both ways along the timeline. At least to the point at which you introduced your Intervention."

"That's around now. But I can' go back further?"

"Yes, but you can't do that at present," answered Drake. "This is all about gain."

"Gain for you, and gain for Earth," said Tomales.

"We thought long and hard about a candidate for this," said Limantour. "It had to be you."

I could see they had me surrounded.

I said we Watchers didn't have superpowers. That's correct in the conventional sense. We can't fly or throw cars around. But we've all got some special features. Mine is to bend gravity.

I don't know why. I can see the forces of gravity and can manipulate them very precisely.

Not strongly, but enough for it to be interesting. For example, I cannot make everyone heavier or lighter, neither can I fly, but I could manipulate gravity as a weak force, with surprisingly big effects. I never use the power. It seems to me to be more trouble than it is worth.

Take it to the limit

I looked toward Drake. He was agitated.

"We need to move; we've been in this place nearing the maximus."

The maximus was the longest time we Watchers could stay assembled in one spot. We all knew it as a quarter revolution of the Earth, but we used 5 hours to provide a safety margin. Failing that, we'd be dispersed and thrown forward to random locations. It wasn't a showstopper, but reminded us all that we were not in control.

"We'll need to continue this somewhere else," said Limantour. I looked along the beach. 'Leave no trace' was an understatement. The barbecue and its trappings had already been flipped into another chaotic metaverse by Limantour. I could envisage some unsuspecting human finding it at the back of piles of rubbish in their yard.

"Tread lightly upon the Earth," said Tomales, smiling towards me. An obvious cliche. I could tell they thought they had me ensnared.

"Where then?" I asked.

"We'll stay with the ley lines but find another point of maximum convergence."

"Le Mont-Saint-Michel," said Limantour, "In France,"

We all knew it. A tidal island and mainland commune in Normandy. The commune's position—on an island just a few hundred metres from land—once made it accessible at low tide to the many pilgrims to its abbey, but defensible. An incoming tide stranded, drove off, or drowned would-be assailants. Nowadays, I realised that outside of tourist hours it was an ideally isolated place. The abbey had been used regularly as a prison during the Ancien Régime exactly because of its isolation.

But we must go back further.

The Mont occupied dry land in prehistoric times. As sea levels rose, erosion reshaped the coastal landscape, and several outcrops of granite emerged in the bay, having resisted the wear and tear of the ocean better than the surrounding rocks. Mont-Saint-Michel consists of granite solidified from molten magma about 525 million years ago, during the Cambrian period.

We each had our different ways to transport ourselves to the new meeting point, and I just hoped our varied routes would not be detected by any of the modern-day Earth dwellers. I knew that much of our science was still in another dimension from that discovered by mankind. They were still in their darkened room.

Now we were sitting inside the Abbey, or maybe still outside? The lights played tricks here and there were many concealed courtyards, such that it becomes difficult to know what is inside and outside. The Benedictine monks knew how to play tricks, even from the 8th Century.

"Something I don't understand," I began, "Is how we

know what bundle of knowledge shards to return to Earth?"

"Don't worry," said Tomales, "Darnell has explained it to Abbott; He has researched the exact combination. Your role, Farallon, is to create a gravity wave which can trigger the release of the shards. Think of it as a signal."

This made more sense to me. I could use my ability to create gravity waves for both purposes. First, to disturb the Hubble Four by one degree; and second, to send a gravity wave which could be received by distant galaxies and would permit them to send a targeted wave towards Earth. A wave which could flip knowledge from another dimension into the Universe.

"But what form do the k-shards take?" I asked.

"They will look like a metallic rain which will fall upon Earth," explained Tomales, "But once they are discovered and disturbed by humans, they will be able to use the knowledge contained within. It will accelerate their comprehension of certain key matters."

"You are sure that this is pure? Knowledge for good intentions?" asked Drake.

"Certainly," answered Abbott, "And I have asked the same questions of Darnell."

I could still feel misgivings when Abbott spoke, but this break from the ennui of a Watcher was overshadowed by the excitement of the moment.

And now, here we all were sitting in the middle of a medieval monastery on a powerful ley line plotting to execute an Intervention.

PART TWO

To the Ghosts Who Write History Books

To them ghosts that write history books
To them ghosts that write songs
Everyone asks would you write one about me
To them ghosts in the train yard
All them ghosts in my drink
Your money's no good here just write one about me

And when you go, where the winds are strong
When you go where flowers bend
Please take along all the best of my luck
and come back unchanged
Your demons all tamed
Your flowers uncut

And when you go where the winds are strong
Where soldiers carve their stones
Please take along all the best of my luck
and come back unchanged
Your demons all tamed
Your flowers uncut

Benjamin Knox Miller / Jeffrey Carl Prystowsky / Jocelyn
Jager Adams

Modified Newtonian Dynamics

I mentioned my ability to bend gravity. Not the coolest of effects, I admit. And it has a stigma attached.

It relates to dark matter, or the other 80 per cent of the universe. There's a human theory called MOND - "Modified Newtonian Dynamics" which suggests that darkmatter may actually not comprise any particles but could be attributed to the 'odd' behaviour of gravity.

Some human scientists claim that gravity does not fade away as quickly as current theories say it does. This 'stronger gravity' may be the dark matter holding together galaxies.

My clue: It is where the other dark rooms are hidden.

Further, this 'Cosmic Ghost' theory is an attempt to resolve three mysteries of modern cosmology, in one 'ghostly' presence.

Astrophysicists refer to the 'ghost condensate' in their revised modelling. They envisage it to produce a gravity to drive cosmic inflation in the Big Bang; and it may also account for the current rapid acceleration of the Universe, which has been attributed to unknown dark energy. The hypothesis that this ghost condensate could combine into particles of mass and thus could account for dark matter. I guess a few 'Bongs!' need to go with those last thoughts.

It reminds me of talk of Phlogiston by Johann Joachim Becher in the 1600s. It seems incredible that it was so long after Darnell's Intervention to give the Earth life-supporting quantities of oxygen.

According to Becher, substances that burned in air were said to be rich in phlogiston. That combustion soon ceased in an enclosed space was taken as clear-cut evidence that air had the capacity to absorb only a finite amount of phlogiston.

When air had become completely phlogisticated it would no longer serve to support combustion of any material, nor could phlogisticated air support life. Breathing was thought to take phlogiston out of the body. It was another science theory eventually debunked in the 1700s by the discovery of oxygen.

But back to the ghost condensate. Scientists are still using the current mathematical frame of reference and with it define 'Sterile Neutrinos' implying that dark matter is made of conveniently elusive particles.

Human science considers sterile neutrinos heavier than known neutrinos and may interact with other matter only through the force of gravity, making them impossible to detect. Sterile neutrinos could also be essential in the formation of stars and galaxies. Or it could all be hokum as human science toils in its lonely dark room.

To my mind, playing with gravity is toying with a serious and dark force always just eluding human vision. It's the dragon that follows me, always just on the corner of my eye. From Rene Descartes to Morpheus, humankind has always tried to look beyond its walls. For

Limantour's low-key sounding project, I was terrified that I was being asked to steal the fire from heaven.

So here I am, seated in the Benedictine abbey, surrounded by other Watchers. In our timeline we know Earth was once populated with small living creatures, survivors of an asteroid collision and a mysterious wipe-out of the Earth's climate.

I was about to engineer another possibly similar wipe.

"Think of the dinosaurs," said Abbott, "We've all experienced that. Evolution theory may be sketchy for them, although we've seen that dinosauromorphs were small animals, larger than insects but maybe only reaching the size of a house cat.

Limantour added, "The oldest dinosaurs were from Argentina, and they were around 231 million years old. There are several dinosaurs of this age found together, including the horse-sized meat-eater Herrerasaurus, the dog-sized meat-eater Eodromaeus (a distant relative of T. rex), and several dog-to-bear-sized cousins of the giant long-necked sauropods, including Panphagia and Eoraptor."

Abbott nodded, "The fact that so many dinosaurs, with different diets and sizes, lived at this time tells us that dinosaurs were already diversifying soon after they evolved from other reptiles."

Drake added, "But what you've just described are a grouping of dinosaurs any of which would terrify humans. But none of these dinosaurs were giants, and none were at the top of the food chain. Those species would come later, during the Jurassic Period."

Limantour continued, "Yes. 201 million years ago to around 145 million years ago is the time when the dinosaurs developed, roamed the Earth and were then wiped out."

Abbott added, "Every reptile and mammal had something else it could eat."

Limantour again, "Remember that by the beginning of the Jurassic, the supercontinent Pangaea had rifted into two landmasses: Laurasia to the north and Gondwana to the south. The climate of the Jurassic was warmer than the present, and there were no ice caps. Forests grew close to the poles, with large expanses of desert in the lower latitudes. The fauna transitioned to one dominated by dinosaurs alone."

Drake smiled, "The first birds appeared during the Jurassic. Other major events include the appearance of the earliest lizards and the evolution of therian/marsupial mammals. Crocodylomorphs made the transition from a terrestrial to an aquatic life. The oceans were inhabited by marine reptiles such as ichthyosaurs and plesiosaurs, while pterosaurs were the dominant flying vertebrates."

I was still thinking about what Abbott had said, 'Every reptile and mammal had something else it could eat.'

I remembered the collapse of the food chain ecosystem around 66 million years ago. We'd already seen a mass extinction during The Great Dying, and then we had another one. For the first 175 million years of their existence, dinosaurs took on a huge variety of forms as the environment changed and new species evolved that were suited to these new conditions. Dinosaurs that failed to adapt became extinct. But then 66 million years ago, over a relatively short time, the dinosaurs, except for birds, disappeared completely during the Cretaceous extinction.

This was caused by another 15-kilometre-wide asteroid impact, in Yucatán Peninsula in Mexico. A few minutes later and it would have hit the sea. Instead, the crash created the 150-kilometre-wide Chicxulub crater and threw huge amounts of debris into the air and caused massive tidal waves to wash over parts of the American continents.

Around three-quarters of Earth's animals, including dinosaurs, suddenly died. The asteroid hit at high velocity and effectively vaporised. It made a huge crater, devastating the immediate area. A huge blast wave and heat wave went out, and it threw vast amounts of material up into the atmosphere. It sent dust all around the world. It didn't completely block out the Sun, but it reduced the amount of light that reached the Earth's surface, impacting plant growth and, in turn, right the way up the food chain.

"Okay, so how will we do this, and when?" asked Abbott. He stared across at me. I noticed he was sitting in a shadowy corner and somehow his presence melted into the darkness.

Limantour stood, "Farallon has had time to think about this now, and I guess must have seen this request coming?"

I hadn't seen it coming, and I had no actual idea what to do. Cornered; that's what I felt. I could envisage a different perception if I said 'yes' and wondered if I would even be in the same dimension.

One thing for certain: I knew I'd be unable to continue as a Watcher.

"Don't you see?" asked Drake. "It's the whole point. We are Watchers until one day we are not. If we get involved - Intervene - then we become something else. I guess you, if successful, would be a Wakener."

"But what about if something goes wrong?" I asked.

Abbott answered, "Can it really be any more mundane that a life as a Watcher? Seeing 13.8 billion years of history unspool until Earth and then the Universe run into an end-state?"

Limantour and Tomales were both nodding.

"I need some time to think," I said, aware of how ridiculous this must sound. A Watcher asking for more time.

"Three days," spoke Abbott, "That should be long enough to weigh the options."

"I'll want to talk to a few people too," I said.

Limantour spoke again. "We'll be able to help you. Who would you like to speak to?"

"Two people first," I said, "Lepton and Darnell." Lepton because he'd done something agreed to be successful - i.e., Light, and Darnell because his intervention was more controversial.

"I can assist that," said Abbott.

"And then a third person," they all looked at me.

"Limantour. Alone," I looked across to her. She looked back, then nodded.

"Agreed, although if you speak to me alone the others will probably be able to share it via the back-channel we've created between us all."

Limantour looked around. I could tell that some of the others wanted to honour my privacy with Limantour. But not everyone.

"Okay, we'll worry about that when it happens," I said.

Three days. Three people. All of them Watchers. Two of them already Wakeners.

Ice, ice, baby

I must recognise us crossing a huge bookmark in time. It was when Earth's ice ages occurred. Some might call it climate change, but that understates its significance. Some people think of simply one ice age, and I suppose as a Watcher you'd be able to fast forward from 2.4 million years ago until a mere 11,500 years ago and it could look like a singular occurrence.

However, around fifteen thousand years ago the coldest part of the last ice age was coming to an end, and the climate of the key landmasses north of the equator began to improve. During this time, the Earth's climate repeatedly changed between very cold periods, during which glaciers covered large parts of the world, and very warm periods during which many of the glaciers melted.

I counted 17 cycles between glacial and interglacial periods. The glacial periods lasted longer than the interglacial periods. The last glacial period began about 100,000 years ago and lasted until 25,000 years ago. Right now, we are in a warm interglacial period. These interglacial times played havoc with the development of society and were well outside of any Intervention. Suffice it to say, we are entering the period when history is recorded, whether through writing or passed down stories.

Masamune and Muramasa

A legend tells of a test where Muramasa challenged his master, Masamune, to see who could make a finer sword. They both worked tirelessly, and when both swords were finished, they decided to test the results.

The contest was for each to suspend the blades in a small creek with the cutting edge facing against the current. Muramasa's sword, the Juuchi Yosamu "10,000 Cold Nights") cut everything that passed its way; fish, leaves floating down the river, the very air which blew on it. Highly impressed with his pupil's work, Masamune lowered his sword, the Yawarakai-Te "Tender Hands"), into the current and waited patiently.

Only leaves were cut. However, the fish swam right up to it, and the air hissed as it gently blew by the blade. After a while, Muramasa scoffed at his master for his apparent lack of skill in the making of his sword.

Smiling to himself, Masamune pulled up his sword, dried it, and sheathed it. All the while, Muramasa was heckling him for his sword's inability to cut anything. A monk, who had been watching the whole ordeal, walked over, and bowed low to the two sword masters.

Then the monk explained what he had seen.

'The first of the swords was a fine sword, however it is a bloodthirsty, evil blade, as it does not discriminate who or what it will cut. It may just as well be cutting down butterflies as severing heads. The second was by far the finer of the two, as it does not needlessly cut that which is innocent and undeserving.'

DAY 1 - Darnell

As agreed, Abbott helped set up a meeting with Darnell. It was immediate and ahead of the session with Lepton, which was not the order I'd hoped for.

I should have guessed that Darnell would want to meet me somewhere theatrical.

He chose the Colonnade in Rome. I'd not visited this part of Rome for centuries. I remembered it as an extensive collection of imposing buildings rising along the hill. Most were now in ruins. The Rise and Fall of a great Empire.

Under the Roman Empire, the forum had become primarily a centre for religious and secular spectacles and ceremonies, as well as being the site of many·of the city's most imposing temples and monuments.

Among the surviving structures I could see the Palazzo Senatoro, various Tempio, and distant Basilica, on the way toward the equally ruined Colleseo.

I'd also forgotten just how crazily busy this part of Italy can become, with the mixture of tourists, small Italian cars and natives of Rome mingling together in this sun-baked area.

Abbott had arranged for Darnell to meet me on the steps of the Capitoline Hill, next to the Lupa Capitolina - the She-wolf, which famously had tended Romulus and Remus. I could tell that Darnell's sense of humour had not been displaced by his new powers.

Darnell was with someone else - a splendid woman, carrying a basket, as if she had just returned from the market.

"I wanted you to meet another one like me," said Darnell, "This is Ceres."

I had to wrack my brain for a moment. I couldn't place the name but knew it was powerful.

She smiled at me, and I felt the air around me crackle. I realised she was some kind of super deity.

She spoke to me in informal Italian, "*Ciao - Darnell mi dice che potresti voler unirti a noi qui. Che tu possa anche fare un intervento, anche se penso che i potenti siano già avvenuti. "* - "Hello - Darnell tells me you may wish to join us here. That you may also make an Intervention, although I think the powerful ones have already occurred."

I found I could switch straight into Italian and realised that there were a few extra useful things which I'd picked up along the millennia. I still couldn't remember who she was.

"You don't know, do you?" asked Darnell. I could see he had triumphantly expected me to fail to recognise her.

"Ceres," he continued, "Ceres, the Roman goddess of agriculture, grain, and the love a mother bears for her child. She is the daughter of Saturn and Ops, the sister of Jupiter, and the mother of Proserpine."

I noticed Darnell spoke in the present tense, and simultaneously realised we were in the presence of Ceres.

Now he switched to the past tense, and I realised that some things about Ceres had been lost in the mist.

He continued, "Ceres, the goddess of agriculture, was beloved for her service to humankind in giving them the gift of the harvest, the reward for cultivation of the soil. Also known as the Greek goddess Demeter, Ceres was the goddess of the harvest and credited with teaching humans how to grow, preserve, and prepare grain and corn. She was thought to be responsible for the fertility of the land."

Ceres spoke, and I once more felt the crackle, "It is close to the time of Ambarvalia, when my powers increase, " she explained, "I guess you, as a Watcher for all time, must be used to experiencing some facets of human emotion?"

"No, never," I said. "This is the first time. Ever, it is a most strange sensation if that is what it is."

Now this was making sense. Other gods occasionally dabbled in human affairs when it suited their personal interests, or came to the aid of mortals they favoured, but goddess Ceres was the nurturer of mankind.

Ceres continued, "It is interesting that so few know of my story, yet I brought humanity one of their greatest gifts. The gift of farming and agriculture. I am the sister of Jupiter, and Proserpine was my daughter. Pluto, god of the underworld, kidnapped Proserpine, to be his bride.

"By the time I could follow my daughter, she was gone into the Earth - the underworld.

"To make matters worse, I learned that Pluto had been

given Jupiter's approval to be the husband of his daughter. I was so angry that I went to live in the world of humankind, disguised as an old woman, and stopped all the plants and crops from growing, causing a famine.

"Whoa," I said, "But you were a god before you gifted your intervention to humanity?"

"Not god; it's goddess, we still use the term around here," she replied. "Yes, but nowadays I'm on the same reality plane as Darnell."

My mind was racing as I thought again of the metaverses and the 'branes created by alternative dimensions unseen to humans. Was the underworld another darkened room?

Ceres continued, "Jupiter and the other gods tried to get me to change my mind to stop the famine, but I was determined. Jupiter eventually realised that he had to get Proserpine back from the underworld and sent for her.

"Unfortunately, Pluto secretly gave Proserpine food before she left, and once she had eaten in the underworld she could never leave forever.

"Proserpine is forced to return to the underworld for four months every year. She comes out in spring and spends the time until autumn with me but must go back to the underworld in the winter. The Romans decided it was why plants lose their leaves, seeds lie dormant under the ground, and nothing grows until spring, when Proserpine and I are reunited.

"Cereals," added Darnell, "Think of Cereals - Ceres..."

Darnell was up to his old annoying tricks, and I could see Ceres dig him in the ribs.

"You forget yourself, " she said, and I could feel a different crackle in the air. One I decided was of a mocking directed toward Darnell.

Ceres added, "Darnell and I are different. He acts from self-interest. I try now to act for the good of humanity. Darnell has a string of plans which he will tell you in a moment, of that, I am sure. I performed one deed, driven by family ties, which led to the gift of agriculture."

Darnell added, "But even that has an unintended consequence."

Ceres smiled. "You mean superabundance? I can never see this the way you do, Darnell. Not ever."

With that, Ceres, lifted her eyes towards the sky and then, as we Watchers do, glitched herself away. I felt an overwhelmingly pleasant sensation drain away as I was left with Darnell.

 "Don't let her fool you," he added, "Ceres is probably still listening to what we have to say."

"You don't believe in half measures," I said to Darnell, "In the heart of the Roman Empire talking to a Roman super-god."

"Goddess!" said Darnell.

"Goddess, " I quickly corrected.

"The lesson is that one of the biggest interventions ever - agriculture - is hardly even recognised and the instigator of the event - Ceres - is equally lost in the mists. I don't think you need to worry too much, if you cross to this reality."

I'd worked out that Darnell was as crazy as ever. But now there was a new facet. Ceres had mentioned self-interest. I wondered what this could mean.

Darnell looked toward the winding road leading up to the she-wolf statue. Then he waved.

"It's Cardinal; he has also come along at my request."

A gaunt, pale-faced man approached in a long sand-coloured coat which seemed incongruous in the hot Italian weather. I could see that he already knew both Abbott and Darnell.

"Hello," he said - with his gaze directed toward me, "It is best that you don't look directly toward me."

Darnell explained to Cardinal that he was describing Ceres' Intervention, "Ceres had turned things around. Her intervention created a conundrum. People would choose to farm although it did not make for an easy life."

Cardinal interjected, "Ceres intervening to create the rise of agriculture allowed far more humans to be alive. A hunter-gatherer needs about ten square miles of game and berry-filled land to live on, whereas agriculture can produce enough calories in a tenth of that space to keep fifty people alive."

I interrupted, "No wonder Ceres carried a basket overflowing with produce."

Darnell continued, "I'm pleased you noticed the

symbolism. But now for the unintended consequence. The increase in population came after agriculture started, not before. Across the planet vastly more land was inhabited by hunters than by farmers. This is the unrecorded narrative of the Indian forests, the Eurasian steppes, the jungled islands of East Asia and the migrations of the Americas. Most people found ways of not farming. And yet farming was repeatedly invented in completely separate parts of the world.

Cardinal spoke again, "It took nearly ten thousand years from the first attempts at agriculture for the world's population to reach a billion. Now humans are adding extra people at a billion every dozen years. World food stocks, held for emergencies, are tiny. To avoid famine every person needs to be fed by a far smaller patch of land than ever before. It is why I am seeking action."

I wondered what this could mean. Ceres had hinted at self-interest.

Cardinal continued again, "Ceres' interventions happened first in the Fertile Crescent, which curves from today's Jordan and Israel, up to Anatolia in today's Turkey, and then back east into Iraq.

"But then it started in northern China. It occurred in Mexico; and independently in the Andes; then in what is now the eastern United States. Thousands of years separate these breakthroughs, but they are Ceres' handiwork."

Darnell added, "Do you see? Humanity had walked into an accidental trap. An unintended consequence from Ceres' gift. Humanity's decisive step had consequences which could never have been imagined, and from which there was no pulling back."

Cardinal continued, "The trap was that settled farming communities produced bigger populations. Even with Late Stone Age technology, each acre of farmed land could support over ten times as many people as each acre of hunted land. It was not simply about food, either. As we have seen, hunting tribes, always on the move, must carry their children. Once people settled down, the birth rate could rise, and it did.

Abbott added, "Larger families meant more mouths to feed, which meant that farming and herding become ever more important. Herds can never be untethered and returned to the wild. The early farming men and women were shorter in stature and more prone to disease– because parasites and pests settle down as well."

Darnell added, "They cannot stop. Before, they were shaping and taming the plants and animals; now the plants and animals are shaping and taming them, too."

Cardinal continued, "They also had to develop other skills. They had to grind and sift their grain and store it. Their precious domesticated animals, which had to be protected from wild beasts and allowed to wander for food – but not too far – must be exploited in every way. Wool could be sheared and carded and woven. Blood could be drawn off and used to enrich meal. Some farmers developed the habit of drinking the milk of lactating goats and cows – and most of their European descendants remain lactose-tolerant to this day."

Cardinal was on a roll now. "The preparation of hides, the weaving of ropes to help with ploughing, and making baskets and pottery for storing or cooking grain – Think of it - a whole new world of domestic jobs and skills emerged.

"By around eleven thousand years ago, groups of humans realised that by keeping some animals near by–the ancestors of today's sheep, goats and pigs–they could ensure for themselves meat and hides.

Cardinal shifted his gaze toward me, "People had probably been gathering edible seeds for centuries before they planted them, then returned to the same place for the annual harvest of seed-heavy grasses or nutrition-rich peas."

Darnell spoke again. "In societies where men would be expected to hunt further from their settlements, the breakthrough discovery prompted by Ceres was made by women. There are supposedly fifty-six edible grasses growing wild in the world—cereals like wheat, barley, corn and rice."

Cardinal took over, "Additionally, the people of the Fertile Crescent had at their disposal a disproportionate number of the thirteen large animals that can be domesticated."

Then Cardinal's brow furrowed. "They had not only pigs and nearby wild horses but also cows, goats and sheep, plus thirty-two of the edible and grain producing grasses. In the Fertile Crescent, people called Natufians could gather grain around thirteen thousand years ago; and early on—presumably to stay close to the precious grain—they settled down in villages rather than moving around as hunter-gatherers."

Darnell again, "All this gave humankind a surplus of energy no mere predator could hope for. Using it, humans grew from family groups to tribes to villages to cities to nations, allowing them to change much more of the original environment. Humans altered the courses of rivers and dug into the mineral covering of the planet, pulling out coal, oil and gas to provide more power, exploiting ancient vegetable reserves that lived and died long before they arrived."

"And that is where Cardinal's idea came in," said Darnell excitedly.

"Exactly," said Cardinal. "The villages had to come together, to create and then to maintain the complicated system of waterways and dykes needed for agriculture. I showed them a ruling system."

Abbott continued, "Workers had to be organised; the improvements to farming produced surpluses of grain and Cardinal's intervention allowed the introduction of rulers and priests, who developed religions, temples and employed servants to tend them.

He added, "A system of notation and recording developed into a system that could record stories and ideas. Individual families or villages were far too small and had too little spare time to achieve what was needed. Only by combining in large numbers, organised by managers, could they survive. The managers seem to have been priests, or at least to have been based in the temples, from where they oversaw vast irrigation projects."

Abbott added, "Once the system of labour and specialised skills was in place, the managers had access to the brawn to build ever greater temples. The feedback from successful irrigation to the power of those who directed it is obvious: over time, the managers could claim they spoke for, with, and to, the gods."

Cardinal's turn now. I could see they were operating as if a single being, with Cardinal having the main ideas and Bishop running a kind of follow-up explanation.

Cardinal explained, "These priests and rulers handled the settlement's very survival. The original ruling class, high on their platforms, ears tilted to the heavens, had arrived. Below them, totting up the deliveries of grain, beer, meat, and metals they required from the toilers, were the scribes or middle management. You cannot have a hierarchically organised society without the bureaucracy"

Cardinal continued, and I could see where this was leading.

"Priests demand their special places–intimidating, nearer the gods. This required huge numbers of workers and full- time artisans, as well as measuring and planning. That meant detailed notetaking, indeed writing. Then, large tributes of food, beer and raw materials were called for, to keep the building workers alive."

I could see how it was falling into place: Priests of religion. Large-scale building projects. Writing. Taxes. Soldiers. Kings.

The ability to make war.

Darnell, with Cardinal and Abbott, had hastened war's arrival.

But the set of conditions is broader than all-out warfare and all arrived in human history alongside one another, based on the first cities - really the first concentrations of stored wealth, themselves based on riverside farming cultures that needed to work together to tame nature.

This is the shift that is more powerful than the old ties of clan, kin and lineage, and marks the next important moment in human development after Ceres brought farming.

Darnell hastened the rivalry between cities and people, which sped up change, until full-scale war brings catastrophe.

The rise of trained bureaucrats, with their cuneiform writing implements, permitted different people with different languages to communicate; Sumerian becomes the lingua franca for Mesopotamia, and scribes become bilingual. A momentum is under way, which may be lost here or there but which has never stopped since.

The first cities also nurtured a flowering of abstract thought. The ruling class of kings and priests had time to speculate, not least about the mysterious world of winking lights and movements overhead that had also obsessed the builders of Stonehenge.

It is no surprise that Mesopotamia gave us mathematics, both the simple sums to tally trade and taxes and the more complicated ones used to track the stars. Looking up, the Sumerians and Babylonians wondered about this nightly message, with its shapes and regular patterns. If the gods were able to send messages back to them, were these the divine writing? Was there a pattern, which could then be imposed on the rhythm of human life?

Reading the stars required measurement of angles. The Sumerians plotted the movements of the five planets they could see—Mercury, Venus, Mars, Jupiter, and Saturn—and named a day after each. They then named one day after the Moon and another after the Sun, giving them a seven-day week. They regarded seven as a perfect number; and the Sumerian week is of course still the human week, its days still named in the Sumerian fashion.

The Sumerians also developed a counting system based on the number sixty, which is divisible by eleven other numbers and so particularly handy for Bronze Age accountancy. From this derived the 60-second minutes, 60-minute hours, 360-day years and 360-degree circles.

All of this is remarkable enough, but the first cities also brought a flowering of art and design, with alabaster carvings and mosaics and useful stamp-seals for parcels of goods from Uruk, plus inlaid gaming-boards, musical instruments, and delicate gold jewellery from Ur—even before the carved reliefs of the Assyrians and Babylonians.

Today, thanks to the habits of nineteenth-century archaeologists, the loveliest of these things can be found in Berlin and (controversially) London, not in Iraq. Each Mesopotamian city had its own gods, culture, and reputation. Uruk was famous not only for its huge ziggurat and sky-god but for its sexy female deity Inanna, who was associated with all kinds of fertility and whose rites shocked one Babylonian writer: 'Uruk . . . city of prostitutes, courtesans, and call-girls - the party-boys and festival people who change masculinity to femininity.'

These first cities are among the most important sites in the human story. Successive floods have reduced many of them to gritty stumps, and obliterated others. Neglect, war, and the lack of interest of later cultures followed by aggressive, treasure-hunting Victorian archaeology has meant that while some of their greatest carvings and other artefacts are in European museums, the sites themselves are often dusty disappointments.

This is tragic, since the achievements of the Sumerians, Akkadians and early Babylonians were huge, and in some ways much more impressive than those of the better-known Egyptians. Their city culture was bureaucratic and clearly in some ways oppressive, weighing heavily on farmers, requiring payment in return for the canals and wells that kept their fields so fertile.

It allowed the emergence of kings with enough muscle to go to war against one another, and to carve out the first empires, along with the misery that early mass-killers such as Sargon of Akkad brought to the land. But these first cities were also places of beauty, intellectual advance, wonder and a great deal of not very innocent fun.

It was Abbott who delivered the ominous summary, "Making people pay tributes as taxes would not have been pleasant; force would have been needed. All the accumulating wealth would be a temptation to robbers and ultimately to rival cities. Therefore, walls were built and some men given the job of full-time protectors. A warrior class emerged. Nothing has advanced technical progress faster than war. "

It left me to wonder how knowingly Darnell, Cardinal and Abbott had provided the worst kind of Intervention. That of warfare.

But this was where it ended. Like a fairground toy that has run out of money, the scene suddenly dimmed. I realised I had consumed my first day. Never had I experienced the feeling of wanting a day to last longer.

DAY 2 - *Lepton & Lekton*

I didn't have long to think about what had just happened. I was suddenly in a dingy apartment reeking of decaying fish. I was in China, in a makeshift wooden room above a market. I'd been sent here by Lepton, who was the second person I'd asked to consult about creating an intervention. Originally, I'd hoped to meet Lepton first, but Limantour was working her chaos as usual.

I looked around. The building was rudimentary, and there were baskets and plastic crates in the corridors. As I looked through the clear plastic lid of one crate, I could see large, rough-skinned red fishes, still alive and gasping through mouths the size of small plates, revealing searing blue tongues.

Then I noticed the next crate. I thought it was rats at first, but then realised it was agitated bats, trapped inside the crates.

Just like when I was in Italy, I could read and understand the Chinese writing. This was Huanan, or Southern China, and all the crates seemed to have shipping labels, outbound from Hankou Railway Yard and then to varied destinations. I wondered about refrigeration and food safety, when, with a flickering of the lights, Lepton and another stranger appeared.

Lepton beamed when he saw me. "Hello, Farallon, it has been a long time! You asked to see me and to witness for yourself that I have continued to enjoy a good living beyond your metaverse."

I looked around and the stranger was also smiling, although his smile looked as if he had been told to look happy in my presence.

"Hello," he said, "My name is Lekton."

I could see what Lepton had done. Lepton was the originator of one of the earliest interventions - the provision of light for the universe. Lekton was almost the opposite. Watchers knew that Lekton's intervention was to bring warfare to humanity. I was standing with the bringers of Light and Darkness.

I decided to deflect matters whilst I considered my options.

"Well, it is an honour to meet you both, although I am not sure of our location?"

"This is the workshop of Scheppach," explained Lekton, "She lives outside of the law around here"

"In what way?" I asked, realising that I could not be in any danger from the human world.

A new voice cut in. Heavily accented, not of China, but of Germany. A slender woman with blue-tinted spectacles appeared from the shadows. She was wearing a form-fitting light-brown well-worked leather apron.

"Farallon, it is my pleasure to meet you," she said, "I am Scheppach, and I import the finest quality knives, which can be used in the wet market below."

I noticed in her short blonde hair; her earrings were formed as two precise tiny daggers. She also had the slender, controlled hands of a concert pianist.

I remembered something about Bladerunners. They were the outlaws who smuggled micron-fine-bladed surgical knives. These knives were used for off-grid operations on humans as well as to augment the combat skills of the warrior classes. The importation was more commonly referred to as smuggling and the kind of person would be correctly thought of as an arms dealer.

I'd been in the room long enough to get acclimatised, and I could see other bulky black cases stacked up beyond the wet fish. I realised from their markings that they were gun safes. I saw Scheppach notice my reaction.

"Yes, you'll see I have a supply of armaments ready for export,"

I knew she meant for trafficking.

Scheppach continued, "This is a useful location to keep others' eyes away. They can't abide the stench from the market and, well, you can see that the way the market operates is barely legal.

"Once we have access to the construction details for the weaponry, we can easily ask the Chinese factory towns to take on the production of as many units as our clients need. The RPG-7 is a case in point. We have supplied 80,000 of these 40mm grenade launchers to Iraq and Afghanistan battlefields. They are badged as 'Made in the USA' and machined out of ordnance-grade steel instead of castings in the way of the original Russian designs from the 1960s. Now we are designing factories to build copies of entire warplanes, based upon stolen

American designs."

I decided there was more darkness than light in this room.

Scheppach added, "Oh yes, and we produce Katana swords too, but with modern steel instead of the pig iron of the Samurai. Masamune would approve if the technology had been available in his time."

I could see outside of the apartment through the grime-caked windows. I wanted to be on the other side of the glass, away from what I could only think of as psychopaths.

Lekton spoke, "You should see now that whatever happens, you get placed in our metaverse after your intervention. I'm sure you have seen that for Darnell, Cardinal, Abbott, Lepton, and myself, it becomes a rewarding journey.

I looked toward Lepton, deciding he was the most beneficent of all of them. I could see a sadness in his eyes. "I've still to visit most of the 84,000 edges of the universe," he said, cryptically.

I realised he was describing the Chinese Courts of Hell - after the Chinese Diyu stories or the Japanese Yomi. It left me with another feeling washing over me like when Ceres had been close, except this time I realised I was experiencing dread.

Lepton continued, "We decided to meet here but our plans call for a journey to Knossos. But first you needed to see the wet market. A precursor symbol of what is happening, used as an excuse. The excuse used to mask the creation of a global pandemic. Humankind resolves it but take no heed of the implied consequences. The huge and unsubtle warning to humanity does not work. We know, we can see forward. We don't need the powers of a scryer. We can run forward on the rails of time."

"Except they stop, " said Scheppach, "Very soon. A few hundred years into the future."

"That's right," said Lekton. "For all of us. We can't see past the event horizon."

"What event?" I asked.

"That's what we need to discover," said Lepton. "I brought light, but from that point in the future all we can see is darkness."

"So why do we need to travel to Knossos?" I asked, "And why is this market a precursor symbol?"

With that, I could feel a ground rush, as if we were in a minor earthquake. The shakes subsided and a bright, hot daylight suddenly bathed us all. I realised we had moved to Crete, and the air filled with the scent of jasmine and oregano, replacing the rankness of the fish market in Huanan. I noticed how all of us, even Scheppach, took it in their stride. I had wondered about Scheppach, because she was not a Watcher, but had become implicated through Lekton.

Scheppach answered, she was smiling, "For all of your time here, you don't seem to have learned much about the ways of the Earth. How it tries to self-heal,"

"And Knossos here. It has been the hub of civilisation," answered Lepton, "The Minoans were the first European civilisation from around 3600 to 1160 BC. Their island of

Crete lies in the far south of the Greek peninsula. They were trading and seafaring people, whose pottery turns up in Egypt and whose art was influenced by the Egyptians."

Lekton added, "Their art and architecture are instantly attractive, giving an initial impression of a tranquil, female-dominated society whose palace walls ripple with dancing dolphins. Amid the fat red columns are images of a little bull-dancing here, a moment of saffron-gathering there. But the Minoans are useful as a warning about history and how we romanticise it."

"But I remember Santorini's earthquake blasting a hole in that part of the Mediterranean?" I said, "Isn't Knossos a reconstruction?"

Lekton continued, "Correct. Knossos only dates to between 1905 and 1930. It is a reconstruction of a Bronze Age palace, filled with fake pictures. It was the lifetime achievement of a rich British archaeologist, Sir Arthur Evans.

"Evans supported the ruined buildings he was excavating with wood and plaster, and then slowly began to 'improve' them with the flexible and useful recent invention of reinforced concrete.

"The extent to which his re-imagining of the Knossos complex is an accurate and reasonable guess, or merely a modernist fantasy, divides even the experts. Evans commissioned modern artists to retouch ancient wall paintings so comprehensively that they produced new ones.

"So, from this rubble, what can we know for sure about the people we call Minoans? Their civilisation lasted for around thirteen hundred years and survived a whole series of natural disasters, including a hugely destructive earthquake and two volcanic eruptions.

They traded tin, very well made and painted pottery, as well as a wide range of foods, oils and other staples. Their agriculture was sophisticated, and it seems that their religion was dominated by priestesses and bull-worship.

"But what warning from the Minoans?" I asked.

Scheppach was shaking her head, "Don't you see? There is a darker side to the Minoan culture. Lekton's intervention pushed them into war and taught them to protect their units with citadels and defensive walls. I could only assist with my infinite supply of increasingly technological weapons. Back then it was spears, Right now we are moving into the era of rail-guns and satellite enforcement systems."

Lekton continued, "Far from being a society of peace and love, wafting about in gossamer and admiring the dolphins, the Minoans became bloody but calculated with it. As the first civilisation in Europe, they combined beauty, human sacrifice and military planning with a class system and fighting elites."

Lekton added, "You could say, with my intervention, they invented war."

Scheppach continued, "And then, through Lekton's interventions, Earth suddenly had a way to correct imbalances of humans. For me, it has been immensely profitable."

"But you are not a Watcher?" I asked, looking towards Scheppach.

"No, correct. I am a mercenary- a soldier of fortune. With Lekton's technologies, I can profit greatly from Earth attempting to rebalance - and Lekton has provided me with longevity, so my operation spans from Cyrus the Great and his Persian Empire conquests at the expense of some 100,000 souls."

She looked wistfully into the air and then towards Lekton. I couldn't believe I was looking at the two people responsible for the slaying of so many. We had been walking through the ruins of Knossos. I was struck by the paintings of bull-leaping and varied figural frescoes, but there were few pictures of conflict. Just like Knossos lacked warfare depictions, Lekton and Scheppach had been suppressing the imagery of warfare. One immense edit.

Then Lekton relaxed a little, "It makes the campaigns of Alexander the Great look small, with around 150,000 deaths, particularly before the Punic Wars killed 1.5 million. Put another way, ten times as many to go to their deaths."

Scheppach added, "And don't let Knossos and the Minoan flower-children fool you. Their peace and freedom came at the price of great conquests. I helped them design long swords and shields. They even devised a signet ring to show they were from the fighting elite, much the way that a current day US Marine wears their ring with pride.

"But how could anything good have ever come from all of this fighting?" I asked.

Lekton looked toward Scheppach again, "If my original intervention was about protection and security, we learned that the unintended consequence was around greed, profiteering and control."

"But it was too late?" I asked.

"The situation accelerated," said Lepton, who had been quiet until this point, "The Three Kingdoms War in China killed around 40 million in the period between the foundation of the state of Wei in 220 and the conquest of the state of Wu by the Jin dynasty in 280, some 60 years later.

Lekton added, "Yes, and there were other big wars in China, like when the Tang Dynasty China and Islamic Empire fought the Yan state in 755, causing a further 13 million deaths in the An–Shi Rebellion."

Scheppach added, "Then there were the Mongol invasions and conquests which took place during the 13th and 14th centuries, killing 40 million and creating the Mongol Empire, which by 1300 covered large parts of Eurasia. Historians regard the Mongol devastation as one of the deadliest episodes in history. It is when humans learned about biological warfare, as the Mongol expeditions spread the bubonic plague across much of Eurasia, helping to spark the Black Death of the 14th century."

Scheppach continued, "And it brings us back to the agrarian economy - back to the farming intervention of Ceres. The Mongol Empire was a land power, fuelled by the grass-foraging Mongol cavalry and cattle. Most Mongol conquests and plundering took place during the warmer seasons, when there was sufficient grazing for their herds."

Lekton added, "Yes, every soldier tended four armoured

horses for interchangeability and speed. No small wonder that the rise of the Mongols was preceded by 15 years of wet and warm weather conditions that allowed favourable conditions for the breeding of horses. This which greatly assisted their expansion - although Scheppach did not profit so much in those days of Genghis Khan's violent cavalry."

Scheppach replied, "I'm not so sure. They had gunpowder and could build bombs. Someone had to show them the way. They also set chains across the harbour, to wreck ships and used firebombs to set fire to all manner of things. Then they got the level of nitrate correct and were able to create the early hand cannons which fired devastating shot across the enemy."

Lekton added, "Just another 300 years later and we saw, in 1519 to 1632, the Spanish cutting their way across the Aztecs in Mexico, then through the Yucatán and the Incas in Peru. Another 12 million casualties from those three related campaigns.

"And then the French religious wars, totalling some 4 million casualties which pales into insignificance contrasted with the further Chinese Wars in 1616–1683 as the Qing China vs. Ming China vs. Shun dynasty, totalled some 25 million deaths."

Lepton said, "It makes the American Revolutionary War seem minor with its 37 thousand deaths."

Scheppach added, "And the Chinese hadn't finished yet. The Taiping Rebellion between 1850-1864 saw another 70 million deaths as Qing China fought the Taiping Heavenly Kingdom. As a context, the American Civil War between 1861–1865 saw around 1 million deaths."

Lekton continued, "We'll jump forward to World War I, which saw 16 million deaths plus a further 24 million through the Spanish flu pandemic, which was exported to Spain from a Kansas war-room barracks. And then approaching 85 million deaths from World War II."

Scheppach added, "But we are seeing a shift now. Earth was able to keep itself balanced through increasingly vicious warfare. Now we are seeing a mixed effect. Part of it is through warfare, but this has been hybridised by the onslaught of other methods, such as the biological one. 230 million cases of COVID and around five million worldwide deaths from the disease."

Lekton spoke, "Earth rebalances, seeking a new equilibrium."

Lekton was stuttering, like a poor-quality video, and I knew I had run out of time. That's a feeling I'd never experienced until this week.

Declare independence!

Declare independence!
Don't let them do that to you!
Declare independence!
Don't let them do that to you!

Start your own currency!
Make your own stamp
Protect your language

Make your own flag
Make your own flag
Raise your flag (higher, higher!)
Raise your flag (higher, higher!)

Declare independence!
Don't let them do that to you!

Damn colonists
Ignore their patronising
Tear off their blindfolds
Open their eyes

With a flag and a trumpet
Go to the top of your highest mountain and
Raise your flag (higher, higher!)
Raise your flag (higher, higher!)

Declare independence!
Don't let them do that to you!
Declare independence!
Don't let them do that to you!
Raise the flag!

Björk Guðmundsdóttir

DAY 3 - *Limantour*

My third day, and I was looking forward to meeting Limantour. I'd saved her until last because I wanted to amass some facts before I saw her to discuss the Intervention.

I could see the lineage. Ceres had pressed agriculture upon the unsuspecting humans. They evolved from hunter/gatherers to farmers and created such a food surplus that they could safely populate newly established towns and cities.

Then an elite had formed; people who wanted greater power, and Lekton had given them the tools and weapons to achieve this. Intriguingly, the first established civilisation - the Minoans - had swept their fighting under a carpet, at least if we are to believe the reconstructions of Evans, the explorer. I believed Scheppach, who had explained to me about long swords and other forms of assault weapon, evolving much later through rocket propelled grenade launchers towards rail guns.

It had also taught me that writing a history of the Earth is a ridiculous thing to do. Everything gets filtered. The victors write the history books and cast themselves as the righteous. It really was the ghosts writing history books.

Limantour arrived on a skateboard, with a new haircut bobbed white and with light blue streaks which matched her powder blue tee-shirt. She had swapped her shorts from two days ago for tight black trousers, with some kind of block logo running up the legs. She was turning heads. We were in Santa Barbara, just along from the pier. I wondered if she had made her way along the Pacific coast, as we were only about 350 miles from where we had first met at Point Reyes.

Limantour smiled, "It becomes about living in wide time and living a long time. I guess you've realised that unassailable information becomes too vast for any individual to absorb. A doom-scroll with the likelihood of immense errors?"

She was right, I'd seen plenty in the last two days but had to piece together my truth.

"C'mon," she said. She flipped her board, grabbed hold of me and whisked me toward Stearns Wharf - the pier structure in Santa Barbara.

"You don't wear any of that body armour?" I asked.

"And the point of that would be what, exactly?"

"As a uniform? So that you don't look - well - unworldly?"

"No, why do you think I wear this hair and these clothes?" she replied, "unless it is to be noticed?"

We were on the boardwalk part of the pier; several cars were driving along it to park. I noticed the slogan on the back of Limantour's tee-shirt, "Off The Wall" and a picture of an anatomical heart.

"Payin' my dues," she said, "Lizzie Armanto - the board rider," she explained. I reminded myself that Limantour was the mistress of chaos theory. She gestured toward a fish restaurant, which looked out onto the calm blue waters of the bay and further towards the Pacific Ocean.

Cormorants with outstretched wings were basking in the sunshine, perched upon the pier's timbers.

Limantour continued talking, although my mind was in a confused meltdown, "We may regard the present state of the universe as the effect of its past and the cause of its future. An intellect which at a certain moment would know all forces that set nature in motion, and all positions of all items of which nature is composed. If this intellect were also vast enough to submit these data to analysis, it would embrace in a single formula the movements of the greatest bodies of the universe and those of the tiniest atom; for such an intellect nothing would be uncertain and the future just like the past would be present before its eyes. Yeah," She grinned at me.

"Don't you see, Farallon? Laplace's demon? Step outside the world to observe the world fully? - Take your brain to another dimension!"

Limantour laughed and said, "Chaos theory may well prevail, where so many minor events can tilt everything, that we are mindful as Watchers of the golden rule never to interfere."

"Is that what you want to prove to me?" I asked.

Limantour answered, "Even in a current event, there are too many sides and perspectives to gain a genuine sense of the truth. Watch politicians swear the sky is any colour they feel like. Or the little self-aggrandised spokespersons sent in to distort and confuse everyone with specious counterarguments."

"I've just listened to a whole series of spokespersons," I said, "Darnell, Cardinal and Abbott, then Lepton,

Lekton...even Scheppach. None of them seemed fully balanced. Even Ceres had a rough time of it, with her daughter partially caught in another dimension."

Limantour continued, "Hold that thought...It should be better that we understand how rulers lose touch with reality, or why revolutions produce dictators more often than they produce happiness, or why some parts of the world are richer than others. It should make it easier to understand how things work."

She added, "But now I must tell you something. About me."

She continued, "I've also, as a Watcher, got a special skill." She was eating a platter of ceviche with sliced jumbo shrimp, jicama, mango, pico de gallo and fresh avocado. I wondered if it was something to do with food.

"I can channel hop. From this human plane of reference to other 'branes. Into other metaverses. And I can, in a limited way, roll backwards and forwards along the timeline."

"Wow. You've never told us, or me, about this?"

"I've told Tomales. She can also jump to other reference planes, but she can't go backwards at all along the timeline. We agreed it was best to keep quiet about it, because otherwise we would be asked to run all kinds of errands."

I could understand this. I'd told very few Watchers about my gravity bending ability. And that was partly because I couldn't think of anything useful to do with it, except party tricks, which were not advisable.

Well, you know about LIGOs, of course?" asked Limantour.

Yes, I knew about LIGOs. LIGO stands for 'Laser Interferometer Gravitational-wave Observatory'. They comprise two enormous laser interferometers located 3000 kilometres apart. LIGO exploits the physical properties of light and of space itself to detect and understand the origins of gravitational waves.

Though the LIGO mission is to detect gravitational waves from some of the most violent and energetic processes in the Universe, the data LIGO collects may have far-reaching effects on many areas of physics including gravitation, relativity, astrophysics, cosmology, particle physics, and nuclear physics.

And, since the "O" in LIGO stands for "observatory", there are unique aspects to its function. LIGO is blind, it is not round and cannot point at a particular part of the sky, and it is rare for a single detector to make a discovery on its own.

The US had built two LIGOs, In Hanford, Washington State and in Livingston, Louisiana. Then a couple of industrial giants had built them, in Bodø, Norway and somewhere in Alaska. And not to be outdone, the Russians had one somewhere to the north of Moscow. Singly, they were of no use. It was only when paired that they could detect anything useful, using triangulation, so it was assumed that either there were secret deals done between LIGOs, or that there were even more in existence.

"But how do you know that any of this intervention will work?" I asked Limantour.

"Have you heard of Holden?" she began, looking at me closely.

"Only as an Australian car brand," I answered.

"Holden is something of a two-edged sword in all of this," replied Limantour.

"Holden has been working to seed a cross-over event between two of the 'branes."

"So has he managed to influence you and Tomales?" I asked Limantour.

"You should meet him," said Limantour. She took both of my hands.

"I'm going to take you across to Holden's metaverse," she said.

I didn't have time to protest, and was suddenly in a white room, decorated with 20th Century Earth furnishings. I knew this was Artificial Intelligence based, and I tried to think of something else to add to the room. A child's rocking horse. Now I could see its head on the other side of the sofa. And maybe some wooden toy train tracks. They were underfoot.

A deep voice spoke, "You are playing with the environment!"

I almost laughed out loud, because Holden seemed to have styled himself on 'Voice of God' television announcers.

Limantour interrupted, "Yes - you have guessed correctly. It is Holden. He wants to take human reality and then - as a sidebar - to bring back knowledge that will allow humanity to self-start its own escape route from the impending disaster."

"How can Holden know about the disaster with such certainty, or about the solution?" I asked. I realised I was speaking as if Holden wasn't there.

Holden answered, "You are correct to speculate that I am a cypher. I operate in another dimension. I can jump from your dimension to other ones and even slide along the timelines."

"With such powers, surely you can get whatever you want?" I asked.

"Precisely," answered Limantour, "Holden has the wish to become a ruler. If the Earth continues upon its current path, then there won't be anything to rule. Holden has seen that, and he knows that the sight lines forward are blocked in a few centuries, as Earth spirals ever further into dystopian decay."

I could see the plot that Holden was creating. Save the world, but then seize control of it.

Holden continued, "I have bypassed the end of the world by riding into an alternative metaverse. I have used my powers to engineer a situation which the human metaverse could comprehend. I have positioned one energy escape route for Earth. It is a series of new knowledge fragments which human minds can use to save Earth."

I was beginning to understand. Holden had created some heavy hints for humans related to scientific knowledge. It would take a discovery of a new source of energy to allow the parts of the original jigsaw fall into place. I was seeing my part in all of this.

"But what if this doesn't work?" I asked.

Holden spoke, "Think about it. Earth is on the road to destruction in any case."

"But how to get the knowledge from your alternate metaverse to the human one?" I asked.

"I have done all of the heavy lifting. By positioning knowledge and even some working examples on the alternate metaverse at roughly the right co-ordinates for a crossover event to cause them to drop into the human metaverse."

"Why can't you just bring them across?" I asked.

"It needs a gravitational distortion to allow the meta verses to cross-over. That's where you come in. You'd bring over the knowledge shards and possibly some other residual 'wash' from the other metaverse. You would do it while you diverted the Hubble Four."

I was thinking of how waves wash up on a beach. The random debris of seaweed and occasional other things that got left behind. It was how I'd imagined the start of the universe, rather than the way the scientists had said.

"Think of it," said Holden, "You'd be restarting everything! Universe 2.0"

"With a few gifts from Pandora included!" I said.

"But most importantly, Earth will have the new knowledge and the means to deploy it," said Holden.

"But only just in time," I added.

"Yes, but Earth is a survivor; you bring the pieces and Earth will know what to do with them."

"And when is this all supposed to happen?"

"We are right on the event horizon," answered Limantour, "you'll need to act almost immediately."

"But I'd need the data if I'm to create a gravity wave, let alone two?"

"I have the co-ordinates for what will be two gravity waves. One for the Hubble redirection and then a second for the crossover event. You'll need to use significant energy for the second event."

I knew there was more to this than Holden and Limantour were telling me.

"So, what else?" I asked, "Like where did you, Holden, get the knowledge?"

"I found another planet, in an alternate metaverse. It exploits magnetomics to create energy and to drive power trains."

"Magnetomics?" I queried.

"Using magnetite to create energy," explained Limantour, "I didn't know about this either."

"But I thought magnetite was found in rocks. That it was iron in origin and fairly magnetic."

Holden answered, "You are right, but that pre-empts one of the devices to make this work. I am altering humankind's understanding of magnetite to mean an additional element which can create extreme energy and can be used to build power-plants, batteries and small motors capable of powering vehicles, with no detriment to the Earth."

I gasped as I realised, "It's because you are bringing something otherworldly into the Earth's metaverse. Something that doesn't belong. It's like tilting the universe."

"Correct," said Limantour, "And that is Holden's genius; except, well, it is you that will be bringing this to the Earth."

"And how will this get to Earth?" I asked.

Holden replied, "I have already prepared packages, 'gifts' if you will, which will land close to major scientific research institutions. The plan is that several places will make simultaneous discoveries and that this will speed up the magnetite discovery program."

I said, "But this could be like da Vinci all over again; where no one will take the lead on any of his discoveries."

"The Cnidarian metaverse I selected is one with interesting properties. I am using its apex predator to spread the knowledge through knowledge shards. Although apex in its current water-bound world, it is still insubstantial on Earth. Think of it like a jellyfish; a member of the Chrysaora. A big mushroom cap, long tentacles along the centre and smaller trailing ones from the cap!" - only larger than any jellyfish known to Earth."

I could feel an inward scream of terror.

Holden continued, "They have a device called nematocysts, which can fire out the knowledge payload - in normal use these are like the barbed stings from the jellyfish. They would propagate the knowledge of magnetomics widely and can also suppress counter-thought."

I kept thinking of the unintended consequences of letting mind-controlling jellyfish with a barbed secret payload loose on Earth.

Boundary Condition

I could hear Holden's fusion rate boosting after the last few moments of interchange. He was hoping to outrun our thought processes, by speeding up his own. Much like a fly sees a human with a swatter as if in slow motion.

Holden spoke, "You are thinking about what happens if we can't control the Chrysaora? We thought of that. It is why we have already introduced DAARQ monitoring. Distributed, Artificial Intelligence, Augmented Reality and Quantum computing. These military systems will monitor for sophisticated intrusion detection and self-protection mechanisms if the Chrysaora - or anything else for that matter - get a foothold. The principle of is that of a Doomsday sensor and it is already used in my other reality. It has been running ever since the US scientists built the first two LIGOs, back in the mid 1990s. The humans have simply not realised how it can be used,"

"In intersecting two meta verses, we knew we must use similar safeguards here on Earth. All we need to do is identify if a boundary condition is crossed."

I was confused, "Boundary condition?"

Holden answered, "Yes the conditions that create triggers for other events."

I asked, "Is that what the LIGO was about? - A condition?"

Holden answered, "The LIGO is watching for a certain condition here on Earth. It can check that the condition of Earth is still viable. If it detects a problem three times in a row, it will invoke the Apex."

"The Apex?" I queried.

Holden answered, "Yes, there's a set of conditions that re-establishes a dominant strain on Earth, on an 'if all else fails' basis. And the apex will function until it has reasserted an equilibrium that ensures the continuation of Earth. Why do you think the dinosaurs became extinct? No true Apex meant that Earth's environment was overly chaotic.

Limantour smiled, "Exactly. it is supposed to be an Earth countermeasure if the planet is attacked, or a new war is creating mutually assured destruction."

"But then there would need to be the second trigger?" asked Sam.

"Correct," said Holden, "That explains the LIGO. It listens for the end of the world, so that it can be triggered as a one-time event. It listens for a pulse - like a heartbeat from Earth. When it's received, all is good, otherwise expect Kratos to exert itself and generate a new Apex.

I thought this was 'a damned if you do and damned if you don't' situation.

My options seemed to be:

A) Let the Earth sputter out;
B) Intervene but set up an Earth self-defence using LIGO which could also terminate everything.
C) The slimmest chance that Limantour's plan might work, even if it was guided by a megalomaniac.

I reviewed the options in my mind. These had not been inspiring days.

Day One, I'd met Ceres who seemed like a breath of fresh air, introducing farming to the world. Darnell, Cardinal and Bishop had exploited this situation to incorporate groups of dwellings which could be controlled and taxed.

Then on Day Two, I'd met the promoter of darkness, Lekton, who had decided to introduce warfare to the world.

Day Three, I'd met the chaotic Limantour, and observed the madness of Holden, who had the created the Intervention Plan, but with many downsides. I'd also discovered that, for some of the Watchers, it was possible to bridge across to another metaverse.

I wondered why I was not getting 'end of meeting' jittering like the first two days.

Limantour sensed this and said, "You won't get end of day phaseouts from me. Remember where we are. Santa Barbara is on the same ley line as the one we started in Point Reyes. Indeed, there are several things about Santa Barbara which are kept special.

She pointed back from the Wharf toward the town.

"The Mission is, of course, on El Camino Real, between

Mission San Buena Ventura and Mission Santa Ynez. And the roads in Santa Barbara are not always a true grid, because they have been realigned to Native American Chumash monuments. Further toward Santa Monica, the Burro Flats pictographs also include predictions of winter and summer solstices and the billenial calendar there incorporates suggestions not of Earth's end, but of Earth's reboot.

Limantour decide to throw me a rationalisation, "Farallon, you have to remember there are no abstract forces in history. Everything that brings change is natural. Climatic shifts, volcanoes, diseases, currents, winds, and the distribution of the plants and animals have shaped humanity.

But most history has been made by human choice and human endeavour. Made by individuals, acting inside their societies. This is where we come in now. To make a choice for the humans. To protect Earth by providing it with the blueprints and access to the material that will release it from the shackles of organic dependency.

I felt the walls had closed in. I would need to flex my powers to engineer this next intervention. It was as if the mad gangster Holden had made me an offer I could not refuse, via the smooth talk of Limantour.

Limantour looked me directly in the eyes. I could sense that her power must include some kind of soul search.

"You are ready?" she asked, "To do this thing? Tomorrow?"

I nodded.

"No," she said, "You have to answer me directly, with

words...Will you assist the Intervention? Tomorrow? From Paris?"

"Yes, I'll help you tomorrow," I said, "and from Paris, but you will need to tell me where."

PART THREE

Got to keep on moving

Got to keep on moving
To understand both sides of the sky
You got to keep on grooving, yeah
Good grooving
Because you got your God and so do I

Jimi Hendrix

Paris

Limantour and the others - Drake, Tomales and the still unexpected Abbott met me in Paris the next day. We were in Jardin Des Tuileries. There's a 136-pace wide ley line there which links to the Louvre. It is one of the strongest ley lines, and is on a tourist-travelled white gravelled path. A popular author once wrote about ley lines in Paris, but missed this one, focusing instead on the double pyramid inside the Louvre.

But Limantour knows about ley lines and their adaptations, including the use of crystals to bend their directions. Underneath the Arc de Triomphe de Carrousel, which is the more famous Arch's smaller sibling, there are many buried crystals which divert and split the ley lines.

"Are we here for the ley lines?" I asked.

There was a crackle and a 'voice of God' spoke out again. Holden was back.

"Yes, in a manner of speaking," he boomed, "The configuration here gives us the greatest chance that your gravity waves will be transmitted in the correct directions."

"Correct dimensions!" added Limantour.

I was clutching the single sheet of paper onto which the two sets of co-ordinates had been written by Limantour. Unlike a normal map reference, these were three dimensional. I didn't let on that these were easy for me to manipulate, nor that they lacked the precision that I was used to. Instead, I asked, "Are you certain that this is all I will need?"

To my surprise, Abbott answered, "Yes, we have cross-checked them numerous times. They lack the fine-grained precision you are used to because we didn't know exactly when you would be here. It means you'll make a slightly bigger splash with gravity than is actually needed to accomplish the task."

I sat down on the grass, cross legged.

I would do this thing.

Chaos absorption

I was aware of Limantour as I closed my eyes to set the co-ordinates. She was holding my hands.

"It is my intervention," she whispered, "I want to ensure that you are safe, and that we don't revert."

I could feel her pulse and her breathing. I was getting a similar sensation to when I'd met Ceres; a warmth and comfort which I didn't ever experience except in these last few days.

"I'm guarding you from Chaos," she explained, "If that occurs, we lose everything. Any Chaos will be absorbed into me."

Suddenly it made sense. I was diverting the Hubble Four and about to ripple gravity for the second time. If our plan was to work, we could not let original Chaos re-enter the scene. We need to know that the Earth's dwellers could continue as before and that the plan handed from Holden would unfold in its most linear sense.

I remembered that in the classical Greek world, Chaos was the origin of everything and the very first phenomena that ever existed. Limantour was my guardian across a primordial void - shielding me from the empty unfathomable space at the beginning of time, from which everything was created.

In performing our reset of the Universe, we did not want to send ourselves back to a start condition.

Limantour was whispering to me: "Just think of the Greek gods: Gaia as Mother Earth, Tartarus representing the underworld, Eros for love, Erebus for the darkness and Nyx as night."

She continued, "In the beginning, Chaos was a state of random disorder existing in primordial emptiness, and later a Cosmic Egg formed in its stomach, hatching and producing the first gods into the darkness.

"My Chaos was a space that separated and divided the Earth, where mortal humans lived; from the sky, where the gods lived.

"Remember that Zeus was the Olympian god of the sky and the thunder and was the king of all other gods and humans. He was the son of Cronus and Rhea and was unfaithful to his sister and wife, Hera.

"Zeus spawned many children including Athena for wisdom, Apollo for divinity, Artemis as the hunter, Hermes as messenger, Dionysus for wine, Heracles for bravery, Helen of Troy to protect sailors, Hephaestus for fire, Hebe for youth and Ares for war."

I realised Limantour had seated next to me and talked to me through the entire time when I was sending out the gravity waves. I wondered if anything was different.

I looked to Limantour, she was breathing heavily and I realised it looked as if she had been in combat, all the while protecting me from Chaos. She handed me something. It was five dice.

"Throw them," she asked, "Quickly."

I did, and they landed as five times sixes.

"Again," she asked.

This time all twos.

"Again!"

Five times four.

"Keep throwing them," she asked. She was now gripping my other hand tightly.

This time I threw them and one kept spinning longer than the others. A four and four fives.

Again. One dice bounced away. I had to retrieve it.

Then I threw them all again. Two, two fours, a five and a six.

Limantour's breathing slowed. She smiled and stood.

"Good," she said, "I have absorbed the interdimensional chaos and we have a normal condition here on Earth again."

Limantour had protected me from importing new chaos from the metaverse with which we had intersected. I wished she had told me about this aspect.

"What? to deter you even more?" she asked, reading my thoughts. Limantour was fully recovered.

I looked across toward Abbott.

"Has anything changed?" I asked.

He remained silent.

"Abbott?" asked Limantour. Tomales and Drake looked across to Abbott's position.

Suddenly, Abbott stood.

"I must go," he said, "To check with Holden."

The air around him glitched and he disappeared.

"I guess that means we'll need to find out for ourselves," said Tomales, "Hey, great work, you two!"

Tilted

Je commence les livres par la fin
Et j'ai le menton haut pour un rien
Mon œil qui pleure c'est à cause du vent
Mes absences c'est du sentiment

Je ne tiens pas debout
Le ciel coule sur mes mains
Je ne tiens pas debout
Le ciel coule sur
Ça ne tient pas debout
Le ciel coule sur mes mains

Ça ne tient pas debout
Sous mes pieds le ciel revient
Ils sourient rouge et me parlent gris
Je fais semblant d'avoir tout compris
Et il y a un type qui pleure dehors

Sur mon visage de la poudre d'or
Je ne tiens pas debout
Le ciel coule sur mes mains
Je ne tiens pas debout
Le ciel coule sur mes mains

Ça ne tient pas debout
Sous mes pieds le ciel revient
Nous et la man on est de sortie
Pire qu'une simple moitié

On compte à demi-demi
Pile sur un des bas-côtés
Comme des origamis
Le bras tendu pareil cassé

Tout n'est qu'épis et éclis
Ces enfants bizarres
Crachés dehors comme par hasard
Cachant l'effort dans le griffoir

Une creepy song en étendard
Qui fait

J'fais tout mon make-up
Au mercurochrome
Contre les pop-ups
Qui m'assurent le trône"

Je ne tiens pas debout
Le ciel coule sur mes mains

Je ne tiens pas debout
Le ciel coule sur

Ça ne tient pas debout

Le ciel coule sur mes mains
Je ne tiens pas debout
Sous mes pieds le ciel revient

I'll die way before Methuselah
So, I'll fight sleep with ammonia
And every morning, with eyes all red
I'll miss them for all the tears they shed

I'm actually good
Can't help it if we're tilted
I'm actually good
Can't help it if we're tilted

I miss prosthesis and mended souls
Trample over beauty while singing their thoughts
I match them with my euphoria
When they said, "Je suis plus folle que toi"

I'm actually good
Can't help it if we're tilted
I'm actually good
Can't help it if we're tilted

Nous et la man on est de sortie
Pire qu'une simple moitié on compte à demi-demi
Pile sur un des bas côtés comme des origamis
Le bras tendu paraît cassé, tout n'est qu'épis et éclis

Ces enfants bizarres
Crachés dehors comme par hasard
Cachant l'effort dans le griffoir
Et une creepy song en étendard, qui fait

I'm doing my face
With magic marker
I'm in my right place
Don't be a downer

I'm actually good
Can't help it if we're tilted
I'm actually good
Can't help it if we're tilted

- Héloïse Letissier

144

Safe Mode

Now although I haven't really mentioned it much, we Watchers are able to keep up with technology. We can tap into human communications, although we don't have too much need of other services like transportation.

We decided we'd better keep a check on NASA comms traffic to see if they had noticed the Hubble deflection or arrival of knowledge shards. Normally Drake would do this, but I noticed Tomales was on the case.

Tomales spoke, "So far I'm not convinced that they have even noticed the telescope deflection, let alone the imminent arrival of the knowledge shards. Look, would you believe it, the Hubble telescope has entered so-called safe mode. She flipped a report across to each of us.

Hubble Instruments Remain in Safe Mode, NASA Team Investigating

NASA is continuing to investigate why the instruments in the Hubble Space Telescope recently went into safe mode configuration, suspending science operations. The instruments are healthy and will remain in safe mode while the mission team continues its investigation.

Hubble's science instruments issued error codes at 1:46 a.m. EDT Oct. 23, indicating the loss of a specific synchronization message. This message provides timing information the instruments use to correctly respond to data requests and commands. The mission team reset the instruments, resuming science operations the following morning.

At 2:38 a.m. EDT, Oct. 25, the science instruments again issued error codes indicating multiple losses of synchronization messages. As a result, the science instruments autonomously entered safe mode states as programmed.

Mission team members are evaluating spacecraft data and system diagrams to better understand the synchronization issue and how to address it. They also are developing and testing procedures to collect additional data from the spacecraft. These activities are expected to take at least one week.
The rest of the spacecraft is operating as expected.

"Unbelievable," said Limantour, "The one time we need it, the telescope is being repaired."

"That timing is before I generated the gravity wave," I said.
"But look," said Tomales, "It says the rest of the telescope is still operational. Would that be enough?"

I felt the air crackle as Abbott suddenly returned.

"It's deliberate," he said, "The outage was all part of Holden's plan. The change in trajectory of the Hubble won't be detected, although it will still be able to transmit it's diagnostics and findings. It is so that what Farallon has done will not be accidentally course-corrected by the human operators."

"But what about the knowledge shards?" I asked, "I sent the gravity wave exactly as directed."

Abbott smiled, "The second gravity wave is almost irrelevant," he replied, "The reason it was necessary was to provide some apparent causality for the sudden appearance of meteor showers around the Earth. Watch tonight, there will be quite a show. The humans will retro-fit the explanation to the gravity wave."

I realised Holden was playing a deep strategy and had revealed only a part of it to us. I wondered why Abbott seemed to have gained his trust.

I started to wonder if I was becoming trapped in Holden's system. Whether he inhabited a dark room and a science incomprehensible to Earth -bound beings.

They say that there is nothing new under the sun.

Ecclesiastes, the originator of the statement complains frequently in his writings about the monotony of life. The entire passage (1:9) reads,

> *'The thing that hath been, it is that which shall be;*
> *and that which is done is that which shall be done:*
> *and there is no new thing under the sun.'*

Even Shakespeare in his Sonnet 95 was of a similar mind:

> *'If there be nothing new, but that which*
> *Hath been before, how are our brains beguil'd,*
> *Which labouring for invention bear amiss*
> *The second burthen of a former child.*
> *Oh that record could with a backward look,*
> *Even of five hundred courses of the sun,*
> *Show me your image in some antique book.'*

Of course, Holden's twist was that his sun was a different one from ours. He had now successfully created a linkage from his world to ours. A Trojan horse.

Homer's 5th Century BC tales of the Iliad and Odyssey with their myths of Helen's abduction and the Trojan Horse dug their way into the world's imagination, with influence extending from Roman generals to the poets of Shakespeare's England and modern filmmakers.

But it was the Virgil's Aeneid which described how, after a fruitless 10-year siege, the Greeks at the behest of Odysseus constructed a huge wooden horse and hid a select force of men inside, including Odysseus himself. As a war trophy, it was pulled into Troy, from whence the warriors released themselves to sack Troy.

I was anxious that we had not somehow created history repeating itself, as our unintended consequence, with Abbott lurking with black wings, able to bridge between Holden's powers and the missteps of humanity.

Not only are Homer's stories of a saturation bloodshed war, but they are of an unusually convincing one, about an army whose leaders are petty and sometimes mutinous, where disease stalks the camp and wounds are frightful and the enemy is to be admired, not merely hated.

And in which the good guys die. The Iliad glories in violence yet was written by someone who found the human lust for war stupid and bitter. Homer was deeply conflicted about warfare, a deathless poet of the human condition.

These are the centuries when humankind's core civilizations moved (with Scheppach's commercial zeal) from bronze weapons to CMC-steel ones, and from oral tales to written stories. The role of war as a dark driver of change has been unavoidable. Advances in metalworking, wheels, horsemanship, sailing, mathematics and counting, architecture and religion, have been driven by confrontation – in China, India and the Mediterranean.

I always considered the Iliad an ambiguous story. Greece is a good place to start it, both because of what will happen there and what had happened just before the Iron Age, when we get a tantalising glimpse of a better future that would be snuffed out.

That is what Holden has now promised. A better future for Earth, if only we follow his instructions. But he has kept back some of the key parts. He wants control of the

Intervention, just as he wants control of a surviving Earth.

But I knew the rest of the Iliad's story. Across the Mycenaean Greek world of Homer's heroes, a dark shadow would soon fall, scattering the people, destroying the palaces and cities, until even the ability to write was lost. The Greeks who followed, using Homer to recall their identity, blamed war for their predicament.

Humans do not know quite what happened. Watchers know that around 1000 BC a string of disasters hit the eastern Mediterranean, causing a dramatic depopulation. But that was not all. Invasions of Dorian tribes from the north came upon Greek statelets weakened by local conflicts and wiped them out.

The natural disasters—just like climate change and a series of terrible earthquakes which provoked local wars of mere survival. Who owns the village, the food, the water supply? The answer? Whoever wins the fight for it.

All of this was troubling me now. My first act had been to create the gravity waves. From my new vantage point after that effect, I could see a whole chain reaction could take place, meticulously planned by the power-hungry Holden.

Before the Mediterranean's disaster, the Bronze Age world was booming. Shipwrecks yielded much of the evidence for modern-day historians, illustrating the wealth, sophistication and a cosmopolitan culture that disappeared.

Holden would need to guard against this happening again. He needs to preserve whatever has developed over millennia and not to plunge the Earth into a total restart - prevent an Initial Program Load, if you will.

Limantour interrupted my thoughts.

"Welcome to my world of Chaos," she said.

Just hangin' around

"Don't leave me hangin'," became a stock phrase in the early 21st century. It referred to how people greeted one another with a five-finger elevated splayed handheld forward. The other person was supposed to do the same, resulting in a 'high-five.'

If the recipient didn't return the 'five', then the originator was left hanging. It was how I felt now I'd executed the gravity wave for Holden. I was left hangin'.

Drake had been monitoring comms again.

"Yes, it is being reported as the onetime event known as the meteor shower Comet 24P/Djamma This new meteor shower gets its name from the Iceland-discovered Comet Djamma and there might be a hidden pocket of dust from Comet Djamma that will collide against Earth years from now, but astronomers are still unsure about how, if at all, it could appear then."

"Good," said Limantour, "They have spotted the k-shards."

Drake continued, "They say that a single-shot meteor shower is possible because a dust stream's orbit changes shape and size. Because material enters space after cascading off a comet, the debris cloud's original path through the solar system is the same as the comet's orbit."

"I guess it is useful because once someone has broadcast

about it, then most observatories will want to track the event," said Tomales.

At least part of the plan was working.

Are we going to Bodø? Asked Limantour. "You'll see the Northern Lights and the new meteor shower!"

We checked a location and then flickered across to Bodø, Norway, located just north of the Arctic Circle. Because of atmospheric refraction, there is no true polar night in Bodø, but because of the mountains south of Bodø, the sun is not visible in parts of the municipality from early December to early January.

Drake explained, "It features among the strongest tidal currents in the world, with water speeds reaching 22 knots, is Saltstraumen, about 30 kilometres southeast of Bodø.

Drake continued, "Bodø has a long history with the Norwegian Armed Forces, and especially the Royal Norwegian Air Force (RNoAF). The Norwegian Armed Forces Joint Operational Headquarters are at Reitan, east of Bodø. Parts of NATO air forces attending the annual Cold Response exercise are stationed at Bodø Main Air Station (MAS)."

"So, it is a major military base?" I asked.

Drake continued, "In a manner of speaking. Bodø is also home to the Research and Development arm of Raven Holdings, a major international conglomerate whose subsidiary Brant Industry has created a base for the development of Artificial Intelligence. This includes supplying Bodø MAS with trial equipment."

"Scheppach would be pleased," I exclaimed, "The further computerization of warfare?"

"And conventional force too," continued Drake, "Bodø MAS is a major Norwegian military air base, housing two-thirds of Norway's F-16 fighter force and two of RNoAF's SAR Sea Kings. Bodø, competing with Ørland and Evenes, is a candidate for the Northern Air Base in the new RNoAF system."

He added, "Bodin Leir located near the air station was an RNoAF recruit school, including Norwegian Advanced Surface to Air Missile System personnel and a national response unit. The base was central during the Cold War because of its strategic location and proximity to the Soviet Union.

"It would have been vital in the build-up of NATO air and land forces to defend Norway, and thus the entire northern flank of NATO, in a war with the Warsaw Pact. It could also have been used as a forward base for American bombers to strike targets in the Soviet Union.

"Now Bodin Leir is a camp to house military personnel for The Norwegian Joint Headquarters and Bodø Main Air Station."

"Okay, " I said, "So does this mean we are sending knowledge shards to a war factory?"

"A major military base, run by NATO and next to a huge Raven Holdings Research and Development facility," answered Limantour.

"What could possibly go wrong?" I asked.

Bodø Travbane

The location that Limantour had given to us all was an oval racetrack track about 10 kilometres outside of Bodo. It was called Bodø Travbane and the small building there showed pictures of horses pulling carts around the racetrack.

I could see that Limantour had selected the spot because it was away from people and gave a sensational view of the sky, only interrupted by distant mountains.

Drake had disappeared and now returned with a selection of food items in carrier bags marked with Coop Prix Storgata.

"I thought we'd blend in better if it looked as if we were here for a picnic," he said.

Tomales was already looking through the bags, settling for a banana and bottle of fruit juice. We checked the time. An hour until darkness.

"I got this, " said Drake, pulling a small pair of binoculars from another of the bags.

"Four altogether," he elaborated, "That way, if the shards are small, we'll stand a better chance to see them."

Drake busied himself with checking the communication

system again.

"They are saying the comet tail is one of the largest they have seen. It almost passes as a meteorite collision."

Then we saw it. Suddenly. A huge green light coming down from the sky. We already had the green from the Northern Lights, and now we had this meteorite with Holden's payload about to hit the earth.

To indicate size, it now looked like a second moon, only moving much faster. It was still silent. Then we say a violent flash and the moon shaped object broke into smaller pieces, all speeding toward earth.

Then we heard the rumble, which just kept getting louder. We were experiencing the delay to the sound caused by the distances.

"Wow! Said Tomales, "some display!"

"Not just here," said Limantour, "But also in several other important locations."

Drake nodded. "Yes, the comms is picking out the NASA chatter. As predicted, the k shards are targeting other research and development facilities. Now we just need Hubble to spot the magnetite bearing asteroid and it will be 'Game On'."

"It'll be a waiting game," observed Limantour.

The others nodded.

Cargoes

Quinquireme of Nineveh from distant Ophir,
Rowing home to haven in sunny Palestine,
With a cargo of ivory,
And apes and peacocks,
Sandalwood, cedarwood, and sweet white wine.

Stately Spanish galleon coming from the Isthmus,
Dipping through the Tropics by the palm-green shores,
With a cargo of diamonds,
Emeralds, amethysts,
Topazes, and cinnamon, and gold moidores.

Dirty British coaster with a salt-caked smokestack,
Butting through the Channel in the mad March days,
With a cargo of Tyne coal,
Road-rails, pig-lead,
Firewood, ironware, and cheap tin trays.

John Masefield

She has that razor sadness that only gets worse

So far, I've only remarked obliquely about the gods. Humankind seemed to sense them all by itself, but with a remarkable consistency. Watchers knew we were not gods or super-deities. Our common special power was the avoidance of boredom over exceptionally long periods of time.

Limantour looked wistfully on: "I think we can wait here for a while. The observation systems cannot ignore this much cosmic activity. " She looked deep in thought and flipped the wheels of the roller-skates she'd arrived here on. Drake and Tomales continued to run monitoring of the science station.

I looked at Limantour, "Hey, I thought skates had four wheels?" Take our mind off the wait.

"Not these. Three large in-line wheels. Ideal for the boardwalks and rougher surfaces but still fast on tarmac," explained Limantour.

I decided I couldn't keep up with Limantour's daily change of sport activity.

"But why are you looking like that?"

"Like what? I was just thinking..."

"About...All of this?"

"Yes. I just hope Holden and Abbott haven't played us."

Drake and Tomales looked around, Drake spoke, "Yes, I don't want to be played like an ice puck."

Limantour lowered her voice, "I've sometimes wondered if we were gods, but that we didn't know it."

I smiled. I knew this conversation and where it led from talking to others.

"If we were, would we let it get to this?" I asked, "Earth's endgame being rescued by knowledge from another dimension!"

"I know, it does all seem pretty crazy," answered Limantour, "But you know that is how I survive. Chaos is everything for me."

I answered, "Yes, it is strange that out of this chaos comes so many belief systems that seem fundamentally the same."

She nodded, "I wonder that too, sometimes. After the Hebrews tipped human beliefs to monotheism. A single universal god who has a personal relationship with everyone who believes in him."

I smiled as I remembered a human movie with a comedy star who had become the recipient of global prayer emails to a single being.

Limantour added, "The gods of polytheism, in all their buzzing, boisterous confusion, were within the universe. They were subject to nature. They did not create it. The

Jewish god gave life meaning from outside, and allowed a new politic of the Covenant of a people pledging themselves to one another and to the common good. This new way of understanding would bind people together with a new intensity. "

Tomales called over, "Sadly, it would divide them with a fresh ferocity, too."

We'd all seen this, of course. The Jews spending around four hundred years in captivity in Egypt before breaking free under a leader with an Egyptian name - Moses - and trekking to the Promised Land where they ousted local tribes and settled down. But there are no Egyptian history books to show this, nor any archaeological evidence, and the Old Testament story was written down only some seven centuries later.

We had seen that the Hebrew god was not alone in his universe. El (as in Isra-el) was the father god - like the Zeus - of a divine family. His wife was Asherah, his children were the storm-god Baal, who also brought fertility, and his sister Anat. These all had a significant overlap with the Nordic god-king Zeus, his wife Erah (think ash-Erah) and thunder god Thor.

 Of course, there were plenty of stories to accompany these characters, as we see their legends handed down through word of mouth before the written word became commonplace.

The idea of 'god' as a Greek-style being walking on the Earth, speaking and intervening personally in human life faded in favour of a more transcendent, obscure and alarming presence.

This took centuries and is traced by scholars of the oldest

parts of Jewish writing in what Christians now call the Old Testament.

The Assyrian empire, with its huge capital at Nineveh, produced a series of hugely successful warrior kings, who carved out most of the Middle East as their fiefdom through a mixture of intimidation and raw terror-tactics.

Their army was by far the most professional and well-equipped of the age and their punishments for anyone who stood against them included decapitation, flaying alive, impaling and deportation all as depicted on clay tablets and memorialised with stone-slab carvings. War propaganda intended to intimidate visitors to Nineveh.

The Assyrians were eventually to meet the formidable Babylonians, led by their king Nebuchadnezzar. There were two phases to their attack. In the first, the Jewish king and ten thousand of his people were taken into captivity. But this did not finish Judah off. There was a revolt, led in part by the prophet Jeremiah.

The Babylonian army came back in 586 BC for a fearful siege of Jerusalem. After many months of being driven to starvation, the inhabitants were overrun, and the city almost destroyed.

A further twenty thousand people were taken off, not to Nineveh this time, but to Babylon. The Temple where Yahweh had resided was almost obliterated. The famous 'Babylonian exile' during which, by the waters, the captives lay down and wept, remembering Zion, had begun.

For the people of Jerusalem, led east to Babylon, it must have been an exceptional spectacle. It was one of the world's great centres, a melting-pot of Middle Eastern peoples, mingling under its enormous gates, by its dramatic stepped ziggurats, in its temples and hanging gardens. It was a glittering spectacle of blue- and yellow-glazed tiles, statues of bulls, lions and dragons, and great processional roads. Here, sensible exiles would adapt and conform.

The Jews refused. Their scribes and priests consulted written scrolls and decided that Yahweh had not been destroyed with his Temple. Instead, he followed his people like a giant shadow and was with them in their exile. He was with them, however, for only so long as they observed purity laws, which had originally been just for the priests. They must keep themselves apart from heathens.

Slowly, it strengthened into a branding exercise, with different religions putting their stamp on things. The Christians with their cross, itself a symbol of a torture instrument, and other power bases such as the Catholics trying to gain ascendancy.

"Or," said Limantour, "There's Buddhism from India in the sixth century BCE. I wonder sometimes if our state as Watchers is that a kind of Nirvana?"

She paused, saw I was looking confused, and then continued. She could do razor sadness well: "They say that space, time and nature were created naturally, have cycles, survives for a set time, then is destroyed and remade.

"Buddhists believe everything depends on everything else, and everything is interconnected. Present events are caused by past events and become the cause of future events. Buddhism is human-centred and states that existence is endless because individuals are reincarnated repeatedly, experiencing suffering throughout many lives."

Limantour was on a roll, "Only achieving liberation, or 'nirvana', can free a human being from the cycle of life, death and rebirth. Buddhism has six realms into which a soul can be reborn. From the most to least pleasant, these are:

"Heaven, the home of the gods (devas); which is a realm of enjoyment inhabited by blissful, long-lived beings. In this realm of nirvana, life is a continual round of pleasure and enjoyment, with no suffering, anxiety or unfulfilled desires; and where human beings are rewarded for many past good deeds."

She smiled and then continued, "Then there is the realm of humanity. Although humans suffer, this is considered the most fortunate state, because humans have the greatest chance of enlightenment. In this realm, passionate and perceptive human beings experience many states of mind and have the most opportunity to free themselves from the cycle of death and rebirth, and hence progress to the heavenly realm."

Limantour flicked the wheel on one of her skates, "Now we reach the realm of the Titans or angry gods (asuras); these are warlike human beings who are at the mercy of angry emotional impulses."

She looked around, "Then the realm of the Hungry Ghosts (pretas); these unhappy human beings are bound to the fringes of human existence, unable to leave because of particularly strong attachments. They are unable to satisfy their craving, symbolised by their depiction with huge bellies and tiny mouths."

She grimaced, "Then it is the Animal realm. Undesirable because animals are exploited by human beings and do not have the self-awareness to achieve liberation. This is a life of ignorant complacency and dullness, in which one does not look beyond avoiding pain and seeking comfort."

Now she looked as if it was the end of what she was describing: "The Hell realms. People here are horribly tortured in many creative ways, but not forever, only until their bad karma is worked off. This is a claustrophobic place of extreme hot or cold in which human beings cannot escape the torment of their own intense anger and hate."

"It's a conundrum," I said," deciding where Watchers are among these realms."

"No," answered Limantour, looking tormented, "Remember, we are outside of all of this. Outside of human imposed belief systems. It is through this that we can investigate what is happening to the Earth and the Universe. We Watchers come out of nothingness to enjoy a spectacle which remains quite indifferent to us."

I knew what Limantour meant, living outside of Earth's systems, but I was also aware of the actions being developed by Holden and coat-tailed by Abbott. To exploit the end of everything, via a manipulated reset.

Deviate

We'd been at the Travbane for a couple of hours by now. We didn't want it to go over five hours with us all together, because we'd exceed the maximus and be randomly re-distributed around the World.

Suddenly Tomales called out, "It's happened! They have seen the effect of the Hubble deviation. They are piecing it together but believe that the Hubble has found magnetite in that asteroid, and that it originates from Jupiter's moons."

"Now they need to realise it can only be from Ganymede," said Drake, "This is where the knowledge shards come into play."

Then, Abbott reappeared, like spectre. He was still all in black and could easily have been mistaken for a ghoul.

"It's okay, " he said, "The Chrysaora landing has worked. They have fired their nematocysts, containing the knowledge payload. There are plenty of scientist walking around now who already know how magnetomics works."

I looked at Limantour. It was as if humankind had just received an injection from another dimension. I worried it was from a dark room which humankind has steadfastly avoided.

Abbott continued, "Of course there are some details to this knowledge finding. First, it will take Earth maybe 20 years to develop the technology beyond its first iteration. However, some of the regular space missions from Earth can be repurposed. Instead of launching new satellites for better surveillance, they can be re-engineered and provide the first missions to Jupiter. It will take around eleven earth years to reach Ganymede, and the only option is for countries of Earth to work together to make this possible."

Tomales asked, "Is this Holden's plan? And did you know about it all along?"

Abbott answered, "I know it now, but I don't think I knew it before I reached Bodo. There is a net being drawn around parts of earth and it inhibits certain memories, and certain types of thought."

Limantour and Tomales looked concerned by this revelation. Tomales asked, "So an unintended consequence of the Chrysaora-based knowledge shards is some kind of neural shield?"

Abbott grimaced, "I wish I could say it was unintended, but I think Holden knew all along. He had told me he needed to focus minds on creating the space-trains to bring back magnetite. That it would require some level of control of minds to achieve this."

I could also feel a numbness affecting my memory. I wondered whether it was a side-effect from the neural shield. Alongside I could feel new ideas developing. Small motors capable of driving large craft. Batteries that required almost no recharge and yet whose charges did not deplete their capacity by more than tiny amounts. Then it flipped to other topics. Blood management in humans. Nano-engineering on the human body and a myriad of smaller ideas.

"Is this affecting you?" asked Limantour.

I nodded, "Yes, I'm being affected. Maybe I'm being infected. The nematocysts from the Chrysaora are doing something. I'm learning about new things, but I fear I am also forgetting some others. It is like there is someone probing around in the back of my consciousness."

Tomales nodded, "Me also, and I'm wondering if it can effect Watchers this much then what effect will it have on humans?"

Drake answered, "I've still been monitoring the normal airwaves. They have gone crazy."

He continued, "Raven Holdings have already announced that they will support a new blood management system to protect humankind from various predicted illnesses. They say they have been working on something for years, which is now ready to be launched. Cartridge based tropus which will both manage blood heath and which can be used to introduce nanoparticles into a body to provide healthcare services."

"And that is not all. They are calling it a Great Leap as several companies have announced new space craft which would be capable of reaching Jupiter."

"Finally, there have been some great strides in Artificial Intelligence, allowing vastly improved cybernetics to be included on the spacecraft."

I wanted to suspend disbelief that these items were occurring together, but I could feel that my critical faculties had been numbed. I could only accept what I was being told at this firehose speed.

Limantour spoke to me quietly, "We have been infected in the same way as Earth. I'm guessing that we will be further manipulated as this goes forward."

I nodded. I noticed Abbott didn't seem to be at all concerned by any of this. Then I realised Holden must have taken control of Abbott earlier.

Limantour took my hand in her right hand and Tomales with her left hand. She whispered to Drake," Stay here dear Drake, we will be back before you know it."

Drake nodded his assent.

I felt the vibration of us moving to another location, still in the northern hemisphere. We were in the United States again, but this time I could see the car tags mainly said The First State - Delaware.

We were looking out to a placid river and Limantour told me we were in Wilmington.

Limantour spoke, "It's where much of the gunpowder and then ships and other war-related goods were created. You could say that the city prospered from the surrounding wars."

"But why bring us to Delaware?" I asked. Tomales was also looking enquiringly toward Limantour.

"I've looked at the next few decades," answered Limantour, "An area - the so named New Delaware - becomes a hub for much of what happens."

"I thought the visibility was blocked. Like it marked the end of the world?" I asked.

"It was, but ever since those gravity waves, things are clearing. I think you brought about a bridging between the universe's metaverse and another one brought to us by Holden."

"I think I've been able to cross some artefacts from that other place into our universe," I corrected.

I was thinking of how a gravity wave was like a wave of water washing the shore and bringing debris with it. Except this debris was from another dimension. It only crystallised that my theory of the beginning of the universe and the Big Bang was more plausible than the 'Bong!' laden one created by human scientists.

"You remember I said I could go forward and back along the timeline?" said Limantour, "I suspect you can too, now that your Intervention has happened. If we hold hands then I can protect you. We will both be able to hop forward, or else I'll be blocked whilst I am holding on to you."

This fascinated me because I felt no different and could otherwise have taken many decades to realise, I had a new power.

300

"Okay, Limantour, you'll have to explain?"

Limantour looked at me. "I said I could go forward, that I could channel hop to another dimension?"

I nodded.

She looked at me again, "Tomales can also channel hop, and just like I can hold your hands to take you with me, so can Tomales. It is how we found out about the ways to bypass the End."

Limantour continued, "And now, I am sure that it has given you some of the Wakener powers since your Intervention. You might not know it yet, but you should be able to slide forwards and backwards along the time-line."

"But only backwards to where I was given the powers?" I asked.

"Yes, to when you Intervened. But it is why I want to hold you when we go forward the first time. To protect you and to ensure we can get back."

I understood. Although I didn't feel any different, like after a human has a flu jab, I now had extra powers.

"Shall we do this thing?" asked Limantour, she held my hands ever so gently and I felt the familiar fast-forward of a time jump.

I looked around and could see that the area was different from the original Wilmington, Delaware. There were none of the familiar automobiles, and there seemed to be new transportation provided by Pod-like structures which seemed to run on guided paths.

The cityscape was also different, with slab-sided glassy tower blocks replacing the low rise of the older city. I realised that moving forward in a developed economy was vastly different from moving forward across older times.

Limantour spoke again, "Interesting. We have now passed the old endpoint. Your Intervention must have created a new path for Earth."

"How far did we jump?" I asked.

"Oh, maybe 300 years and still at the same location. Far less than we've been jumping in olden times."

I felt the floor shift very slightly. I'd always been aware of my gravity sensitivity, which was something like a human child's sense of balance on snow-skis. I was aware that I was making a very slight balance correction, beyond the normal one I'd associate with the pull of the moon.

Limantour looked at me, "What are you thinking?" she asked.

"I don't know, I said, but it feels almost as if there are two moons, both exerting some kind of gravity onto Earth."

"Yes, there are," answered Tomales, "The second moon has been created by humankind, and is used to launch space missions outside of Earth's atmosphere."

By now it was early evening, and she pointed into the

darkening sky. "See," she said, "There is the moon and to its left in the sky is Moon 2, smaller but also in a geocentric orbit. It is much closer than the moon, but is on a similar rotation to the moon, in the Outer Van Allen belt."

But I thought that was dangerous?" I asked.

"Not if the Moon 2 is positioned in the safe slot, " answered Tomales,

"The gap between the inner and outer Van Allen belts is caused by the Very Low Frequency (VLF) waves, which scatter particles, and which result in the gain of particles to the atmosphere. Solar outbursts can pump particles into the gap, but they drain again in a matter of days, " answered Limantour.

I could see that Limantour and Tomales had this worked out.

Tomales approached me, "We have something else to do here."

"But I thought my Intervention was complete? I mean, I sent out two gravity waves?"

"It is," said Limantour., "Any of us can do this next part. Tomales has obtained the number that we need to pass on."

I was intrigued. We had somehow leapt forward into a new zone, called 'New Delaware,' and the effects of the knowledge shards appeared to be already in full swing. What could we possibly need another code for?

Tomales showed me a hand communicator. It was from

the place we had left and I wasn't sure if it would still work, realising that it would need legacy technology to communicate with anything else.

Tomales had already pressed a transmit button and I could hear a sound like a faint bell ringing.

Then I heard a click and someone answered, "Hello," said a faint female voice, "who are you?"

"You will need to hold it to your ear," explained Tomales loudly.

"Ah," said the voice, "It is vintage. Like an old telephone."

Tomales had put her device into speaker mode, and we could now all hear the distant voice more clearly.

"Hi Cindy," said the Tomales, " I know you are with Sam now. The reason we are using this old device is because it operates on frequencies that are no longer used for routine communications. It gives us an advantage in that we can talk without the risk of being monitored."

I was completely out of my depth now. I did not know what Tomales was talking about, nor to whom she was speaking

Tomales continued, "We would like to meet you to discuss what you have found in the envelope. It is from two of your distant colleagues. Like you, they have identified some anomalies with the work on Ganymede. There are some major forces in play."

Cindy asked, "So how do you know about us, and how do you know about this? Have you been following me?

Have you been listening to my conversations?"

"Only in so far as it is in your best interests," said Tomales, "We don't want to snoop, but there are some things we will need to find out."

"You also said I was in danger?" said Cindy. "How do I know the danger isn't from you?"

"You're right," said Tomales, "But the truth is we think you have discovered something, and we need to follow it up as a lead toward a bigger situation."

"But you seem to imply that we made some mistakes?" said Cindy, "Other people have been taken away from their workplaces because they found similar anomalies to me but then extrapolated the wrong conclusions."

"Possibly," said Tomales, "But we think they are probably closer to what is happening than even you are."

Cindy answered, "If you expect me to cooperate, I will want Sam to also be involved."

"That's fine," said Tomales, "We wanted you both."

"And what assurances do I have that if I talk to you further face-to-face, that it won't get me into other difficulties?" asked Cindy.

"I can't give you any guarantees," said Tomales, "But I think our approach so far has been trustworthy. As I said, we want to understand what is happening on Ganymede."

"So where are you taking this conversation?" Asked another voice, which I took to be Sam.

Tomales remained silent.

Cindy repeated the question, "Where are you taking this?"

Tomales answered, "Sam, you need to remember that this is an old communicator, and it limits the voice sensitivity unless you are very close to its microphone.

"We want to meet with you, and to show you some things about Ganymede. There is some risk, so we will need to take precautions in how we meet you."

"Just a moment," said Cindy, "I need to check with Sam about this."

"They want us to go somewhere with them," I could hear Cindy speaking to Sam, "They say they want to tell us something about what is happening and why they have taken the information about the ship."

There was a pause whilst Sam considered this.

"Okay, let's say we go with you," he said. We could hear him pick up the communicator and the sound suddenly became louder.

"Let's say we go with you; will we meet somewhere fairly close to here and how will we know you're not putting us into danger?"

Tomales answered, "All I can say is that I think you will be in less danger by working with us than if you try to continue as if nothing had happened. "

Tomales continued, "You should look for a Sven

Mallinson. He'd like you to meet you now so that we can continue this conversation."

"Okay," said Sam, "If we agree, how would we do this?"

"I can get a pod to you where you are," said Tomales, "It will be unmanned but can bring you to a safe location where we can talk."

"It really is in your best interests," Tomales added, speaking softly, "We can walk away from this now and leave you to run with this alone. Working with me and the others, you stand a better chance."

Sam spoke to Cindy, "They want us to go with them now for a meeting. They are prepared to send us a pod to take us there."

Cindy replied, "There seems to be more downside for us if we don't follow this."

Sam spoke back to the communicator, "Okay, I guess you know where we are right now. Send your pod and we will come to meet you."

"Look over to the pod bays. The pod flashing, XTZ 564 is yours. Just get in and it will bring you to us. And bring this communicator with you."

We could hear Sam walking toward the pod bays. Then the sound of climbing into a pod. The communicator had acted as a digital key and the door opened as he approached. Then we heard the pod door closing, and the pod manoeuvring. According to Tomales, they were speeding towards north New Delaware, towards the boundary zone that separated the space zone from the rest of Amerika.

"Okay," I said, "How did you know those people, and to do that? How did you set it up?"

Limantour answered, "Tomales has the ability to look forward through time. She has already visited this area. She knew that the two known as Sam and Cindy would be here and they would need an escape route to get them to a certain other location. We simply had to join the dots. It is where Pod XTZ 564 came in. Tomales had pre-programmed it to go to the next location to support Sam and Cindy. They are both being hunted by some ruthless people, but we can help them. Look, I'll show you around this area."

She gestured toward the nearby transit system, and we all clambered into a pod. I had never used this form of transit before and didn't know what to expect.

"It's okay," said Tomales, "Best to go in this rather than be seen out in the open."

The pod had blanked its windows as we started the journey. By human standards the pods travelled fast, although to a Watcher it was still very slow. Now the pod had slowed right down, and the glass cleared. We all looked out.

"It's horrible," I said, "Like some kind of Armageddon."

"Yes, and the pod has slowed right down to traverse this area," answered Limantour, "It can only mean it must run extra surveillance."

We could see industrial rubble stretching away as far as the eye could see. Some tall posts rose through the rubble, and atop them were video surveillance systems.

A few drones hovered over the landscape, which looked arid and hot.

We passed several gantries with cameras pointing inside of the pod.

"I see, they have partly slowed down to scan us incoming," said Tomales.

Razor wire had been draped around and there were several laser-triggered alert systems.

"We are most definitely heading for the Scratch," said Tomales.

Once we had flickered into this time period, our memories had been backfilled with relevant information. We all knew that the Scratch was a zone just outside of the New Delaware boundaries. It extended for around 10 km in a ragged line around most of the exit points from New Delaware. Like the ramshackle towns that had developed beyond many areas of military installation, the Scratch population centres were at the gates exiting from New Delaware.

Right at the border, there were many layers of security to stop people from getting into New Delaware. Just outside of this was the zone where many people had tried to get in but had then stopped and instead tried to make an opportunistic living around the borders.

The area had become known as the Scratch. It was thought the name came from the phrase 'To scratch a living'. This close to the border was a rough area and didn't have many people passing through unless they were bound in or out of New Delaware. Nowadays it was a label on a ramshackle microclimate filled with rough-necks looking for ways to make a turn on what was happening inside the space zone.

Uniformed new Delaware residents were usually safe because everyone knew they had full identity tracking and anyone interfering with them would be rapidly traceable.

The New Delaware Security Force would swoop into the Scratch at the first sign of trouble affecting New Delaware residents and bring the full force of their law to bear.

This almost 'take no prisoners' policy meant that the Scratch residents would keep a distance from New Delaware residents.

The same didn't apply to those from the rest of Amerika who passed through the Scratch. NDSF security forces also paid little attention to anything affecting Amerikan residents entering the Scratch. The assumption was that these people would only go forward knowingly.

It suited the New Delaware Security Forces to have this buffer zone because it acted as insulation for the Space Zone.

I noticed the pod had progressively dimmed its windows during this latter part of the journey. This has happened faster than the change in daylight, but I knew that the pod had also put up its protective shields across its windows.

I could also feel a very slight vibration from the pod on the last part of the journey. The pods were self-levelling but on extreme terrain it was possible to sense that there was an un-made route being used.

"We've left the main transit route," said Tomales.

"Yes, yes, I know," said Limantour, "We are going deeper into the Scratch."

Suddenly there was a click and the door on Tomales' side of the pod cleared. A light appeared, and then the door swung overhead as the pod settled into a docking bay.

"Amber conditions," said the pod.

"Pod explain," asked Limantour.

"Amber condition, uncharted territory, zone alert for outside of New Delaware. Repeat. Uncharted territory, zone alert for outside of New Delaware."

"Okay," said Tomales, "I'll go first."

She climbed past Limantour and out of the pod. We were all suddenly in a clean and modern complex; not at all like I was expecting.

Inject

"Okay, " I said, "I need some explanations!"

Limantour spoke, "It's like we discussed back at Point Reyes. Earth was reaching its endpoint. We were helpless to watch. Then Tomales went forward and used her ability to hop into another dimension. She met Holden, and he showed her that there was a way forward. She brought back the information, and we set up the Intervention.

Tomales added, "Holden's plan, supported by Abbott, had two parts; to show a discovery of magnetite from Hubble Four and then to bring in knowledge useful to Earth to exploit it. Holden retrieved the knowledge shards from another dimension and then your gravity wave caused them to wash up in sight of Earth."

I nodded. So far, I had understood.

"But how has the knowledge spread so fast and how is it I already know some much about this time?"

Limantour continued, "I can't say exactly how the knowledge of magnetomics spread so fast throughout the scientific community. Holden told us it was using nematocysts from Chrysaora - jellyfish stings injecting knowledge to the humans."

Tomales added, "Yes, Holden had brought these knowledge artefacts from another dimension. Their use on Earth is unpredictable. I didn't know that it would affect us as Watchers, for example."

She added, "Your other knowledge gain does not differ from when you have done forward jumps before. It's just that there is so much more that has occurred this time. The Great Leap they are calling it."

Limantour continued, "Now we are in the newly created and saved version of the Earth. Remember, we have jumped forward three hundred years. Earth now understands magnetite and has created a mining programme which runs between Earth and Ganymede, a moon of Jupiter. They are using partially machine-based robots - androids if you will - to pilot, operate and mine the distant moon. There has also been a technology shift between Earth and Ganymede, which is now considerably ahead of Earth."

"But what about Holden?" I asked, "He was trying to take over and run things?"

"It is a partial success," answered Tomales, "Earth has formed three super-powers which together operate an Earth Council. They are America, Eurussia and Sino-Nihon."

Tomales continued, "On Ganymede and its Earthside infrastructure support, there are zones run by different closed communities. The Eurussian zone, the Amerikan Mafia zone, and an area operated by a mix of Chinese and Japanese called Sino-Nihon and run by the Japanese Yakuza. We are currently close to the Space Zone of Amerika."

"But that suggests that gangsters run the entire mining operation?" I asked.

Limantour answered, "Mining, Manufacturing and Infrastructure augmented by robotics and nano-systems. Here, Earthside, the entire New Delaware facility is run by Torus industries, once known as Raven. They had seen through the acceleration of the space program to support Ganymede. They were a consolidation of several other companies, including Biotree, which had developed much of the nanotechnology prevalent on Ganymede and Earthside.

Tomales added, "There are two other equivalent huge corporations operating in other parts of the world. AlfaCorporatsiya (AlfaCorp) for Eurussia and Kǎxīmǔ gōngyè (Cassim Gongje) for Sino-Nihon."

"Is this what they call an Unintended Consequence from my Intervention?" I asked.

"Kinda," said Limantour, "But remember that there wouldn't even be an Earth without our intervention."

I also realised I was now too close to this unfolding situation. As a Watcher I should remain detached - on the outside - but I felt as if I was inside of these events.

Arbitrary

"We must go!" said Limantour, suddenly," We'll exceed the maximus here in this version of Earth."

I looked at Tomales, who was nodding, "Yes, we need to get back to our familiar unspooling of events - we are in a metaverse which could be unstable. If we get arbitrarily shifted to new locations, it could now be as far away as Moon 2, or even Ganymede"

Tomales continued, "The good news is that you'll keep the knowledge of this outcome, if we go back to the time of your Intervention, Farallon. And when we go forward under normal conditions, it should lead to this future. One we are now prepared for."

With that, Tomales took my hand and that of Limantour and we jumped back to the time and place just after the Intervention. We were back at the Travbane Racetrack, in Bodø, Norway. Drake looked over.

"You've only been gone for a matter of minutes!" he said.

"It's okay," said Limantour, "We've reset our time here, we'll be good for a complete maximus again. Another 5 hours together here in Bodø. But Drake you'll need to leave soon."

"Only after you tell me what you discovered," he said.

"So, we've just visited a new version of Earth's timeline," I said, "Where our Intervention has worked."

Tomales continued, "Yes, and that is because of your gravity wave, the second one, which washed the results of Holden's preparations across from another metaverse into our one. Earth has gained the science to make small powerful motors, tiny, long-lasting batteries from magnetite, advanced healthcare via the Tropus and nano-engineering and an apparent uplift in robotics knowledge."

"It's a lot, but is that everything?" I asked.

"No, I don't think so," said Tomales, "I've travelled further along that timeline that we were on. The Intervention transferred some mind manipulations as well, and some powerful new weapons. It seems that Holden's plan for control of Earth has also been to block part of it. A large part of the Southern hemisphere will be partially erased from Earthside memory. Some entire countries will go. An exclusion zone will be created and at the boundary, some high-powered weapons installed to prevent transgressions."

"Has this happened yet?" asked Drake, looking like he was keeping up with the conversation.

"No, but it happens quickly," answered Tomales, "First the tropus and a new cartridge delivery system for all humans to use. These cartridge systems become a proprietary money spinner for Brant, then Raven and finally Torus. Everyone on earth needs the tropus and it must be renewed every month. Think of the income stream."

Drake added, "Yes, the comms I have been monitoring are already talking about them. Apparently, some of the major Research & Development for the programme has been run from here in Bodo."

Limantour added, "Once the cartridge system is in place, it can also be used as a delivery system for the nano-engineered machines which become responsible for certain health care maintenance. And of course, there are degraded versions too - below standard clones that can be used as cheap substitutes for the real thing."

"You knew all of this?" I asked Tomales, "Yet you still let me go?"

Tomales replied, "No, actually I could see none of this until you Intervened and broke through the blockage which was preventing us all from looking into the future. You can see it all too now that you have moved from Watcher to a more proactive status."

"But what about Abbott?" I asked, "Would he have known about any of this?"

"Probably, but he chose not to share it," said Limantour, "I guess he wanted to stay on the right side of Holden."

On cue, Abbott reappeared. It was as if I only needed to say his name and he would appear.

Swirling chaotic waters

Abbott spoke, "You have all played your part now. Holden is pleased and may seek you for other roles later. You see what this has done?"

I answered Abbot, "No, beyond bringing the new attributes across from the other metaverse, I can't see anything else."

"Well, you have answered one of the 13-billion-year-old questions. That of creation. Most religious systems have positioned this. As an example, the Egyptians had several creation myths."

Abbott continued, "All the Egyptian ones begin with the swirling chaotic waters, called Nun. Amun-Ra, the supreme Sun god, willed himself into being, and then created a hill. Amun-Ra possessed an all-seeing eye. He spat out a son, Shu, god of the air, and then a daughter, Tefnut, goddess of moisture.

"These two gods were charged with the task of creating order out of chaos. Shu and Tefnut produced Geb, the Earth with all its natural landforms and life, excluding humans, and Nut, the sky and celestial bodies.

"Geb lifted Nut above him and gradually the world came into order, but Shu and Tefnut became lost in the remaining darkness.

"Amun-Ra removed his all-seeing eye and sent it to search for them. When Shu and Tefnut returned, thanks to the eye, Amun-Ra wept with joy.

"Where the tears struck the Earth, humans began to form."

"But there is only good news from this story?" I asked, "No Heaven and Hell?"

"That's right," said Abbott, "You have to look further to find the darkness."

Like in the Quran?" asked Tomales.

Limantour answered, "Even the light and darkness of night and day came together with good will."

Abbot nodded, "Yes, in the Islamic religion from the Middle East in the seventh century CE, the Quran states Allah created the Sun, the Moon and the planets, including the Earth. Allah also created the night and the day.

"Originally, the heavens and the Earth were joined as one unit, before they were ripped apart. Following a big explosion, Allah 'turned to the sky, and it had been as smoke.'

"He said to the sky and the Earth: 'Come together, willingly or unwillingly.' They said: 'We come together in willing obedience.' As a result, the elements and what was to become the planets and stars cooled, came together, and formed shapes, following the natural laws that Allah established in the Universe.

"The first human beings, Adam and his wife Eve (Hawwah), appear in the Quran, which states that they were created from clay, and were brought to life by the blowing of the soul into their bodies.

"It's the same all over the planet," said Limantour.

"The Sumerian Mesopotamian creation myth involves a struggle of the younger god Marduk, against the chaotic water gods; the male Apsu, representing fresh water and the female Tiamat, representing salt water. A cycle of violence then erupts, resulting in Marduk, the aggressive upstart, leading the gods to a final decisive battle against Tiamat.

"Marduk defeats Tiamat and splits her body, creating heaven and Earth. The moral of this Mesopotamian myth is that the human being is an insignificant part of a much larger struggle within the natural world; and that there is a monumental struggle between order and disorder."

"Order and disorder. Now we are getting somewhere," I said.

Abbott continued, "Then there was the Zoroastrian religion of Ancient Persia, in which the world was created by the deity, Ahura Mazda. The great mountain Alburz grew for 800 years until it touched the sky. From that point, rain fell, forming the Vourukasha sea and two great rivers.

"The first animal to be created, the white bull, lived on the bank of the river Veh Rod. However, the evil spirit Angra Mainyu, killed it. Its seed was carried to the Moon and purified, creating many animals and plants.

"Across the river lived the first man, Gayomard. He was as bright as the Sun. Angra Mainyu also killed him. But the Sun purified his seed for forty years, which then grew into Masha and Mashyanag, the first human mortals. There's more cannibalism in this story, but do you see, Farallon, conflict leads to a stable outcome.?"

Tomales asked, "I'm not sure what you are trying to prove here? Do you imply new conflict before the Earth settles again?"

Abbot answered, "We can look at Central African, Maori, Viking, Japanese and Aztec origins and they all show humans born out of a suffering or conflict. As an example, the giant pale god, Mbombo, a white-coloured figure who had been ill for millions of years in Africa but then brought the Sun the Moon, stars and nine animals, from which all other descend; Or the Vikings Odin,Vili and Ve, who killed the primordial frost giant Ymir and made his body into parts of the Universe; Japan with Izanagi and Izanami gliding down the rainbow-striped floating bridge of heaven, stirring the water but Izanami dying after giving birth to the fire god Ho-musubi. In Aztec beliefs, Huitzilopochtli, the God of War and the Sun, sprang from his mother, Coatlicue, fully grown and armoured. He attacked Coyolxauhqui, killing her with the aid of a fire serpent and so appeared the Moon."

" I could go on but the pattern is a coda of illness and death followed by the recognition of the Universe and the dawn of humanity. We all know that the Universe goes back much further than these humans, and so their beliefs are surrounded by what they know. The main symphony finishes before they are allowed their embellishments."

I thought Abbott was speaking some sense but wondered about a punchline.

As if on cue, Abbott continued, "So, to make an omelette, you have to break a few eggs."

I groaned. Next, he'd be quoting that revolution is not a dinner party. But deep down I knew he signalled that the restart of Earth might be surrounded by some form of conflict. Conflict introduced by Holden's wish to attain control.

I looked over to Limantour and Tamales. Their expressions were like mine. I think they had both been duped and I could see they were trying to think of ways to remedy the situation. But of course, the world was entirely different now that a slice of another reality had entered.

Complexity and entropy

I began to realise that by crossing some artefacts from an alternate reality into our own metaverse, we had just thrown another layer of complexity across the Earth.

All societies need to create some meaning for the natural world around them and their existence. Without scientific evidence, they develop mythical creation stories traditionally passed down orally from one generation to another. An earthquake, storm, volcanic eruption, famine or pandemic pestilence, was often believed to be the sign of an angry god.

These creation stories were powerful and acted as the glue which embedded the people with a common purpose and within their respective cultures.

Then, with the passing of time, we see the added thresholds of complexity.

This may seem at odds with the second law of thermodynamics in physics, which states that the total entropy or disorder in the Universe will increase over space and time.

If that's true, and the Universe is getting constantly more disordered, then why do we see ordered things like galaxies, stars, planets and complex life, like humans? In physics there is a difference between complexity and entropy.

Entropy refers to the number of ways matter and energy can be re-arranged within a relatively stable system, in this case, the Universe. It is possible for something to grow in complexity and yet become less disordered at the same time.

Typically, as entropy increases, disorder also increases, reaching a peak and then decreasing again. But within this large disorder are pockets of ordered complexity.

This is where we will find galaxies, stars, planets and life in all of its complexity and diversity.

In the beginning, the Universe had little entropy and was very simple and homogenous. As the Universe expanded, it became more complex despite the greater disordered entropy.

At the end of its existence, the Universe will become simple again, as all the galaxies and stars exhaust matter and energy, and disordered entropy becomes extremely high during this process of the Universe contracting and "dying."

But the wash from the alternative metaverse means there will still be a long time for this to unspool. It felt like we had just added some new glitter into the equation.

Navel

In Hinduism, which also originated in India sometime between 2,300 and 1,500 BCE, it is believed that the world is created many times, repeatedly, and not just once and for all.

Hinduism also states that this Universe is one of many multiple universes in existence, with other forms of life abounding and existing on different planes.

In the Hindu religion, the creation story from the Vishnu Purana states that Vishnu, while laying on an ocean of milk on top of the serpent Sesha, sprung a lotus from his navel that contained the god Brahma.

Having been sprung from Vishnu's navel, Brahma created all living things including humans, as well as the Sun, Moon, planets and several other gods and demigods.

Now that's a far-sighted idea and seems to organise some of the thinking. Multiple planes of existence. The 'branes and their metaverses. Humans, animals, and other entities. Maybe even Watchers?

PART FOUR

Versatility at the Crossroads

I could tell we were at a crossroads. This is where adversity favours the versatile.

I thought back through the times I had experienced. From two million years ago, as tree-living African hominids attempted to live on two feet after cold, dry weather attacked their forests.

The open grasslands that resulted made it imperative to run and hunt and see into the distance, which resulted in *Homo erectus*, an important early version of humanity, with a brain around two-thirds the size of a 21st Century brain. The versatile walk tall.

Homo erectus, ranged far out of Africa and evolved first into the bigger-brained *Homo heidelbergensis*—people who were hunting and making axes in England half a million years ago, and had a brain not so much smaller than ours, around 1,200 grams compared with our 1,500.

This picture of human development is a brutal simplification. Scientists name and slot them into neat divisions, and assemble evolutionary trees, but the truth was messier.

What most needs to be grasped is that modern humans were not just a single super-bright, planet-conquering ape, who leapt as if by magic from an earlier world belonging to dim ape-men.

Those earlier species, including the famous Neanderthals, and in Asia the 'Denisovans' also survived dramatic changes in climate and pushed into new territories as pioneers, equipped with cutting- and killing-tools.

They decorated themselves, had a language, and interbred with the newcomers, *Homo sapiens*.

Humankind, as we know them in the 21st Century, had arrived.

Peer inside

This is how it works
You're young until you're not
You love until you don't
You try until you can't
You laugh until you cry
You cry until you laugh
And everyone must breathe
Until their dying breath

No, this is how it works
You peer inside yourself
You take the things you like
And try to love the things you took
And then you take that love you made
And stick it into some
Someone else's heart
Pumping someone else's blood

And walking arm in arm
You hope it don't get harmed
But even if it does
You'll just do it all again

Regina Spektor

We'll always have Paris

Now we are in Casablanca, Morocco, sitting in an old courtyard-style mansion built against the walls of the Old Medina. The seafood restaurant and piano bar is filled with architectural and decorative details: curved arches, a sculpted bar, balconies, balustrades and beaded and stencilled brass lighting and plants that cast luminous shadows on white walls.

This was Limantour's idea again, and I knew that she'd chosen somewhere that resembled a film set. I sensed she was about to make another suggestion.

The air rippled, and Tomales appeared.

I wanted to say, "Of all the gin joints in all the towns in all the world, she walks into mine," but Tomales hadn't walked in. She'd rematerialized.

Instead, a "Hi Tomales!" but Tomales replied, "We'll always have Paris!" so I guess there is some justice in this human world.

"Will Drake join us?" I asked.

"Any moment," answered Tomales, and then, with a quiet crack, Drake appeared and was almost instantly sipping a cocktail. "Cheers, " he said, then looked at me, he raised his glass, "Here's looking at you, kid."

I worked out that they must have had a pre-meeting about whatever we were to discuss.

Limantour looked at me.

"Farallon," she said. Limantour speaking my name again could only mean that something intense was to follow.

"Farallon. This is how it works. You've successfully created the Intervention. Both parts. The seen part, and an equally important unseen aspect."

She paused for effect, then continued, "Holden has engineered for unseen artifacts from another dimension to accompany the ones helpful to humankind. The knowledge shards also contained something else, which we are sure has something to do with his attempt to win power."

I sensed the dark-side dragon slither across my body, like the time we were on the beach at Point Reyes.

She continued, "Abbott will assist him and we think there are others, too. Tomales has gone forward along this timeline and seen that it could be as long as three hundred years before this plays out."

Limantour spoke, "And if you say to me tomorrow, 'Oh what fun it all would be.' Then what's to stop us, Farallon?"

I could see what she was doing.

Tomales spoke. "Yes, and there seem to be a couple of future outcomes as well. Heavily influenced by what is and what should never be."

I spoke, "But the wind won't blow and we really shouldn't go. It only goes to show, we must take our time."

Drake added, "Yes. I have checked communications and Earth gets Trigax weapons, blood management, nano-engineering and new cyber robotics. It learns to mine magnetite and how to craft new miniature power units to save the earth from a climate catastrophe. There are wars along the way, including a terrible Klima War, which decimates the Earthside population. Shortly after this, Holden gets a foothold towards power."

"What do you want?" I asked.

They looked at one another, then Limantour spoke. "We think that we four can divert this from a catastrophe."

"How?" I asked, fearing the worst.

"We are now all Wakeners, still with our special powers. We have stepped beyond being Watchers and are now in the proactive world of Wakeners. It means we can move along the timeline. It also means that we can adopt a human persona."

"I'm confused," I said, "How can this be?"

"In the new world with its intersect with an alternative reality, we can adopt Personas from another's Presence. Here is our proposal."

My mind was whirring. This was becoming too much to comprehend.

Limantour continued, "Drake has been running an

analysis of the players in the next section of this. People who directly or indirectly interact with Holden. He has made a list of Candidates."

"Candidates?" I asked.

"Yes, people or other points at which we can insert Wakeners to take over the actions of the original actors."

I wasn't sure what I'd just heard. Did it mean that as a Wakener, I could slip into someone else's personality? Become them?

"It is supposed to be a technology for use with cybernetics," explained Tomales, "It is not even due to be discovered for another 60 years, but because we went forward 300 years and have retrieved the residual knowledge, we know about things that are yet to be. However, this cybernetic transformation can also work with Wakeners, who can add themselves to anyone, not just cyborgs.

Limantour added an explanation, "Wakeners know to exploit the darkness when humans sleep. In 5 hours of their darkness, we can exchange considerable memories with them. We need five nights to completely exchange everything."

"What does it mean?" I asked, still confused.

"It means, after five nights tuned into a sleeping human, we will have enough power to completely inhabit them. They will still have everything they had previously, including free will, but we will examine everything from their perspective and in extreme situations give them insights from our vastly more complete knowledge."

"But that sounds like the jewel wasp," I said, "the one that paralyses cockroaches and then enslaves them to support the wasp's own life cycle?"

"Not at all," said Limantour, "This is benign - we are using the host human as a means to an end, sure enough, but only in the best interests of humanity. They won't even be aware that they are being supported."

I realised that since we had washed the alternative metaverse fragments into Earth, things had taken an altogether unpredictable turn.

I asked, "You say 'we'. Who exactly do you mean?"

Limantour answered, "All of us. Tomales, Drake, me and you."

"We've worked it out. Here," Limantour showed me a fragment of paper.

I read it:

- *Farallon becomes Scrive Mallinson.* An experienced ex-military freelancer based in London
- *Limantour becomes Chantel.* Madcap adventure-seeker London socialite.
- *Tomales becomes Charlie.* Gambler and mercenary based in New York.
- *Drake becomes Nathan.* Security specialist in Bodo, Norway

There was not much to go on.

"How can this possibly work?" I asked.

Limantour answered, "Don't you see? You have turned from a Watcher into a Wakener. You'll be able to act now, and not only that, but you'll also carry the knowledge of the future from your trip forward with Tomales and I. Your mind has been loaded with the next 300 years of developments."

"But what is the downside?" I asked.

"1 You'll be linked into a specific human. They operate slowly, so you'll have to get used to that, although you can help them at our normal speed of thought and knowledge."

"2 You'll need to get used to travel at a human rate. No hops to another position on the earth."

"3 You will need to take care of your human. They are not immortal like us, and our persistence is interrupted if they are killed. You'll still be able to get back to the Wakener dimension though."

"4 You'll need to resist some of the human emotional traits. It can be like a massive sensory overload when you start."

"5 All of your Wakener back-channels will go, although you will still have a powerful affinity for each of us and we will gravitate toward one another."

Limantour paused. I could understand each of the points. To be honest, it felt like a useful break from the billions of years acting as a Watcher. I was ready for this, even if it tied one hand behind my back.

Tomales reminded me, "You will be on a 300-year journey, but you will need to experience all of it. It won't be like now when you can jump forward to avoid some tedium."

Limantour added, "Yes, but nearly the entire time will be packed with human-level adventure and decisions. It won't be dull!"

"It is tonight," said Drake, "When the humans are sleeping, we will enter their minds. Our Personas will interact with their Presence. They will awaken, feel little different but have all the augmented knowledge and speed that we bring to them."

"Do we have to come to a special laboratory?" I asked, "Like some kind of movie scene?"

"No, " said Drake, "You'll be linked by precognition. A perceptual ability that allows the acquisition of non-inferential information arising from a future point in spacetime. As Watchers we have been bestowed with the Multiphasic Model of Precognition (MMPC). In the physics domain, this addresses the question of retro-causation and how it is possible for information to traverse from one spacetime point to another."

I'd temporarily forgotten that Drake was a geek.

He continued, "Then, the second method is from the neuroscience domain, which addresses the acquisition and interpretation of retro causal signals across three stages:

"(a) perception of signals from an information carrier, based on psychophysical variability in a putative signal transducer.

"(b) cortical processing of the signals, mediated by a

cortical hyper-associative mechanism; and

"(c) cognition, mediated by normal cognitive processes, leading to a response based on retro causal information."

To my surprise, my brain was keeping up with this, like it had received an upgrade during its 300-year-round trip.

Drake continued, "If I put this simply, we just need to target the right individuals and start the transfer, which will end after around five human sleep periods."

In five days and nights, I would be completely operating from within my target. In this case, someone named Scrive Mallinson.

"How will it work?" I asked, "...How do we start this?"

"Oh, It's already started," said Limantour, smiling.

Scrive

So now I was Scrive. I clicked a new cartridge into place in my forearm and felt the cold rush snake from my arm to burst somewhere inside my head. These were entirely new sensations to me.

Next, I checked the small plexi-inspection window briefly and could see my blood already changing from a bright red back to orange, and I knew that within another twenty minutes it would again be the safe yellow colour.

I seemed to know that red blood spelled danger, and I had been careless to let Scrive's system deplete its supply of the tropus for so long.

I could feel the pulse bubbling on the left side of my head above the eye-line. I knew this was my body regaining its equilibrium. I squeezed both hands into a fist shape the way I, Scrive, had been taught and used my two middle fingers to massage the fleshy areas below my thumbs while the system adjusted.

This cartridge was having a vastly stronger effect on my sense of being, compared with the transfer of my Farallon Persona into Scrive's Presence. I could know that I, Farallon, was now operating from within Scrive. That I, Farallon, could bestow great knowledge and powers onto Scrive.

Another five minutes, and I was walking across Chelsea Bridge to the Tube station. I was operating on Scrive's internal logic, living as a human. Scrive lived less than ten minutes on foot from the nearest transit stop, and his ride to today's meeting was around fifteen minutes. I could feel the cartridge working, and a relaxed acceptance of the day's tasks was already returning.

Scrive looked briefly toward the sky. A jagged spark had flicked across. Now gentle vapour trails were crawling behind what had been a brief tear shooting along the path of the River Thames.

Others walked at a similar pace towards the station, although Scrive ducked to the right into a quieter street that also cut a corner and missed some traffic crossings.

I glanced as I prepared to cross the diagonal into the station and glimpsed someone I recognised.

She had a petite almost boyish build, dressed in black, dark hair in a black band. Scrive had noticed her for three days now, at the same spot, the same pace, and the same appearance. I knew she would look up and I'd see the small tattoo by her left eye. At least I assumed it was a tattoo and not consistently applied daily make-up. As she passed, I thought I could hear her gently humming a tune. Maybe from a streamer, but I couldn't see any signs of her wearing one. I wondered if this was another Watcher or Wakener.

I descended into the TfL transit. My new cartridge meant I had a good range on my transceiver and could access the transport system without overtly waving my arm over the sensor. It would take some time to get used to these human forms of transportation.

Most travellers referred to the sensors as 'oysters' although this was a reference to a long-defunct technology, much as the Tube itself was merely a reference to the shape of the original tunnels that formed the original wheel-based transport system.

Scrive used the moving floor system to get to the so-called 'high-speed' transit level and stood for a moment waiting for the next transit pod. I clipped myself into a TPOD seat and punched in my destination. The system was pretty fool proof. My cartridge provided the principal co-ordinates for routine travel, and a short, personalised menu of options had appeared on the screen.

Of course, I could go to other points within my regular routes or pre-authorise other destinations in advance, from the HomeLink system. Today was ordinary, though, or at least that was what I needed to suggest, despite what had happened with the transfer of Persona into Scrive.

Chantal

Chantal was meeting Janie at Canada Water. It was across the River from their planned rendezvous with Lars but would give them a chance to plan how they would handle the session before they met him. It was also a five-minute trip to their planned meeting place so they could be 'in position' quickly.

Chantal arrived in a polka dot bodysuit. Three different colours; shorts, top and jacket. There was little point in being a mistress of chaos if Limantour could not mix it up when cloaked in a human body.

"I took the ears off," Chantal said as she hugged Janie. "And I brought a black topcoat as well in case we need to wear disguises." She pointed into a bag where a black compressed micro-fabric nestled amongst some fashion headgear.

Janie's mood briefly lifted at Chantal's appearance. They'd party together sometimes, and Chantal would usually go for something extravagant. Janie's appearance was more conservative, and today she was wearing a dark business suit.

"Thanks for agreeing to cover this", she said to Chantal. "I'm a little bit worried that this is all getting rather weird."

"I wouldn't miss this for anything. You're about to meet a dishy Swede or something," replied Chantal, with Limantour relishing the new freedom of expression she found as a human.

"He's Norwegian, or says he is," replied Janie. "And it is something to do with the firm and the way they are operating."

"What do you want me to do?" asked Chantal, "Hold your hand or what?"

"The main thing is to cover the situation and to follow my lead if I say anything. I may want him to think that we've spread the word about this, so there's no point in trying to keep it to just the two of us."

"Does he look as if he would be violent?" asked Chantal, making like a karate move.

"I don't think so, but put it this way, I think he could outrun us if it came to a race."

"Okay, so we'll listen to what he has to say and what he asks. I suggest we move from the first location to somewhere else to keep him on his toes, too. If we take him to the end of the walkway, we can grab a taxi and then go almost anywhere".

"Good plan. I suggest we go to Westminster. It's full of police and security so we can get out there and if needed we could easily attract attention."

Janie looked at Chantal.

"I don't think you'd have any trouble doing that,

anyway".

They giggled briefly. Limantour enjoyed the sensation.

"Okay, that way, he can tell us his story while we are in the cab. It will only take a few minutes and put him under pressure to get to the point."

They nodded and walked the short distance to the transit point to Canary Wharf. They would be at Smollensky's on time.

Charlie

Charlie's previous history had a few dubious moments when she was raiding high roller casinos by reprogramming the payout software. It had to be undetectable to the house, and Charlie had created a small device to fire the software changes into the system. Tomales secretly approved of this method and the gains that Charlie had made.

Scrive knew about Charlie in Vegas. He helped her get out of trouble when she was in danger of being caught. He'd shown her how to mess up the electronic trail, which had previously led back to her. Now I was linked, already as Scrive, to Charlie as Limantour I this strange layout of the known world.

Charlie spoke, "I took the Casino injector and produced a sub-scale version. It's a lot easier now than when I built the original because of the improved scanner resolutions available. I could use laser light to improve the sensitivity and then fire the new software into the bots, in much the same way that the old version fired software into the Casino devices."

"NSA hadn't seen anything like this before, and they've asked me to show them a breakdown of how it works. I'm heading for CERN to show them in detail. It is my comeuppance for the damage I caused when I was testing the bots I'd been building, and they kept failing and destroying one another.

She pointed to a small titanium box with an elaborate digital lock. "There," she said. "the Nano injector is in the box. It's loaded with 'nanoreductives'. They are my attempt to rebrand the dud nanobots which self-destruct one another!"

"Its' not a bug, it's a feature," smiled Scrive. He looked quizzically at the lock.

"Birthday?" he asked.

"Yes", replied Charlie, "Plus 666 - I wanted to add an edge to this devilish device."

Scrive knew the box would destroy its contents if the wrong number was entered. But knowing Charlie, there would be a twist.

"Yes - birthday will open it but destroy the loader logic. Birthday plus 666 will open it and keep the device intact"

"What about anything else?"

"Three goes and thar she blows," smiled Charlie. "I thought I'd give a pirate hacker a sporting chance."

"Are you getting paid?"

"Yes and no," came the reply. "I should be, but I'm sort of paying off the cost of the destruction I created at Biotree with my hack. But now it's your turn. What are you working with at the moment? I assume it is still with Biotree?"

"Direct at the moment. And big. The biggest probably.

Corny as it may sound if I tell you then you might be in some sort of danger.

"Yeah, right," responded Charlie," As if that would stop me, but first, another refill of the wine?"

Scrive reached across, picked up the wine and gently poured.

"You know what? You might want to help with this. I can split some fees and it is worth it to me to have someone alongside who isn't known to the client. But wait until you hear before you answer."

"I'm intrigued already," said Charlie, ...and fees would be beneficial at the moment."

"Okay, so I've been asked to find out the origin of the cloned Chinese nanobots. There's a leak somewhere and the impact of the copies has collapsed the Biotree share price. I'm supposed to track the leak and pass the information back. Sounds simple, but we are talking about government level conspiracies. Hence the potential danger."

"Wow. This assignment sounds great. I assume the fees are in keeping with the magnitude of the challenge?"

"You know something; I think I could have named any price. I've gone high. The thing that worries me is that they seem almost too happy to accept my terms. It makes me concerned they won't honour the agreement."

"What can you do?"

"I already did it", said Scrive. I negotiated the money in two stages. "all of it now and all of it again when I complete. If I don't, they will come for me and take half back. I don't want that to happen."

"You've already been paid and can get the same again?" asked Charlie, somewhat incredulous.

"Plus, expenses...I said they didn't seem to care about the amount. And don't worry - it's a lot. Come in with me, and I'll split the second amount. You'll be my safeguard that we get paid based upon our successful outcome."

"How much?" asked Charlie.

Scrive told her. It was a huge sum.

Charlie nodded and started to feel like a big prize lottery winner.

Nathan

Nathan was at home with Sheri in Bodø. He looked out of the window, across snow toward some distant hills. Beyond them were mountains. He had grown to love the Norwegian scenery. The water, the sparkling ice and snow. It was like every cliché he'd heard about the place, an extraordinary land where bad vistas were not permitted. Nathan knew that Sheri felt the same. It made the already enjoyable work even better. Sheri had an excellent location, good job, good money and most of all, she was now sharing it all with Nathan.

Although they both worked for Biotree, their roles were very different, and they didn't see each the during the working day. The campus was vast, and they would usually go to work separately, and because of the distance and security inside the facility, it was best to stay out of contact throughout the day. There were ways to communicate, but it was much the same as if they worked in different towns.

Sheri was reading a report on her handheld. It was related to engineering advances in India and the possibility of some breakthroughs in molecular design. Sheri could sense immediately that this wasn't a plausible article. There were holes in the logic, and the approach was one she already knew to be flawed.

Nathan was preparing some food in the kitchen and

called through from time to time to inform of progress. She could tell by the variety of hissing sounds, aromas of onions, garlic and the tell-tale sound of a bottle cork being extracted. Drake, whose Persona was inside of Nathan's Presence, was secretly pleased that Nathan was already a great cook. The evening meal was almost ready.

Sheri knew she had the more demanding and specialised job of the two of them. Her work was at the (almost literally) cutting edge of the design of the nanomachines. Originally, there had been a set of standard assemblies that worked well together. Most of the incremental designs were based around these pieces.

When she explained it to others, she likened it to a car. Four wheels, one on each corner, some seats, an engine, steering and brakes. The nanobots had a similar basic construction kit.

At the small sizes of the machines, it was the use of protein as a fuel and the same effect that makes cream in coffee eventually spread through the liquid that created a bot power source. It also provided a significant physical limitation to the small machines and their deployment. The 'coffee-cup' Brownian motion made everything shake dramatically at this small scale. Nothing was every static and the trick with the scientists was to harness this motion as a power source.

The building blocks that Sheri used were like the components of the car. There were components for movement, components for sensing, components to join things together - the so-called fixtures - like the chassis of a car and elements to provide grip and contact between the machines - the end effectors.

Then, as Nathan referred, "JASMOP" - or "Just a small matter of programming" to create the operating systems for these small devices. Just a small matter was an interesting point. The technology of Scanning Probe Microscopes used to view the assembly work had never really scaled itself and relied upon clean, secure environments and mega-voltages. It was not surprising when a speck of dust would be like throwing a planet at some of these machines.

Sheri's work was within the so-called 'exotics' division. As expected, there were pictures of palm trees and Pina-Coladas stuck to the walls inside, but fundamentally this was the area where Sheri tried to outdo her maker.

It was the place where new elements were designed. Original elements to create the missing shapes of matter needed to extend the constructor kit of parts for the nanomachines. The pieces that God forgot. There were practical physical limitations to how they could be used. Apart from atomic forces that would blast structures apart, the continued jittering from Brownian motion and the protein fuel consumption of the tiny devices, there were still some basic components that were proving impossible to construct. It was like the car but with only a few degrees of steering and no gears.

Sheri and the team were attempting to build the new shapes. The missing piece parts that would extend the nano constructor kit.

Nathan entered the room, triumphant. "Dinner is served!", he quipped and gestured towards their dining table. They would still eat together whenever possible because the nature of the work often meant irregular hours, and this would give a chance to spend some time chatting. There was an inverse luxury to 'dining in'. Most of the time, workers in the facility would avail themselves of the vast eco-system of restaurants and cafes that had established around the complex.

Pretty much all cuisines were catered for, from fast-food Americana to the fanciest French or Japanese food. Most evenings they would eat out, sometimes alone but often in company with others from the facility.

Being alone in their living quarters was an excellent time for decompression, even if Sheri appeared to have started the evening with a scientific journal article.

"I can't believe it's nearly two years that we've been here." stated Nathan. "I know it will be after your birthday that it's officially two years for you and about three weeks later for me."

Sheri alerted herself as Nathan started this line of conversation. A meal at home, talk about how long... would this be leading to a discussion of 'them'? She decided to see where it was going, but Nathan moved in another direction.

"I hope tonight's 'dish of the day' is okay?" he inquired, "I had to scratch around for some ingredients".

Sheri relaxed. She was keen enough for a talk about their future, but tonight it didn't somehow seem to be the time. She was just too strung out on the current work.

A few tonal changes were creating some new upsets. Makatomi's business plans were at odds with Sheri's personal beliefs. Instead of the Biotree being about healthcare and the future, it seemed to move towards more ominous goals. They had recently brought some contractors into Sheri's department who seemed rather more lackadaisical about their approach to safety systems.

"I've been working on some new secure perimeter systems this week," called Nathan," It looks as if Biotree are getting even more paranoid based upon the recent share prices and business news."

Nathan worked in another area of high technology, but rather than being progressive and forward-facing, like the pure R&D that Sheri conducted, this was more related to the protection and security of the Biotree complex. Beyond obvious physical defences, there were rings within rings of security measures that could both give an impression of a relaxed environment but could become extremely strict in moments if something inappropriate was detected.

Nathan worked on the improvements to this world. A guardian role that also meant he spent more time around the whole complex than Sheri.

They had initially met just after Nathan had joined. Sheri had been out for a weekend skiing in the adjoining mountains, with a couple of new-found girlfriends when they had run into an 'induction team' part of which included Nathan.

Sheri was snowboarding at the time and noticed that Nathan seemed similarly adept, and they'd broken away from the group to try a particularly exciting route. At least that was what Nathan had said and - on reflection - Sheri had also thought the course unexpectedly delightful.

They'd been together ever since that first encounter. After a short time, they had moved into what was considered one of the better apartment areas. Their facility looked out to the sea on one side and hills and distant mountains on the other side.

Tonight, Nathan had placed candles outdoors in the Norwegian tradition, and Sheri could see a distant twinkle from boats on the sea and stars in the sky.

But no, tonight wasn't the one to have deep conversations about the future.

Aborigine

After what seemed to them like an eternity, the Sun Mother peeked her head above the horizon in the East. The Earth's children learned to expect her coming and going, and were no longer afraid.

At first the children lived together peacefully, but eventually envy crept into their hearts. They began to argue. The Sun Mother was forced to come down from her home in the sky to mediate the bickering.

She gave each creature the power to change their form to whatever they chose.

However, she was not pleased with the result. The rats she had made, had changed into bats; there were giant lizards and red fish with blue tongues and feet.

The Sun Mother looked down upon the Earth and thought she must create new beings, to not anger the All Spirit.

She gave birth to the Earth, Water, Fire and Air. And then to humans.

Pulse

Follow the timeline for the next 300 years.

Pulse
The immediate sequel.
Features Scrive, Chantal, Charlie and Nathan

And then, 300 years later
Edge
Edge, Blue
Edge, Red

PULSE

Ed Adams

Author's Note

Pulse was written ahead of the series of novels Edge; Edge, Blue and Edge, Red. Collectively they discuss Earth after a major series of dystopian catastrophes.

Pulse features as a pre-history of the series, set some 300 years earlier than Edge. To a reader of Edge and particularly Edge, Red, it will become increasingly apparent that Pulse set in motion events with a very long arc.

I hope you enjoy!

Ed Adams

PART ONE

1 uni-thought

Where all were minds in uni-thought
Power is weird by mystics taught
No pain, no joy, no power too great
Colossal strength to grasp a fate

David Bowie – The Supermen

Scrive

Scrive clicked the new cartridge into place in his forearm and felt the cold rush snake from his arm to burst somewhere inside his head.

Next, he checked the small plexi-inspection window briefly and could see his blood already changing from a bright red back to orange, and he knew that within another twenty minutes it would again be the safe yellow colour.

Like everyone, he knew that red blood spelled danger and he had been particularly careless to let his system deplete its supply of the tropus for so long.

He felt the pulse bubbling on the left side of his head above the eye-line. He knew this was his body regaining its equilibrium. He squeezed both his hands into a fist shape the way they were taught and used his two middle fingers to massage the fleshy areas below his thumbs while his system adjusted.

Another five minutes and he was walking across Chelsea Bridge to the Tube station. He lived less than ten minutes on foot from the nearest stop, and his ride to today's meeting was around fifteen minutes. He could feel the cartridge working, and his relaxed acceptance of the day's tasks was already returning.

He looked briefly toward the sky. A jagged spark had flicked across. Now gentle vapour trails were crawling behind what had been a brief tear shooting along the path of the River Thames.

Others walked at a similar pace towards the station, although he ducked to the right into a quieter street that also cut a corner and missed some traffic crossings.

He glanced as he prepared to cross the diagonal into the station and glimpsed someone he recognised.

She had a petite almost boyish build, dressed in black, dark hair in a black band. Scrive had noticed her for three days now, at the same spot, the same pace and the same appearance. He knew she would look up and he'd see the small tattoo by her left eye. At least he assumed it was a tattoo and not consistently applied daily make-up. As she passed, he thought he could hear her gently humming a tune. Maybe from a streamer, but he couldn't see any signs of her wearing one.

He descended into the TfL transit. His new cartridge meant he had a good range on his transceiver again and could access the transport system without overtly waving his arm over the sensor.

Most travellers referred to the sensors as 'oysters' although this was a reference to a long-defunct technology, much as the Tube itself was merely a reference to the shape of the original tunnels that formed the original wheel-based transport system.

He used the moving floor system to get to the high-speed transit level and stood for a moment waiting for the next transit pod. He clipped himself into a free TPOD seat and punched in his destination. The system was pretty fool proof. His cartridge provided the principal co-ordinates for his routine travel, and a short, personalised menu of options had appeared on the screen. He'd just tapped his planned destination.

Of course, he could go to other points within his regular routes or pre-authorise other destinations in advance, from the HomeLink system. Today was ordinary, though, or at least that was what he needed to suggest, despite what had happened yesterday.

Janie

Janie was exhausted. Not from the morning jog, which had been one hour at a fast pace. It was because of the idiotic requests that she was subjected to in the workplace. If anything, the morning run had boosted her mood, but it was from a pretty low starting point.

"Hi," she said to Karin, as they sat together for a morning coffee, "How's today?" Karin threw a knowing glance toward Janie.

"Not great, we are still going downhill, I think. But the coffee is good."

Janie's work colleague was Karin, and they'd been friends since just after Janie had started. Karin had been in the company slightly longer - just enough time to mean she could show Janie around and warn her of any subtle office hazards. The machine coffee, the inefficient procurement process and ways to circumvent it and the slightly sleazy Leonard who worked in accounts.

Karin seemed to be able to operate around most of the recent chaos of the changes without being as perturbed as Janie. Even her functional move just after the new management arrived didn't seem to have affected her spirit. But Janie also noticed that Karin could be somewhat different if they went out together for an after-work drink or occasional cup of coffee.

"I don't know how you do it," Janie exhaled, intensely watching Karin scraping some of the foam from the coffee.

"I asked for a flat, and they've given me a latte," Karin replied,

"Considering I'm in here most days I'd expect them to get it right by now."

"No - I mean about the firm," continued Janie, "It's beginning to drive me nuts. They are constantly changing things at the moment, and each time they do so, a few more people seem to disappear."

It was over the last two months that things had changed. There had first been rumours that the company was in some financial trouble. Then a set of new people had arrived, superficially polite but rapidly asking for increasingly ludicrous changes to the way that they were supposed to operate.

It was supposed to be about boosting profitability, but Janie had seen several of her colleagues summarily dispatched, some to overseas and a few to leave the company.

The ones remaining had been instructed in no uncertain terms to refrain from contact with those that had moved away. It was officially because of privileged information, which was supposed to remain secure, but Janie was far from convinced that this was the real reason.

Janie's unit had remained mainly unscathed except when Mayer and Nikolai, who were two of Janie's bosses, moved to the USA. Replacements were new people from an external consultancy firm, and Janie understood that they would be temporary so-called 'interim' management while the operation was revised.

"I tried to contact Nikolai, after the swap around," said Janie.

"It was a personal matter - I'd forgotten to return a couple of items before the move. I sent an email to check whether a small package I'd sent through the internal post arrived. I was surprised when the mail returned with a non-contactable message."

"Not just an out of office then?" asked Karin.

"No, I don't think so. It landed me in trouble with the new management, who explained that the 'no contact' protocol was rigorous. I was told not try to reach anyone that moved, and if anything were to be forwarded, then the managers would handle it. "

"It doesn't surprise me," said Karin, "So much is changing. Even the email system itself. My workstation was upgraded to a new model and now needs a biometric scan of fingerprint and retina before I can use it."

Several of them had made macabre jokes about this because the technology was in some ways more traditionalist than merely using the proximity detector built into their tropus arm cartridges. It was slightly irritating that they could use the cartridges to access phone systems, the transport system and most types of door access but now had to revert to bio scans for something as simple as browsing the infranet.

Biotree

The Biotree company they worked for was a producer of biotech equipment. It had developed several of the nanotechnology-based products which had created a renaissance for British industry. The most famous was the Aport, which could be used within a bloodstream to manage the walls of veins and arteries. It had revolutionised healthcare since its originally controversial introduction and development into a range of products which could manage blood flow, cholesterol build-up and some aspects of the cleansing of contaminated organs. The Aport ran as a series of nanobots, which inserted into a person's bloodstream via the same type of cartridges used to manage general health.

The company made its fortune from the devices and the sophisticated software that was required to make them run successfully and without error.

London was still the global headquarters for the company, with other administrative locations in most major countries. The tentacles from the company spread wide, and the product base was routinely customised to markets.

The huge secretive manufacturing plants for Biotree's core nanotechnology resided in several locations around the world. Nevada, US; Toulouse, France and Shandong, Eastern China.

Research and Development had been moved to Bodø in Norway as a strategically safe location. Just within the Arctic Circle, it still had good infrastructural connections including fast land transit, extensive seaborne links and the small matter of a major NATO airbase nestled within the town. The origins as a strategic base went back to annual shows of strength known as the Cold Response, which still occurred under the less obvious title of CORE.

It had other advantages. A local population with their own language, while also possessing excellent English language skills for handling the incoming scientists. A university base developed extensively as part of the run-up to the creation of the research faculty.

The location also had appeal for the people stationed there, who were attracted by world-class research, the best facilities, no practical budgetary limitations and a premier lifestyle during their term. Many tried six months and then remained for much longer.

Additionally, the Norwegian government had been particularly understanding since the changes in global energy policy because they had needed to re-provision from the decline in North Sea oil and natural gas. They had granted the area a special status as a world economic development zone, and it had boosted the relative ranking of the still sparsely populated Norway to a top fifteen economy in terms of its economic freedom.

The subtext was the immense security that surrounded the environment and the commitment of those employed to maintain the secure nature of their work. Bodø was also small enough to mean that unusual activity would be quickly spotted and with the added incentives of the Norwegian kriminalitetsforebygging (KRÅD) - the criminal intelligence organisation providing added rewards for useful intelligence.

In its heyday, Biotree was simply a money machine as the demand was pretty much world-wide, and the patents and manufacturing processes locked down during the prototyping cycle.

Therefore, the employees of the company were routinely subjected to heavy screening before they joined, were provided with extensive benefits and the equivalent of 'golden handcuffs' making it exceptionally undesirable to want to leave.

That had been the case until when a Chinese manufacturer had started to produce the first clones. Strictly, they were not clones at all. They were a different way to provide the same outcome. It was evident that some brilliant people had somehow reversed engineered the 'bots and also the operating systems and now created something remarkably similar in its function, but at what worked out to be one-tenth of the price.

That had tipped the market and the little nest egg of un-vested shares that Janie and Karin had received when they joined the company was now worth less than one-tenth of their original value. These changes had heralded the management changes and the new people that walked the corridors.

It was understandable that the company was now jittery and that many of the longer serving associates were beginning to look at the job sites again for new roles.

Janie and Karin continued their coffee.

"I think we are still at the very beginning of something," replied Karin, "I won't be surprised if the new management also gets replaced within a month or two."

"What, just a revolving doors management style?" smiled Janie.

"No, more a double-blind protocol," responded Karin, "Remove the people who know what is happening, replace them with new ones and break the chain. I've seen it elsewhere; it severs the Corporate knowledge before another move is played."

"How come you know so much about this?" asked Janie.

"I'm letting you into one of my secrets when I tell you this." said Karin, "This process is the reason I joined Biotree".

She looked long at Janie. "We're all smart people in this organisation, but some of us have roles that are far enough down the organisational tree not to be a threat. We won't get replaced like Mayer and Nikolai. And that's important because I've been sent here to find out what is happening."

"Why are you telling me this?" asked Janie, "It all sounds a bit far-fetched."

"Tonight," said Karin. "meet me - this coffee bar at six o'clock - and I'll show you something."

Tube

Scrive was travelling across London. He looked around the TPOD compartment. It was a recent model and had the evaluative advertising module that had been creating a commotion in the media. Linked to the occupants, it would select advertising materials with apparent associations with the people in the compartment.

Despite the trials, it had proved something of a disaster. People wanted to see aspirational products like expensive holidays but instead were presented with perspirational products like deodorants. When the adverts were non-selective, it really didn't matter, but when the demographics of who was in the vicinity chose the banners, then it became a question of 'who triggered that one?'. The marketeers had an answer for it all based upon product weighting. Still, most people assumed it was either them or their neighbouring travellers that had created the demand for unpleasant cereal selections or dating agencies for the lonely.

The countdown began, and ten seconds later, they were moving towards their next destination. They accelerated to a couple of metres from the next pod and hurtled through tunnels at blurry speeds towards East London. The scheduling was impressive, with individual pods able to manoeuvre around one another and to take the right branches, guided by a lidar and radar system that avoided collisions.

For most of the journey the windows were black, not because of the view, but to avoid inducing vertigo into the passengers. Everyone knew the buckle-up protocol on the system, and it was frowned upon if anyone fumbled too long getting into their seat.

There was a moment of phase shift which sounded like a suction motor as the pod slowed and stopped suspended on its soft magnetic levitation while a few passengers swapped for the second part of the ride.

Scrive was flicking idly through the touch screen pages, which had been interrupted by the safety announcement before the pod started again. He noticed the newsfeed was referencing the Biotree financial difficulties and the emergence of the Chinese alternative biobots.

It wasn't exactly new news, but the media liked to recycle the same few facts every way they could, and there were now plenty of cartoon simulations of how a nanobot worked and the various components used to make them.

Scrive had been through surgery after a fall which had broken a bone. The medics injected the local area with nanobots to speed the repair. He was astonished at the way they had linked into a structure which effectively fused the bone halves and then as he healed naturally, the nanobots progressively reduced their links and eventually flushed from his system.

He had no idea how it all worked but had been given a monitoring device while the 'bots were working so that he knew how many were operating. It had been several thousand 'bots but eventually dropped away to a couple of dozen.

That had bothered him at the time because he was aware that they hadn't dropped back to zero and he sometimes wondered what the remaining few were doing inside of him. He'd tried the monitor on others. Those who underwent nanosurgery seemed to have residue; everyone else didn't show any readings on the device.

A sharp ping broke his reverie as the TPOD arrived at the second stop. His destination in Canary Wharf. He'd been lucky to catch a fast transit that had only made one stop along the way. Sure enough, about two-thirds of the passengers unclipped and left the pod, back onto a platform surface and then through a series of moving floor-ways back to street level. He would use the underground retail levels to get from the Tube stop to the office.

Pulse

That certain something

Janie's afternoon passed with yet more unexpected company changes. Laughably they were being sent out via email, but the new system had a fault, and so many of the recipients were getting blank messages with just a title. Janie felt this summed up the current situation, all title and no content.

Janie prepared for her evening meeting with Karin. She knew her well enough to think she was serious about something, rather than it being some form of a practical joke, and it didn't look as if it would be about boyfriends or partying. At about five minutes to six, Janie headed into the coffee bar and ordered a drink. She'd not got much of an evening ahead, so a chat with Karin would be an entertaining diversion, whatever the basis.

Another ten minutes passed, and Karin hadn't appeared. Janie sipped on her coffee and looked around. She hadn't prepared for a long wait, because usually the two of them were pretty punctual when they met.

Another five minutes and she decided it was sensible to call Karin's phone. She tapped the number and diverted straight to messaging. She decided to tap in "I'm here" instead of leaving a voicemail and was surprised when an immediate response returned saying that 'the holder of this number is no longer available'.

Janie looked down to check the code. She had used the right number; it was a click from a call that she had made to Karin earlier today. Janie decided to try again, this time by speaking. The number rang, and she heard a click. No voicemail, no messaging at all this time.

Janie pressed her recall facility and ran back to the message from a few moments ago. She saw it pop back onto the display, but then she noticed it rotate and disappear. It had been deleted, but not by Janie.

She flipped back to Karin's contact entry. It had also gone. Janie looked around the phone for the on/off switch. She cursed that she'd forgotten how to turn the whole device off, it wasn't something that one ever did after it had been powered on for the first time. Eventually, she found the switch, flipped off, counted and flipped on again.

A few seconds for the animated initialisation and this time Janie could see that the contact had been deleted. There seemed to be a few other numbers missing as well.

"Hello." said a man's voice behind her. It was quietly spoken as if to soothe. It didn't work because Janie was already turning to confront what she expected would be someone panhandling for money right there in the coffee shop.

"You must be Janie?" he asked before she had a chance to get angry. He was around the same age, clean-cut and lean but wearing a slightly crumpled looking outer sports jacket. The sort of jacket that would be seen at the top of a mountain. In snow.

"Hello...Who are you?" she responded, "I don't think we've met, at least I don't recognise you?"

He smiled as if trying to look like a friend.

"No, we haven't met, but I'm a friend of Karin's"

"Where is Karin, then?" replied Janie.

"I don't think she will be along, in fact, I don't want to alarm you, but I don't think either of us will be seeing her again," came back his response.

"Now you are worrying me," said Janie, "In a creepy kind of way. I will make a fuss if you don't explain yourself right now." The cafe was busy. They were at a table right in front of the serving area. It would be hard to imagine anywhere else more public. If Janie made a commotion now, there would be people intervening in seconds.

"Look - I'm here for a reason. Karin said she would bring something along this evening. That something was me," he began to explain.

Charlie

Charlie could hear the room. Apartment 123. It made two different sounds. One was a buzz, which seemed to be the refrigeration unit. The second was a low growl, which sounded as if someone had left a music system switched on but without anything playing. A kind of low-frequency hum.

She removed the mirror shade sunglasses that she'd worn into the building. Better to be remembered for dark glasses that are easy to exchange.

Charlie had felt that spin-down relief wash over her too. She had been busy and hyper-alert for the last two weeks. An assignment in Milan had almost driven her to distraction.

She'd been driven back to the airport in a black limousine, which seemed to have priority access everywhere. The driver only spoke Italian but had a black liveried look that said to anyone outside the ride, "Don't mess with me."

She'd noticed the relief on the plane. A conventional jet; she'd sat down in her seat, felt the adrenaline leave her system and then blanked for the next 30 minutes that they'd sat on the tarmac before take-off.

Then she'd flown back to New York, but now she was back in Europe again.

At least she'd had a driver pick her up at this end, who was able to whisk her above the traffic to her destination. But now, her body needed to recalibrate.

Charlie kicked her boots off and felt the springiness of the carpet. She dragged her toes through what she perceived was an expensive flooring surface. Scrive had posted her a key. She said she'd be around this week and would it be okay to stay over for a few days?

Scrive had responded immediately with a "Yes". Not exactly a phone call, just a word in an email. She knew that this was his way of saying everything's fine and that they'd talk when they were together at the apartment. Scrive was frequently travelling and had dozens of people trying to get his time, but Charlie knew that she was an insider and would get the equivalent of the red-carpet treatment, even if that was just a simple "Yes".

She checked a few of the cupboards in the kitchen. There was plenty of cereals and a bowl of fruit as well as a random selection of refrigerator foods. She selected an apple and took a bite. It was the real thing. Not a clone or a forced product synthesis. It had a taste as if it had been grown somewhere in sunshine and with rain.

Charlie savoured the fruit and paused to think what to do until Scrive returned, which she assumed wouldn't be for several hours. She noticed the packaging on the edge of the kitchen counter from Scrive's new tropus cartridge. The rest of the room was tidy. Almost perfect, whereas the packaging here suggested the tropus was the last action before heading out for the day. She glanced at the type of cartridge. Standard. Scrive was still operating low profile.

Charlie headed for the sofa, flipped the remote and clicked into a television menu. She'd watch a movie and maybe enjoy some more fruit while she waited for Scrive's return.

Cedar woodland streams

Scrive passed through Biotree's elegant entrance lobby through which permeated a smell which he thought of as a woodland stream. There was a greenish-blue water sculpture between the entrance desks for visitors.

The building divided into different businesses and a couple of the main ones had entrance lobbies on the ground floor. Further up the building were more entrances and some quite ornate environments designed to suggest anything except a world that was many stories from the surface.

His access was pre-defined, and he could walk through the scanners and harmless-looking glass turnstiles and towards one of the elevator banks. He knew about the glass shield that could drop silently from the ceilings and rise from the floors if someone was attempting unauthorised entry to the building.

The security style of this enterprise was muted, compared with offices having steel cages separating inside from the outside. It was as if the security was in inverse proportion to the appearance. The fanciest buildings appeared to have no protection, whereas a corner shop would have the overt grills, chains and bullet and bombproofing on display.

Scrive hadn't worked with the low-end organisations for several years. He'd made his reputation which meant his line of work was recognised with the big players and the related high fee-rates. He usually worked through other organisations but still considered himself a free agent, able to pick and choose and without the hassle of a long-term boss breathing down his neck.

Today he was deliberately operating through another company and had some special idents made so that he could tow their corporate line with his client. It was a kind of unspoken situation between him and his clients. They knew he was freelance but didn't ask. He would appear to represent the agency or company that fronted him but wouldn't confirm that he was a full employee. It worked for everyone because the clients he worked with knew he was one of the best at what he did.

The elevator pinged open on the 63rd floor. A woman's voice politely announced the level and the company departments represented.

"Company Compliance," she announced. Scrive crossed the lobby to another set of reception desks.

"Hi, my names Mallinson, Scrive Mallinson, I'm here to see Makatomi San".

He presented his ident on a handheld device, and it blinked onto the receptionist's screen. He had used the Japanese traditional method to show the ident.

It seemed slightly arcane to grip the handheld between both thumbs and forefingers and to bow slightly, but this was part of a business tradition that swept back through many generations with the Japanese and was somehow updated to cope with current technology.

"Mr Mallinson, I see you already have full access authority; would you like to be escorted to the meeting room, or do you know your way? I will inform Mr Makatomi of your arrival," responded the receptionist. It was all very formal here, Scrive knew he was in one of the places where ceremony and protocol would still be necessary.

"I'd be pleased to be escorted," he replied, realising that although he knew the way, the client would feel more comfortable seeing him in their keeping rather than running loose, at least until he'd kicked off the assignment.

Another person, slender, Japanese and suited appeared from behind a partition. "Hello Mr Mallinson, let me show you to the meeting area - My name is Takuya, if you need anything while you are here, please let me know." He bowed lightly, and Scrive returned the gesture.

Scrive noted the perfect London English of the Japanese-looking person. He wondered if Takuya was his real name or was an adaptation for his role in the company. As they walked, he thought about asking, but then decided to play it low-key until at least until the assignment was underway. He was ushered into the meeting room.

"Mr Makatomi will be along in a few minutes," said Takuya, "Would you like some coffee, tea or water?"

Scrive thanked Takuya but declined any drinks. He would use the few minutes waiting to scope the working environment. Takuya quietly left the room after pressing a small button which lit an engaged signal. Scrive realised that this would probably also start a monitoring process with sound and vision, so he decided it best to play dumb and settle into one of the comfortable leather chairs.

He noted the camera spots on the walls. One large obvious video camera, facing towards the floor and switched off. Three more small pinprick-sized cameras with tracking arranged around the wall. He couldn't tell whether they were activated, so he pulled his handheld and took a silent picture of one of them. He glanced at the result, and there was a small pinprick of light from the infra sensor. Something not noticeable to the eye, but easily spotted by another camera. They were on; he was undoubtedly being watched.

He walked across to the door of the room and glanced at the panel that Takuya had pressed. It was the latest generation operating system and a specification that he had not seen before. This room and probably this entire floor were kitted out with state-of-the art technology.

There was a muffled noise from outside, and then the door clicked open. Takuya re-entered along with two other people.

"Let me introduce Mr Makatomi and Mr Arusen," said Takuya, "and this is Mr Mallinson". They all briefly bowed to each other, and Scrive extended his hand for a handshake.

"Mr Makatomi, Mr Arusen, you may call me Scrive", he smiled.

They both smiled back, shook hands and took seats around the long rectangular table.

"Mr Mallinson, ...Scrive, we have something of a problem and require some extraordinary assistance. I believe you will be able to help us." began Mr Makatomi,

"This is Mr Arusen, from our legal counsel. He is in this meeting as a matter of record. Afterwards, he will have taken notes from our conversation, and they will form a permanent record of what has been discussed today. That will be the only record, and it will be non-deniable. Do I make myself clear?"

Scrive nodded. He realised the situation. The three of them were about to talk off the record. The notes from Arusen would describe an entirely different discussion which had no bearing on the actual conversation.

"Mr Arusen, you had better provide me with a copy of this conversation", he smiled, "You may beam it to me for simplicity."

Mr Arusen nodded, "Actually, I have taken the liberty of already doing so. I noticed your use of your handheld device a few minutes before we entered the room and beamed you a copy just after I wiped out your picture of this room".

Scrive felt the hairs on the back of his neck tingle. He would need to be on top game. They'd just signalled to him that they knew what he was doing and that they could access his well-protected handheld computer. And even wipe out content.

Scrive smiled again, "My test of your systems worked then," he replied, "you detected my scan and were able to remove the content and replace it with your file. That's my point about your current technology. It looks modern, but if you can do that, then so can someone else. It's a situation where the strength becomes a weakness."

Scrive secretly thought that he was the one turning his own weakness into a strength.

"Let's get to the main business", continued Scrive, "beyond the theatrics, you seem to have a real problem. You've already checked me out, long before I appeared here so please can you explain to me the basis of your need for my assistance?"

Mr Makatomi grinned, "They said you would be direct with your discussion. Let me be as direct with you…"

Tract

Lars lived on the Tract. It was an area outside of the Ellipse where the poorer people lived and survived in the old ways. No biospheres, no transport apart from self-powered systems, because access to fuel was prohibitively expensive.

The Tract dwellers had access to fire, water and farming, but the economy was completely separated from the area inside the Ellipse.

Fundamentally it was an economic model similar to something from the Middle Ages. Protectors ruled, workers survived, and everyone paid a lien to the local Protector who would keep order and prevent incursion from neighbouring tribes.

There was no hope for a Tract dweller to legitimately enter the Ellipse. Lars sometimes wondered whether people inside the Ellipse even knew about the Tract world outside.

He and his partner Carolin knew about the outside, because they had both arrived in the Tract from Norway. They had been children when they crossed into the area, both with their families. After one of the meteor strikes they found that the land had become part of the Tract.

There was a large area separating the two environments. It comprised a white band of light and a power source activated by movement. It sparked first as anyone approached and would destroy anything that crossed into the white area. As a child, Lars played dare games and had thrown stones and even small branches into the light, to see them burst and disappear. As he'd grown older, he'd realised the danger of getting too near and nowadays, like most people, he would stay away.

There were plenty of signs of the previous habitation in the Tract areas. Whole towns were visible and many consumer goods but without power sources nor means to make more than token power from fire or through static bicycling, the energy from which was stored in old-style car batteries. There were some wind farms which still worked, but the energy was erratic, and the control systems for routing the energy was mainly destroyed.

Lars and Carolin belonged to the Dals. They mainly spoke Norwegian and English and also had another language amongst some of them from which he'd only ever managed to pick out a few words. His nearest town still had bookstores and libraries, but they had gradually been raided, and nowadays many of the books had been used as fuel. It was still possible to make a living, and the old metal coins were used as a means of exchange. The coins had a higher value than the banknotes, and only a particular type was generally exchanged.

The biggest threat to Lars and his tribe was from various biohazards which would sweep through the

Tract areas. These could be simple ailments although there was still a significant number of conventional pharmaceuticals in circulation, there were also virulent bugs that would strike and wipe out a whole tribe.

Consequently, most tribes stayed apart, and a delicate balance of equilibrium was maintained based upon survival rather than territorial disputes.

Lars had no idea how large the Tract or the Ellipse were. He knew he could walk all day, but the terrain would hardly change. There were markers on the old highways which represented the ends of his domain. They were words describing what had once been county boundaries. It was usually easy to stay inside the boundaries because the main routes all had signs at their edges, which often, ironically, welcomed people into the next area. Some of the county signs picked out something famous about what the county contained.

Along the side of some of the main routes were also advertising for domestic items or even high technology equipment. He had seen one for an author, Jo Nesbo, as he approached a town. He knew Jo Nesbo had written books and had once found a couple of novels that were dismissively supposed to be used as fuel, but which he'd hidden in order to read.

Despite all of this, Lars could still sometimes feel very happy. A warm day, sunshine on his face, a run along one of the highways and he could feel a kind of exuberance. He was careful though because he knew that if he fell and injured himself, it would be another story. Many people who fell and broke bones would never recover — the same with people who suffered from deep cuts. There was something like a fifty-fifty survival chance in this case. And even with the remains of a past civilisation to use, it was like seeing an infrastructure gradually weakening.

He wondered for the next generation, and whether there would even be one beyond that.

Today was a special day. It was Protector's Day, and the Protector would be in town to collect lien via his taxmen and to provide an update on the current state of the Tract. He was heading for the town centre, not to meet the Protector, but because he had been selected for a special assignment.

He knew it was special, because he had been told he would be leaving the Tract, to visit the Ellipse.

Coffee Bar

In London, Janie looked more closely at the person who had replaced Karin for the evening coffee. The first impression of mountain climber was somehow reinforcing itself. He had a tan, but it was a kind of weathered look, more from snow and wind than direct sunshine. His face was angular and with slicked-back blonde hair. The dark jacket had a small logo with a picture of a mountain. And almost irrationally Janie was deciding he looked very athletic.

"Okay," she said," We'll stay here for a couple of minutes while you explain yourself. Then I will leave".

"Okay," he replied, "My name is Lars." Janie noticed the slight accent and then the name. Before she could ask, he continued, "I am Norwegian and had come to London to meet with Karin. I saw her yesterday for about an hour, and she has told me some things that I will share with you. I don't want to alarm you, but Karin has been in some trouble, and I think that is why she won't be along this evening."

Janie looked around. She could still make a fuss if needed and leave without difficulty. She was across the table from this guy. Outside, he could undoubtedly outrun her if anything weird was going on.

"You'll have to tell me some things that stop me thinking you are a weirdo", she clumsily replied," I've never seen you before; you tell me you've come from Norway to see my friend and then she disappears. It doesn't seem right." She could feel her pulse quickening as the gravity of the situation increased.

"I didn't expect you to embrace this situation", he continued," But frankly, I need your help. Karin was sent into Biotree to investigate some things. She has been undercover since she started. I work with her, and we are investigating a serious matter."

He looked down at his plain black coffee and stirred in a small amount of sugar. Janie noticed a small tattoo on the underside of his hand. She thought she had seen it somewhere before, like a skinny 'S' shape.

"We've been looking into the plans for Biotree," he continued, "We're from an organisation that checks civil freedoms. Like an Amnesty International crossed with a Greenpeace, but we do our work without publicity or public profile. It is better that way, or we get targeted and refused access and even work."

"You'll need to show me some proof," replied Janie, trying to think of her best exit strategy. She needed to know whether to trust this person or to find the fastest way to escape. Janie was concerned that he knew who she was, what she looked like, and where she worked.

"Okay," she added "Suppose I believe your story. I still want to know where Karin is, and then when we are both together, I may co-operate a little more."

"As I said", he responded," I don't think either of us will be seeing Karin again. I believe she has been dis-appeared like a number of your company management. Try her number, try her email, I think she will have been removed from the systems already."

Janie knew this was the case but wondered if it was some trick. An electronic erasure done like some conjuring trick to persuade her that this wasn't a hoax.

"Here's the situation", he continued. "I expect Karin has given you something to look after in the last couple of weeks. It will be something minor, a book, some papers, a loan of some kind — That's how it works. The person investigating discovers something, passes it along to someone who doesn't know anything about it and then one of us collects it. A kind of drop box but with a person rather than a location."

Janie was thinking fast. "Okay, but if Karin has disappeared, it could be because of you, in any case. How do I know you aren't trying to get something from Karin by tricking me?"

She stared at him.

"You don't", he replied. "But why would I go to such elaborate ends to talk to you like this? If I was hostile and had done something to Karin, wouldn't I do the same with you? No. I am who I say I am. Here is my ident."

He held his arm towards her, and she scanned him onto her handheld. A Norwegian passport appeared. His picture. Lars Fjelstad. Journalist. It looked authentic and had the little authenticated icon on the top corner when she played it back.

"Look. I can see you don't want to get involved with this. Think carefully about the last couple of weeks. If you can think of something like I describe, let me know and after I receive whatever it is that Karin passed you, I will be gone forever."

"Also, please don't tell anyone about this, or I will probably also disappear, and then you'll probably get another visitor with a similar request."

He looked towards her. Their eyes met. Janie thought he looked both genuine and a little bit nervous.

"Okay", said Janie," I'll think and then contact you tomorrow if I can remember anything."

"We will need a different meeting place," he replied. "You choose it now, and I will remember it. Don't write it or email it".

"Okay..." Janie felt she was getting drawn into this.

"We'll use Smollensky's by the clocks at Canary Wharf. Whichever day, it will be six o'clock in the evening".

He nodded. It was another very public place in a busy area. Easy to find but impossible to guess.

"I'll go first", he said and slipped quietly towards the door.

Janie looked around. It was six forty. Her world had changed considerably in the last thirty minutes.

Chinese theft

In Biotree's offices, Makatomi stared at Scrive,

"We know that the Chinese stole the designs for the Aport nanobots," he began.

"We also know that they didn't want them for the basic commercial cloning that they have been taking to the market. It is highly likely that they have other plans."

"Why would you need me for that?" asked Scrive. "You know what I do and what you describe is little more than the speculation that we've seen in the media. Conspiracy theories and all. China steals trade secret, makes copies, makes something else as well, makes pots of money…"

"All of that might be true," replied Makatomi, "But we need to track back to an alternative source of work that the Chinese are running."

"What is it then, some weapon?" asked Scrive, "That was the typical criticism of the nanosystems, that they could be used for good or evil purposes."

"Correct. In a way." said Makatomi, "There are certainly options that would allow an unscrupulous group to do something bad with this technology.

"The thing is, we believe the Chinese that took the original design have themselves been compromised. They are in a similar position to Biotree, except their design is weaker and their ability to trace what has happened is probably non-existent. We don't even think they know they have been compromised. That, Mr Mallinson, is where you come in."

"You want me to penetrate the Chinese system and then follow the trail to the second set of thieves? Presumably without anyone noticing?"

"You are astute, Mr Mallinson, that is what we want you to do. We will require the geo-coordinates of the second group and will take actions from there."

"We will need to discuss fees for this, Mr Makatomi"

"That can be taken care of by Mr Arusen; it should not be a problem. We will pay you some upfront expenses too; this will involve significant preparation, I am sure. Exceptionally, we will make Biotree facilities available for you as well. I suspect we have some technologies here which you will not have seen before."

Scrive looked at Makatomi San and nodded. Then he bowed. "I'll be pleased to assist with this. I will find the source of the original leak and then trace the second leak from the Chinese."

Makatomi stepped forward. They shook hands. He looked across to Arusen. "Please assist Mr Mallinson to get started. Mr Mallinson, you realize this is covert and deniable?"

"Always the way", replied Scrive.

Scrive made his way back to the Tube station. He'd return on the same system. Low profile. Traceable. His training meant he wouldn't do anything to trigger alerts from the surroundings of what was his new employer.

As he crossed the pedestrian area back to the curved dome of the station entrance, he could hear the recharge cycle of a high-powered transit. A V-Blade. Commercial version. He could tell from the engine note. Two tones were rising through octaves from a rasp to supersonic frequencies in around three seconds.

He listened for the tell-tale bass thump as the systems prepared for take-off and looked upward to the apparent source. It was from his building and almost certainly Makatomi San making his less low-key departure. The private V-Blades were still in the luxury goods category, so he knew it would be someone with corporate leverage and in a hurry. A few seconds later, he saw the flash as the 'blade departed. Makatomi could be back in Japan before Scrive reached Chelsea.

Scrive had used V-Blades plenty of times. Mainly the military versions which didn't need the dual engines. Commercial versions had to provide full system redundancy as part of safety measures. For the military, the reasoning was it was better to have two of the transports, which would anyway be moving into hazardous terrain. He'd used them since the early days when they required occupants to wear pressure suits to handle the G-forces. The later models had a new atmosphere and pressure walls which dealt with the forces as a kind of electronic cradle around the passengers. When he'd first travelled on the commercial versions, he couldn't believe it was the same craft as he'd experienced in the military field.

Scrive had never been 'in' the military but had worked with them much as he had with his commercial clients. He'd been a system engineer by background and had found a natural ability to see past numbers and sequences. They popped out to him almost as pictures, and he could manipulate huge matrices mentally to look for unusual patterns. Some would say he was gifted, but he had also experimented with some of the biometrics that he ran into in his professional life. He'd seen others running upgrades on their metabolisms and computation and decided he'd run the same tricks on himself.

There were practical limits based upon human systems, like maximum pulse rate and breathing, so the approaches used a blend of pragmatic limits and ancillary adjustments.

He knew how to run himself at 'human speed' but could over-clock his thoughts and reflexes to around ten times normal. At a burst, he could go even faster. It made other people appear to slow down when he was using these capabilities. In a stressful situation, it could make a huge difference. The physics of movement were not quite as accommodating though, and although he ran himself to a high athletic level, he couldn't outwit nature for running or strength.

He sold himself and his abilities around the pattern and analytics capabilities. It was because of this that he needed the best physique to get himself out of the occasional trouble.

Scrive learned that a low-key presence coupled with the ability to ramp up his pattern work was the best and most survivable way to earn the kind of money that would get him a V-Blade or any other flying thing one day if he wanted it.

Santa Monica

Makatomi San had further deals to conclude in the rest of the day. That was why he was using the V-Blade. He didn't like the ostentation of the craft, but it was the most convenient method to travel very long distances for face-to-face sessions. Much of what he was doing now required that type of stealth. He was next heading to Los Angeles, to meet with another figure similar to Mallinson. The task he was setting was pretty much identical because Makatomi knew that there was a better chance of success with two people involved. They wouldn't know of each other, and despite the fees, his corporation could easily afford to pay them both, ideally with both of them bringing results.

Before returning to the 'blade, Makatomi had stopped by the vendor floor in his building. He'd bought a disposable paper phone from the automat. Ten minutes of signal, enough for the calls he was to make. It was to a fixer who could provide him with support in Los Angeles. He tapped the number, and the line clicked with a kind of static that didn't happen on the newer circuits. The old 3G cellular structure was like a barely loved base utility now, and most people had moved onto wave plex. He spoke softly to the fixer about the need for support in LA, then folded the phone into his inside pocket. He'd throw it away later.

Makatomi and Arusen had settled into V-Blade seats for what would be a deceptively short journey, considering the distance covered. The craft would be entering space and then re-entering the atmosphere as part of its journey. Yet, the buffeting of 20th-century space travel was no longer a concern since new intelligent moleculars had been used in flight craft design. In effect, the V-Pulse used technology to make it super slippery, blended with its transit environment and consequently didn't suffer re-entry burn.

As they approached their destination in L.A., Arusen looked out of the windows at what was still a night sky, approaching dawn.

"We'll be at their office before they are even awake", he commented. "I doubt it", replied Makatomi, "Don't underestimate the people at Marina del Rey. I expect they are tracking us incoming".

Makatomi was right, Denny and Suze had been watching their scanner for the last couple of hours. Not just watching the scanner, they were also working on a plan for their expected task.

"We don't know where Makatomi will arrive from, but we do know he'll be using a five-wing or a V-Blade for his incoming journey. When he departs, he will probably route scramble. but there's less need on the way in, so I doubt whether they have gone to the trouble." Suze flipped around a few radio frequency spectrums and checked some high incidence angles for unusual propulsion mechanisms.

"I've got this worked out and am picking up Andrews Airforce Base and way into the Pacific", she commented.

"It'll be a fast entry, probably, so we need to keep the monitoring so we can try to work out the reverse trajectory. I'd like to know where Makatomi is flying from."

Danny already had suspicions that Makatomi would be setting more than one team onto this case. He'd had a tip-off from the beam that this was a big situation and the implied fees supported the claim.

"Got something!", said Suze, she fidgeted in her seat and then stood to point at the screen. "Look, I've put it onto the big one", she flipped a button, and a large plasma displayed several arcs of co-ordinates, one moving noticeably faster than the rest. Almost immediately, it slowed to a similar crawl to the others, and then disappeared from the chart.

"It has to be him", said Denny, "Incoming fast, then going slow stealth for the last part. They won't have picked up our monitoring, and we've got a good arc to run back to work out their take-off spot.

"Easy," replied Suze, "It looks the most like London, with side options on Amsterdam or Paris. I doubt if they ran any evasive pattern so I think we can assume Makatomi has just flown from the company headquarters.

"Broadens the field somewhat", said Denny," If he has come from Corporate HQ, we don't really know if he's been with anyone or not."

"Let's ask him where he's been in the last 24," said Suze. "There's a higher likelihood that this is part of a setup if he was only passing through London".

"Yes. We may be working with paranoia here, but it is the best plan at this stage".

Denny and Suze's office was an apartment south of Santa Monica Boulevard. It was one of the areas that had picked up from the bad-old days and was now quite respectable while being on trend with the local fashions.

It worked well as an area to operate from, because the itinerant population and the side orders of tourists always flickering through between Melrose and Hollywood Boulevard.

They could do most things here without getting picked out as unusual, and long gaps or absences were quite commonplace among residents.

They'd arranged a completely different location for the meeting with Makatomi. An office block across town close to the Santa Monica Airport, which had the trappings of business but was clearly a front for what they were doing. It was a call centre, and Denny had called in a favour from one of his friends to borrow a supervisor's office for the meeting.

He wondered why Makatomi would make such a journey in person and decided it must be high profile. He'd seen Makatomi's picture in Fortune and on various media bulletins and was quietly impressed that he and Suze were sufficient 'players' to get a personal visit. He was also aware that there could be some danger around the situation, hence the separate precautions around the choice of location.

2 bad day

A bad day in London is still better
than a good day anywhere else.

- *Unknown*

Scrive's Apartment

Scrive walked the ten minutes or so from the Tube station back to his apartment. The Thames sparkled as he crossed the bridge, into another borough of London.

The sheet glass of the Japanese-developed shopping mall glinted as he made his way towards his building. Two sets of sensors to get inside, across a small security moat that had been constructed as part of the original building. The security was good all around this area. Suspended structures, gated parking and no direct access from roadside or pedestrian walkway. It required the two sets of secure identity to enter, a further set for the elevator and that was all before the entrance lobby for his own apartment's door.

He smiled as he approached. He could see that someone had entered. He knew it would have to be Charlie. He also remembered that he'd agreed that she could visit, but that she didn't have the requisite access.

But he also knew that Charlie could walk through walls. Not actually, but to be able to spoof past most systems.

"You could get into a lot of trouble doing that", he said as he walked into the apartment. Charlie smiled. The plasma was on, but nothing was playing.

"I decided to get a few minutes sleep while I waited for you", she said, getting up and stretching her arms for a hug. She'd changed from her travel clothes and was wearing one of Scrive's T-shirts.

"Hey", replied Scrive, "Hey Hey" responded Charlie. They kissed. "How long will you be in town?"

"It depends. Depends on whether what I've heard is true? You sound as if you might need some help with your latest big business?"

Scrive smiled. Charlie had connections and an uncanny ability to know where the big games played. Scrive would trust Charlie totally. They had a history.

Scrive had run into Charlie when he was solving a middle eastern puzzle. Things had started to cut up rough as he got closer to the truth.

He'd been told that he would be given backup and then Charlie had arrived. He had been taken aback, because he'd expected a small army, rather than a single 'girl'.

He'd soon learnt that Charlie was combat ready because within a half an hour of their meeting someone had tried to drive a truck into his apartment. They were on the third floor, and the truck had been launched from a building across the street, several stories higher, from the roof.

Charlie had shown her reflexes to be able to get them clear of the building and also to take down the skycrane jet that had dropped the truck onto the adjacent roof. It was a messy scene, but Charlie's jacked-up reflexes and remarkable use of weaponry had probably saved his life.

But Scrive had seen that Charlie's talents didn't stop there. She was fast and accurate with weapons, but also had pretty good tracking skills too. Not to his level, but easily fast enough to be a good sidekick. The difference between them was one of degree, with his world more digital and Charlie's a great deal more physical.

They were both effectively mercenaries and generally held to a code of non-involvement although sometimes drifted into a fling.

Scrive knew that Charlie had other male followers but had set some life goals around making a fortune. He knew she'd been trained by a Para-military group, but she remained silent about how this had come about. He had seen her in action with planes and 'blades too and knew that wherever she had trained and for whatever purpose, it made her one of a global elite.

Scrive realised that Charlie knew about his ultra-fast compute abilities, but Scrive also understood that Charlie was probably the only person that could almost keep up. Even when well-apart they kept an active link running, using one another as sounding boards. Scrive also realised that when Charlie had collected enough cash, there was every chance that she would simply disappear, as a way to break all connections.

"Yeah, I've just been to see Makatomi, He's asked me to find out how the Chinese have accessed the Aport, and what they are trying to do with it."

"What beyond profit-making by selling a variant at a tenth of the price?", replied Charlie. "It'll be some government deal - which superpower wants to get something over on someone else".

"Yes, my speculation also, but Makatomi seems to think its a double-cross or something".

"It's time we dug deep?" replied Charlie. "I'm on my way to a small gig on the Swiss border, but this sounds way more interesting. How come you've got the deal? I'd have thought Makatomi would shop around."

"Cheeky! I expect he did!" replied Scrive. "He seemed in a hurry to leave after our meeting. I thought he was heading back to Japan, but I suppose he could be seeding a duplicate operation somewhere else".

"Are you ready to take a look at this?" asked Charlie, waving the remote for the streamer.

"First things first", smiled Scrive.

Los Angeles

Suze and Denny packed some telemetry and other gadgets before they headed for the basement and into a hire car that Denny had randomly selected from the online Hertz desk. They wanted to run this as anonymously as possible, and the car and office they were using should help them stay unobtrusive.

Denny flipped the start button on the car. It whined into life. It was a fuel cell power plant, one of the new types that could run silently. He grimaced at the noisy engine tone that came as a default setting and selected another one, more a solid tone and less sporty sounding.

"Let's not draw attention", he said to Suze, who nodded in agreement.

They took the car up the four levels to the street and joined the slow-moving traffic as they headed towards the Freeway.

"It shouldn't take us long to get to the venue, especially at this time of morning, said Denny. "Don't jinx it", retorted Suze.

Dawn was breaking, in a half-hearted kind of way.

There were signs of the nightlife still in progress, as people had lost their sense of time altogether. Additionally, a few tired looking tourists were wandering around and there was a police situation on a nearby street corner. They moved along through the traffic, within the speed limits and as low key as possible.

Ahead, taillights from other traffic snaked out in as they took another on-ramp for their short ride.

"We'll be early, and prepared," said Denny, knowing that Makatomi was already in town and had probably landed at Santa Monica rather than LAX. Makatomi and his partners could be very close.

Makatomi and Arusen had a short meeting before they met Denny and Suze. They had arranged for some local support because their meeting with the trackers would be taking place away from Biotree company property.

Makatomi hated it that he was asked to meet like this, it was away from his systems and controls, but it seemed to be the only way that he could get to these two specialist trackers.

He's also done his homework, or rather, his fixer had done the homework and established that these people were the best. That was why he was meeting them on their turf. Being trackers, they knew how to hide. That they would break their own cover for this suggested that they were interested in the size of the reward. Makatomi didn't think there would be any more altruistic reasons for the trackers to agree to this meeting and the likely pursuit.

He asked Arusen to make the arrangements for the hired help and also requested that they lose the land cruisers and acquire some more low-key transport. A compromise was agreed, where Makatomi, Arusen and a couple of the hired team would take conventional transports, but that the rest would still provide less than discreet backup in the black-windowed cruisers.

"I hate LA," said Makatomi, "everything has to be so brash and dramatic. This is why we prefer these type of meetings to be in our own properties," said Makatomi to Arusen.

"Too true," nodded Arusen in agreement, "This type of situation can get so untidy."

They looked at one another knowing there was some risk of a trap or that the trackers would ask for some unacceptable terms. It would have to be something extreme because fundamentally Makatomi was desperate to get this resolved and needed the support from the likes of Denny, Suze and Scrive.

Of course, he wouldn't tell these trackers about Scrive. Part of the plan was to have the insurance of two separate investigations.

Janie gets help

In London, Janie was unsure about whether to meet Lars again, she'd gone home but kept looking over her shoulder to see whether she was being followed. It was partly irrational, but she wanted to be sure that neither Lars nor an accomplice would be behind her. She'd used her regular route so that she could go through the systems without having to stop, but then deliberately overshot her destination and made her way back. That way, if there were anyone following, it would be more visible. She wasn't sure what she would do if someone had followed her, in any case, but to her eyes there didn't seem to be any trail.

At home, she pondered about the meeting the following evening. She thought she should go but wanted to take some insurance. Maybe a friend to accompany her? But who? And how much should she say? She didn't think it should be anyone from Biotree, because of the nature of the conspiracy.

Janie lived in a flat with two others. They'd shared for about a year, as a way to get somewhere central in London. London gave them access to the best jobs, but at the cost that none of the three of them could afford a place to live, based upon their salary levels.

Janie decided that her flatmate Chantal would be the best option. They'd been in plenty of scrapes together, and shared thinking to the degree that they would wear each other's clothes. Or more precisely, Chantal would frequently borrow some of Janie's clothes.

Chantal's tastes were a little extreme for Janie. Although Chantal could scrub up real fine when required. Janie knew that Chantal operated with some of the more dubious parts of London and had quite a street scene going.

Janie was more the Ms Corporate and was used to wearing business attire, or when on courses or training, something they called business casual. It was even funnier watching the men. They'd all don polo shirts and beige chinos when it was an 'awayday' meeting.

The thing was, Chantal was out somewhere and had been for the last couple of days. It wasn't particularly unusual but meant that Janie would need to call to check whether she could help out. Janie clicked the number on her handheld and was immediately routed to voicemail. Instead, she opted to leave a text message.

"Need your help when I'm meeting a guy called Lars tomorrow - It's to do with Biotree, and I could do with some support. Fit guy, but this is business".

She thought the message was to the point, looked innocuous but would get Chantal's attention. She pressed send, and it flicked into the network.

Within a couple of minutes, a short reply from Chantal

"Sure ... How fit?"...Followed by an old school emoticon of a lascivious face. Janie smiled. She knew she'd got Chantal's attention and that she'd provide the backup.

Oil field digitizers

Orange LA sunrise, and Suze and Denny arrived at the offices.

"Not bad," said Suze, "Just above seedy and quite beige and sand coloured bland. We have found a great place to look anonymous."

Denny backed the car away. "We'll do better to park in that area the other side of the building", he commented, that will give us a couple of routes out. They could exit through the car park or the Mall. He nodded towards the adjacent Mall entrance which would put them into a zone filled with shoppers.

As he pulled up at what he now considered to be the optimum spot, he looked around.

"I think we should also look out for our friends," he added. They will be unlikely to visit us alone," He grabbed a bag from the back seat and looked through the content.

"Oil field digitizers", he said. "These little units allow us to build a grid to chart the area around here. We need to drop about 20 of them around and we'll have monitoring of the whole zone front and back".

Suze looked at one of the units. "Neat." she said, "self-powered, small transmitter, geo-sensitive and a motion camera included?"

"It's motion-sensing and also controllable", replied Denny. "These are quite new; I heard about them from a friend in the military. Don't worry; they are commercial grade and used by the large oil companies, so they are commonly available - if rather expensive. Let's sprinkle a few around".

They created a small coverage zone on both sides of the office with the units. The units had a grey speckled cover which helped them blend into the background, somewhat like pieces of discarded building material.

"They remind me of rodent traps", mused Suze.

"Surprisingly appropriate." replied Denny. He picked up another bag from the trunk of the car and they walked towards the offices. It was still early enough that they were not expecting it to be too busy.

However, as they entered there were sounds of activity. "It looks like a 24x7 operation?" said Suze.

"Yes", replied Denny, "But I didn't know it would be 'full on' all of the time. They made their way to the office that Denny had been loaned. His friend had provided access and covered off the paperwork, so they were expected and able to get straight to work.

They decided to treat a couple of the chairs to some transparent self-adhesive bugs. "Just don't sit down there," said Suze as she placed a few of the small bugs onto the chairs.

"I can't help thinking that Makatomi will be way above this stuff?" questioned Denny, "What with being into nanobots and all".

"Maybe", replied Suze, "But on the other hand his strength could also be his weakness."

They looked at the time. It had moved from 4 am to 8 am at lightning speed., They were both used to the early morning effect where the brain's processes change pace during sleep, giving the impact of a speeded-up return of the dawn.

"One hour and they will be here. Bang on time, I'm sure" said Denny.

"And I wonder what they have been doing for the last few hours?"

"My paranoia suggests they've been meeting some backup. I doubt whether they are used to operating this kind of thing on someone else's patch. That's why we have extra surveillance".

Denny pressed a button on his computer, and a whole bank of TV monitor screens appeared as small squares across the screen.

"We are ready to watch you, Mr Makatomi".

The Makatomi files

Scrive and Charlie were sipping wine. "Okay, so what brings you here to London?" asked Scrive, "I thought you'd pretty much settled in New York now".

"Work, in a manner of speaking", replied Charlie. She sipped the wine slowly and looked hesitant. "I'll tell you what happened, but you must promise not to say you told me so..."

"Sure thing", replied Scrive smiling and looking down into his glass.

Both Scrive and Charlie worked around the edges of the security environment. They'd known each other since before nanobots had become such big business. The so-called Great Leap, when technology had made a few giant steps forward.

They kept in contact and had paths that crossed haphazardly. Three years ago, they'd briefly been an item when they worked together on an investigation during which Scrive had found out that Charlie had other 'get rich quick schemes' that didn't seem to be working. Scrive had helped Charlie get out of a somewhat illegal situation, and despite growing apart, they'd kept their friendship and trust very tight,

"I've been working with the security agency in New York and Washington. They asked me to look at the rate of change of the nanobots that Biotree have been making – to try to find out if they were being deliberately weaponised.

We all know they went through a big series of upgrades in the first few years and then pretty much stabilised. I had been asked to take a look at whether there was anything else happening, like a secret 'Version 2'. Anyway, it meant that I had to join a team experimenting with the Nanobot operating system, to see if there were any obvious hooks for them to provide updates. We designed a few changes of our own."

Scrive smiled. He knew that Charlie had a swift and creative mind for software and a 'few changes' would be an understatement.

"They let me join their lab, and we started with just emulations of the 'bots. We knew we would need the real devices to try putting sets of them together. The way that their heuristics work meant that it is chains of nanobots that deliver an outcome rather than individual devices."

"It sounds like that old Lemmings game to me!" said Scrive, trying not to laugh.

"Well, it is, kind-of. In Lemmings you have diggers and flyers and climbers and so on. With bots it is the same except instead of each unit being self-contained, they join together to make the device that does the work. A machine might comprise 2,3,4 or even more separate nanobots.

Scrive nodded. He knew the theory of this as well, but whilst he would be good at tracking the originator of the devices and could probably, right now, find a trail in the 'net that led back to Charlie, he wouldn't be able to do the same complex logic that Charlie was describing.

"I arranged for a selection of the nanobots to be provided, and I first ran them as intended. They were designed for cell wall repair and did join together to create chains to achieve this. It was like a ribosome conveyor belt. You know, the way that ribosome can stitch polymeric protein modules together via messenger RNA molecules?

Scrive looked blank," No, I missed that session," he quipped," But I get the idea, nanobots copying biological processes?"

"Yes, that's kinda right; they were pretty good and speedy. Some of their inbuilt logic was to slow the 'bots down, so that normal body tissue has a chance to keep up."

Scrive continued to drink the wine. "So far, so good, but what is the point of it all?" he asked.

"Exactly, our lab wondered whether the 'bots were just copying biological functions. Then we tried a few modifications. They were quite small, and we were mainly checking the ability of the Nanobots to support new command structures.

"Big Mistake. It all went horribly wrong. The 'bots have been designed to include failsafe. If you don't know the keys and secure modes, it looks impossible to reprogram them. They have their own self-checking logic and if they spot a variation, they just stop functioning. In fact, they go one further than that."

"what exactly?", asked Scrive.

"They attack each other and destroy each other. It really is a 'leave no trace'. Quite a sophisticated design really. The manufacturers can modify them, but if anyone else tries then the whole system is rendered useless."

"Reminds me of some consumer electronics," smiled Scrive. "Take one screw out, and you might as well throw the entire device away".

"Yes - although the purpose here is a more security-minded one. To stop tampering and remove the ability to try out different processes."

"So why did that get you sent to Europe?" asked Scrive.

"Well, the thing is, I got rather bored with the programming of the 'bots each time I was testing my ideas, so I created a small handheld unit to punch in the code revisions. It is based on something that I used back in the Casino days".

Charlie's previous history had a few dubious moments when she was raiding high roller casinos by reprogramming the payout software. It had to be undetectable to the house and Charlie had created a small device to fire the software changes into the system.

Scrive knew about Charlie at Vegas. He helped her get out of trouble when she was in danger of being caught. He'd shown her how to mess up the electronic trail, which had previously led back to her.

"I took the Casino injector and produced a sub-scale version. It's a lot easier now than when I built the original because of the improved scanner resolutions available. I could use laser light to improve the sensitivity and then fire the new software into the bots, in much the same way that the old version fired software into the Casino devices."

"NSA hadn't seen anything like this before, and they've asked me to show them a breakdown of how it works. I'm heading for CERN to show them in detail. It is my comeuppance for the damage I caused when I was testing the bots I'd been building, and they kept failing and destroying one another.

She pointed to a small titanium box with an elaborate digital lock. "There," she said. "the Nano injector is in the box. It's loaded with 'nanoreductives'. They are my attempt to rebrand the dud nanobots which self-destruct one another!"

"Its' not a bug, it's a feature," smiled Scrive. He looked quizzically at the lock.

"Birthday?" he asked.

"Yes", replied Charlie, "Plus 666 - I wanted to add an edge to this devilish device."

Scrive knew that the box would destroy its contents if the wrong number was entered. But knowing Charlie there would be a twist.

"Yes - birthday will open it but destroy the loader logic. birthday plus 666 will open it and keep the device intact"

"what about anything else?"

"Three goes and thar she blows", smiled Charlie. "I thought I'd give a pirate hacker a sporting chance."

"Are you getting paid?"

"Yes and no," came the reply. "I should be, but I'm sort of paying off the cost of the destruction I created at Biotree with my hack. But now it's your turn. What are you working with at the moment? I assume it is still with Biotree?"

"Direct at the moment. And big. The biggest probably. Corny as it may sound if I tell you then you might be in some sort of danger.

"Yeah, right," responded Charlie," As if that would stop me, but first, another refill of the wine?"

Scrive reached across, picked up the wine and gently poured.

"You know what? You might want to help out with this. I can split some fees and it is worth it to me to have someone alongside who isn't known to the client. But wait until you hear before you answer."

"I'm intrigued already," said Charlie, ...and fees would be beneficial at the moment."

"Okay, so I've been asked to find out the origin of the cloned Chinese nanobots. There's a leak somewhere and the impact of the copies has collapsed the Biotree share price. I'm supposed to track the leak and pass the information back. Sounds simple, but we are talking about government level conspiracies. Hence the potential danger."

"Wow. This assignment sounds great. I assume the fees are in keeping with the magnitude of the challenge?"

"You know something; I think I could have named any price. I've gone high in any case. The thing that worries me is that they seem almost too happy to accept my terms. It makes me concerned that they won't honour the agreement."

"What can you do?"

"I already did it", said Scrive. I negotiated the money in two stages. "all of it now and all of it again when I complete. If I don't, they will come for me and take half back. I don't want that to happen."

"You've already been paid and can get the same again?" asked Charlie, somewhat incredulous.

"Plus, expenses...I said they didn't seem to care about the amount. And don't worry - it's a lot. Come in with me, and I'll split the second amount. You'll be my safeguard that we get paid based upon our successful outcome."

"How much?" asked Charlie.

Scrive told her. It was a huge sum.

Charlie nodded and started to feel like a big prize lottery winner.

Polka dot bodysuit

Chantal had arranged to meet Janie at Canada Water. It was across the River from their planned rendezvous with Lars but would give them a chance to plan how they would handle the session before they met. It was also a five-minute trip to their planned meeting place so they could be 'in position' very quickly.

Chantal arrived in a polka dot bodysuit. Three different colours; shorts, top and jacket.

"I took the ears off," she said as she hugged Janie. "And I brought a black topcoat as well in case we need to wear disguises." She pointed into a bag where a black compressed micro-fabric nestled amongst some fashion headgear.

Janie's mood briefly lifted at Chantal's appearance. They'd party together sometimes, and Chantal would usually go for something extravagant. Janie's appearance was more conservative, and today she was wearing a dark business suit.

"Thanks for agreeing to cover this", she said to Chantal. "I'm a little bit worried that this is all getting rather weird."

"I wouldn't miss this for anything. You're about to meet a dishy Swede or something," replied Chantal.

"He's Norwegian, or says he is," replied Janie. "And it is something to do with the firm and the way they are operating."

"What do you want me to do?" asked Chantal, "Hold your hand or what?"

"The main thing is to cover the situation and to follow my lead if I say anything." I may want him to think that we've spread the word about this, so there's no point in trying to keep it to just the two of us."

"Does he look as if he would be violent?" asked Chantal, making like a karate move.

"I don't think so, but put it this way, I think he could outrun us if it came to a race".

"Okay, so we'll listen to what he has to say and what he asks. I suggest we move from the first location to somewhere else to keep him on his toes, too. If we take him to the end of the walkway, we can grab a taxi and then go almost anywhere".

"Good plan. I suggest we go to Westminster, it's full of police and security so we can get out there and if needed we could easily attract attention."

Janie looked at Chantal.

"I don't think you'd have any trouble doing that, anyway".

They giggled briefly.

"Okay, that way, he can tell us his story while we are in the cab. It will only take a few minutes and put him under pressure to get to the point."

They nodded and walked the short distance to the transit point to Canary Wharf. They would be at Smollensky's on time.

Strip Mall Call Center

It was ten minutes before nine in the morning on the busy roads of California, and Makatomi was in a regular wheeled luxury sedan following another sedan and trailed by a Land Cruiser. A couple of small buzzy drones circled their convoy.

They arrived at the parking lot by the side of the offices where Makatomi was to meet the tracker. The parking lot was adjacent to a large shopping mall, which had car parking and a service road running alongside the offices where he was due to meet. He could see the signs for the call-center. Low rise, maybe four stories and adjacent to a few strip mall retail outlets. A pizza place, realtors, a financial institution. These were lower rent locations, and Makatomi was confident it was a short-term front used for his benefit.

Two of the heavies his company had hired approached the door of the offices first, and he could hear them asking for Mr Denny Amelung and then calling through to his car to say they were about to be accompanied to the office. He stepped out of the vehicle, along with Arusen.

"Same protocol?" asked Arusen.

"Exactly", replied Makatomi, "We need to keep this to the least people knowing. I don't want the hired help in the room with us."

They stepped into the office reception, and Arusen had a word to one of the men sent in ahead. He nodded, and the two men withdrew. Arusen led Makatomi to the office where Denny was seated.

"Mr Amelung?" said Arusen, looking at the casually dressed twentysomething sitting on a leather sofa in the room. Denny rose and moved to shake hands.

"Hi, I'm Denny, Denny Amelung." He shook Arusen's hand and turned towards Makatomi. "And Makatomi San? I am pleased to meet you." He bowed slightly towards Makatomi, who returned the gesture with a nod.

"We will be keeping this off the record," began Arusen.

"...and you'll have a record of a different meeting for your records?" continued Denny. "I know the protocol."

"We're running a surveillance sweep on this room," said Arusen. "I'm sure you know that protocol too".

"Yes, and you'll find plenty of electronics here. This is a call centre, with walk-around headsets, computers, high-speed comms, video links and satellite coverage. We are highly wired."

Arusen nodded. The small unit he was using to check for bugs was utterly useless in the current room. There was so much interference, and the EMF emissions were off the scale.

"If you prefer, we can delay for a couple of hours whilst you run an isolation sweep, or we can continue, in the knowledge that it is as much in my interest to keep this quiet as it is in yours," continued Denny.

Makatomi interrupted, "We'll keep this brief. I'll explain the situation and the terms. You'll need to make an immediate decision."

Denny was aware that Suze would typically be in the decision, but they had decided that Denny's view would prevail, and it was too much of a risk to reveal Suze's involvement.

"Please, Mr Makatomi, take a seat and do explain."

Makatomi repeated the story of the Chinese cloning and the need to track the leak. He didn't mention Scrive, but he did explain the generous terms.

The situation was pretty much as Denny and Suze had predicted. A search for a leak, an enormous reward. Almost too large, which implied extra danger.

Denny questioned the situation slightly but accepted the terms and a large down-payment from Makatomi.

As Makatomi and Arusen left the offices, they were tagged by the self-adhesives in the manner that Suze and Denny had hoped.

Denny waited for the Makatomi entourage to depart. The small telemetry units in the car park tracked the exit of two sedans and another two Land Cruisers. Denny waited for another twenty minutes before calling Suze back from the main office where she had been waiting, wearing a headset.

She behaved as if a staff member, whilst they started their conversation, aware that there was an equivalent possibility that Makatomi had dropped tags or sensors in this office. It was easier to exit, so they headed for the Mall's sports center, via a sportswear shop.

They both bought T-shirts and shorts as well as tracksuits before heading to the sports centre where they added new swimwear to their collection. Then into the respective changing rooms, change and both a swim and a sauna, followed by a change into the new clothes.

"That should have separated us from any bugs or trackers," commented Denny as they made their way to the car they had parked in the Mall.

"Let's get back to base." They gently edged the car from the Mall car park and took a long route back to the apartment.

"Let's see what we can trace of Makatomi," said Suze as she flicked on a geocentric tracking device and zoomed into the local area. They were using GeoSat to ping the small tracking bugs now on Makatomi and Arusen's clothes. These would give a short timeframe for information, useful as they tried to gather new intelligence about their new employers.

"Yes, we have them," they are heading back to Santa Monica Airport," said Suze and looked at the Map display. I'm patching to the tourist view camera to see whether we can spot their craft".

Two minutes later they had access to the airport's tourist cameras, and sure enough, they were pointed towards the more interesting craft in the airport, which included the V-Blade.

"I'll get its call-sign", said Suze and she zoomed onto an aero-map which showed planes at rest and amongst them the V-Blade which displayed the N registration of an American plane.

"Gotcha,", breathed Suze as she scanned for the registration,"...and you are from the NSA?? That's a surprise, and I thought you'd be registered to Biotree!"

Suze and Denny looked at one another. This was something very unusual — way beyond the request to drill into what the Chinese were doing.

"Are we being played?" asked Denny to Suze.

"We've been careful so far, and we've already received a down payment, so it doesn't make a lot of sense?"

Denny had flipped onto a Chinese web site. He was trying to run probes into the large corporates of China. Simple stingers designed to elicit port responses from the major sites. He wanted to find some loose links that would allow him to dig deeper into some of the principal Chinese corporates and maybe a few government departments.

"Steady," said Suze, "You'll swamp their networks, and then they'll trip intrusion and saturation detection. If that happens, we'll get tracked ourselves."

"Not behind these walls," smiled Denny. "I'm after speed but not being reckless". He made a gesture of a hand rising in front of his face.

"Shields up", he said, as he hit the Chinese and also the Biotree site with port scans and intrusion probes.

3 lookout

Be on the lookout for coming events;
They cast their shadows beforehand.

Fortune Cookie, King's Chinese Restaurant, Odiham

Beijing

The Chinese Ministry of State Security based in Beijing was monitoring traffic from airwaves and networks into and out of the Republic. Less well known than UK's GCHQ or USA's NSA, the Ministry of State Security (MSS) is the intelligence, security and secret police agency of the People's Republic of China, responsible for counter-intelligence, foreign intelligence and political security. MSS considered itself one of the most secretive intelligence organizations in the world.

It used the common triggers of blacklist and whitelist words and additionally looked for unusual activity. In this case, it was coming from a couple of US-based nodes that were repeatedly running interrogation probes.

The initial reports noted that the probes were quite sophisticated, but that they jittered and sparked their way around North America.

Den Xiapau noticed the signals first," It's a range of messages, but they seem to be different orientations than from the usual mischief-makers. It looks sophisticated, but a little bit rough at the same time."

Of course, that was what Denny and Suze wanted. To shake the tree, to see what fell out. In a few minutes, the Chinese would be running countermeasures, because they perceived it as a more worrisome than average attack.

That was when they would drop their catcher link into the system. It could decode the cryptologic that the Chinese were using as a countermeasure and then replicate itself into the core. Suze knew how to make the logic hopscotch over the kernel, and Denny's logic loop was so tight that he could drop this into play before a gnat had time to blink.

In Beijing, Den Xiapau had flagged the hostile to his superiors and was readying a crypto bomb which would take down the ports that were being probed and move them into a sophisticated sandbox.

The hacker would think it was still running its penetration logic, but all it would be doing was getting ever more complex prime numbers to calculate.

It would run out of computing power within ten minutes.

Den Xiapau fired the logic algorithm. The post scanner flickered for a moment and was gone.

"Wow", said Den Xaipau," That wasn't much of an attack, but it also wasn't static".

"Slam-dunk," said Denny, "We're in- they did just what we predicted". Suze and Denny did a quick high-five.

Den Xiapau wasn't satisfied to leave it at that point though. The disappearance of the trace probe had been almost too perfect.

He'd seen something like this before in the literature. It could be a decoy probe and an iceman insertion into the system. He remembered this software usually stayed dormant for extended periods. Iceman was so named because it had to thaw out but then ran everywhere.

He looked around and ran a few port scans of his own. He may not be able to find it, but he sure would escalate it. He punched the code for escalation. "There's something strange happening here - I know it is not right..." he began to explain, and a small chain reaction started.

Probed double

"Did you notice something?" asked Denny. "When we were running that decoy search, there was someone else doing something similar? It was incredibly fast but seemed to be targeting the same area? If we hadn't been running a decoy probe, I don't think we'd have seen it."

Suze punched a few keys.

"I'm going to re-run that last couple of minutes," she said, "from the log, as an e-Discovery."

She set up a couple of monitor screens, and they watched the party piece that Denny had created. Sure enough, a few other probes were running as well. They were both able to see past the usual scam artist attacks and into something more systematic. "It is coming from London", said Suze. And there seem to be two sets running, suspiciously like us.

"But way, way fast," said Denny. They were running their playback at one-twentieth of normal speed, and even then, they were missing sections.

"...it is someone in the business", said Denny, "look at that switching."

"...and look at that second probe," added Suze. "It is going into Biotree."

"I think we've found our doubles," said Denny, "And they seem to be in London. Makatomi has hired someone else to do the same job."

London

Scrive looked at Charlie," I know we are using some brute force techniques to test the Chinese systems. We need to see what kind of response they provide.

"I just want to see how they react. If it is overt, then we can assume they are treating us as normal hackers. If they do anything special, then maybe they are hiding something in the way that Makatomi suggests."

He ran the scans, initially in a quite visible way and received back the equivalent of a perfunctory 'not authorised' signal. The response was typical for the Chinese, who had managed to suppress parts of their systems from investigation for years and mainly with quite simple techniques involving feedback like 'not available' or 'busy'.

"Let's up the game!" said Scrive and he started a second script simultaneously pinging the Biotree site. It was a fairly blatant attempt to be seen 'red-handed', if there was a situation to hide. One hand in China, the other in Biotree.

"It is no different", said Charlie, "look they are still treating us as normal hackers or spammers".

"Yes - they aren't making any connections. Either they are masking their response very well, or there's nothing to hide?"

"Wait," said Scrive," did you see that?" Someone else running a perfunctory digital attack, but it just stopped suddenly. They killed that one in a heartbeat."

"Let us take a look", Charlie adjusted a couple of settings.

"A pro. that was a decoy attack. They've dropped some code into the Chinese system. It's impossible to tell where its originated though. Pretty slick work."

They looked at one another. "Someone else is playing with the Chinese," said Charlie, "Maybe a threat to us?"

"Or possibly an ally?" added Scrive.

Makatomi was back in the V-Blade. He'd had enough of today's haggling with what he considered to be hackers. At least Mallinson had been polite. Mr Amelung (if that was his real name) had been downright scruffy and rather discourteous.

But it was the price to pay for getting these people to track down the security leak. If he could get it fixed, then his company stocks would start to rise again despite the leakage of data from the hacks.

A signal bleeped. Makatomi was ready to sign off for the day and let it go to voice when he saw it was Holden. His boss. The chair of Biotree. He would have to answer it.

"There's a problem", said Holden. "One of those hackers you've been hiring has annoyed the Chinese. They are creating bad ripples. It will damage Biotree further. They say it's something from London. Fix it."

Makatomi couldn't believe it. In less than a day, the so-called professionals were rattling cages.

If the Chinese got nasty, what could happen next? They'd lower their prices further, and the faltering Biotree would go legs up.

"I'm going to have to contact London", he thought, "...and fix Mr Mallinson. A pity".

Smolly's

Chantal knew Smollensky's. It wasn't her kind of place. She knew it had its share of investment bankers engrossed in discussions about the latest fancy cars.

Chantal decided to have some fun. She would walk in first and draw the stares.

She took out her headgear and replaced it on her head. The polka dots, the colours and the fashion accessory that looked a bit like ears would do it. Minnie Mouse meets Mui Mui.

She heard the silence as conversations missed a beat. Even Lars looked. He was with someone too, as Janie moved across to meet again. A slim woman with boyish looks and a short dark haircut. A small S-shaped tattoo by her left eye.

"Thank you for coming here", said Lars. "I've brought Carolin as well this evening; it may give you some comfort that I'm doing all of this for the right reason."

"Okay and I've brought Chantal," said Janie before Chantal could introduce herself. Chantal realised that they hadn't thought about using a different name or anything clever, despite their preparation.

"You'd better explain what this about and what has happened to Karin."

Lars nodded. Both he and Carolin were drinking small glasses of wine. A waiter appeared. "We'll have the same", said Janie and nodded towards the glasses that Lars and Carolin were drinking.

"Maybe a bottle?" asked the waiter, smiling. "No just two glasses will be fine", replied Janie. Chantal knew they would be leaving any moment.

"Okay, let's introduce ourselves more accurately", continued Lars, in quiet tones "We are both originally from the Tract."

"Outside of the Ellipse", said Carolin, "So we're using quite a lot of Tract resource to come here today and to look like Ellipse people.

"We're leaving right now though", said Janie and pulled Chantal from her seat.

Janie looked across to Chantal. It wasn't a normal situation. Janie hadn't met many Tractwalkers, who mostly had reputations as thieves and roughnecks. She had hardly spoken to any of them before, now being with two of them in Smolly's was quite a revelation.

"We're catching a taxi, right now. Are you coming?" They moved outside and across the short pedestrian walkway to a line of waiting black cabs. They were parked on their landing wheels, and

Janie walked to the expensive rank that didn't use the street rail.

They clambered inside and Janie asked the driver to start the run to Westminster, "But follow the rail", she added, effectively forcing the driver to take a slower route.

They sat two across from the other two. Chantal noticed that Carolin looked as if she hadn't been in a cab before.

"I hope you don't mind me asking...but how do you afford to come into a bar like Smollensky's?", she asked.

"We have been selected from the Tract dwellers to make representations because of what has been happening. I think we see things from the Tract that most Ellipse dwellers are unaware of," continued Lars.

"It links to the economy and governance of the Ellipse people by the major corporations".

"You mean since the lifestyle improvements were introduced after the Great Leap?"

"Exactly, so-called improvements since the division between wealth economies and agrarian cultures"

The taxi was running smoothly now; it had taken one of the road tunnels from Canary Wharf's island back towards the City of London.

Chantal knew that Lars's comments were referencing the progressive split as service industry-based economies became wealthier and the land dwellers who farmed and fished became separated and eventually separated. Then barriers had been introduced that made the free passage of people from one environment to the other more difficult, unless for economic sustenance of the Ellipse.

"Carolin and I are Tractwalkers, who are part of a small group who can pass from the Tract to the Ellipse and back. For reasons we don't understand, our bodies don't trigger the 'glimmer' which enforces the boundaries.

We are considered as the servant class, but today you can see we are trying to blend with the Ellipse people."

"Are, are you really from Norway, then?", asked Janie," or was that part of your cover too?"

Lars nodded, "It's true, I am from Norway originally, although, after the introduction of the cartridges by the Nordic governments, I opted out of the healthcare program bracelets, and then found myself opting out of the economy"

Janie looked at Carolin, who was wearing a dark long-sleeved sweater which concealed her arms.

Chantal assumed that Carolin wouldn't have a cartridge and it was evident that Lars didn't. Chantal also wondered why Janie hadn't spotted this, but with cartridges being so commonplace it was like not noticing whether someone was wearing a watch.

Lars continued, "Remember, the cartridges were originally introduced as a response to the N3Ro virus. N3Ro occurred patchily around 20 years ago and very suddenly had reached an inflection point where it voraciously attacked large parts of the population.

"Some were immune, but the level of death had been on a scale greater than world wars.

Janie nodded, "Yes, some in our family were affected by this."

Lars continued, "A Biotree vaccine was already available and was progressively rolled out. As the virus became epidemic, governments shifted through single-shot vaccines and but soon had to deploy the cartridge solution."

"I remember those days of needing to get regular shots of vaccine. The cartridge was a much better solution," said Janie.

"This was because the virus was also self-modifying, and the consequent protection required regular changes. In effect, the medicinal properties of the tropus cartridges could be adapted to combat the new strains.

"That's how the cartridges became the standard form of inoculation. They were fitted to infants at birth, and to the majority of the population that lived within 'The Ellipse'.

Lars waved his arms, "There was a huge part of the planet which had the economic wherewithal to support the ongoing cartridge programme.

"The other parts of the world were referred to as 'the Tract'. There was also a variation of The Tract around the edge of significant conurbations, where more impoverished people who had the resilience to survive N3Ro had moved."

Chantel said, "I always thought of the virus as Nero." Janie nodded, "Yes, we were taught Nero in school."

Lars said, "There had also been a challenge for those with immunity. Ellipse dwellers were distrustful of Tract dwellers for a variety of emotional reasons. Mainly it was the risk of Nero morphing faster than the cartridge immunity could handle, and so the Tract people might become carriers for a new strain.

"This thinking created the disadvantaged status of Tract dwellers and the rumours that they were all roughnecks and thieves. It created a caste system at a global level. It was stronger than anything from the eras of segregation or the castes in India. All driven from fear.

Lars continued, "So you'll appreciate that we are feeling a lot more threatened here than either of you. One word and you could have us removed from the Ellipse and even with our current special status, we wouldn't be able to come back. Please listen carefully to what I'm about to tell you."

He looked across to Carolin, and she removed a small device from her bag. It was a handheld, but it looked several generations older than anything that Chantal or Janie had seen.

"Wow, vintage," said Chantal, looking with some intrigue towards the device.

"It is all we can afford," answered Carolin," And I've brought it to show you a recording." She flipped it on, and Chantal was surprised at its relatively small screen surrounded by some kind of carbon fibre surround.

It showed a few seconds of what looked like a transit station and then a blur of some feet and a three-second image of a man. He was walking on a path that cut across the field of vision. It looked like Sloane Square, thought Chantal.

Carolin paused the device and made the hand gesture to rewind it a few frames. She paused it on the face of the man.

"This is Scrive," said Lars, "He is our key to a different way of things in the future."

He pointed to the face, and Carolin also looked intently at the freeze frame picture.

"You are still talking in riddles," said Janie, "You'll need to be more explicit. I'll give you another five minutes and then we are out of this cab."

"Okay," continued Lars,

"Some big secrets are being kept by Biotree. The company has always projected itself as very kind and generous, but we are sure there is another agenda at play. Every time we have got close, our friends have disappeared, and we have not seen them again.

"They have all been from the Tract, and that has made them relatively easy to spot, even if they have cartridge implants added."

He put his hand into his pocket and pulled out a cartridge kit. Chantal recognised it as an after-market graft, the sort of thing used by people who wanted a boost beyond the level given by the health service.

"Yes, it's a booster," said Lars, "not a straight cartridge implant, but once it is in, you need to look carefully to spot the difference."

"We have information about some kind of attempt to penetrate Biotree, not by the Tract, but by a National government. Someone is hunting for power. But we think they'll get more than they bargained. The reason it is so important is that we believe Biotree is effectively the planetary power broker."

"What makes you think this is the case?" asked Janie.

"It is simple," replied Lars. "No one dies from Nero. We haven't seen a single death from it in the last 15 years."

Chantal looked at Janie, not sure whether to believe the conversation. "You'll need to prove that," replied Janie.

"It is easy to show you that," said Lars as Carolin also nodded, "...but people are still being affected. We don't know how, but people simply disappear. Much like Karin, but from the Tract it is a regular occurrence. That's where Scrive comes in.

"He's been asked to investigate something by Biotree. We want to adapt his agenda. He is one of the few people globally that can get to the bottom of what is happening. But we can't pay him, and he works strictly as a mercenary."

Carolin nodded again. Janie looked at Chantal.

"I'm ahead of you," said Janie. "You want me to help this guy named Scrive."

Lars nodded. "I do...We do. And it's because you are not from the Tract, you knew Karin, and you work at Biotree. It is the area where you work as well."

"...And in return for this, I get 'disappeared' like Karin?" asked Janie, "It doesn't sound like much of a deal."

"I agree; we have no leverage in this situation. All I can say is that our mission is trying to put things right. To rebalance the world and remove some injustice.

Janie looked at Chantal. Chantal thought that neither she nor Janie had ever stood up to anything at a political level. As far as they could tell the world was kind of 'sewn-up' about politics. People just got on with their lives.

Chantal was always slightly more edgy than Janie, but that was because she diluted her cartridges. She did something highly illegal, which was to siphon out about half of the cartridge and sold the content on the black market. The half shots were called 'boosts' and for certain parts of the population, they were a form of additional rather pleasant relaxant.

Chantal always thought she was taking risks by doing this; she saw her own blood through the plexi, and it ran close to a red colour, rather than orange or yellow. It meant she was usually on a very low dose of the tropus that everyone was supposed to take. But it made her feel more alive.

Others did this, but it was frowned upon in much the same way as 21st Century class A drugs were considered harmful. Chantal wasn't addicted to the low dose lifestyle, but it certainly gave her more freethinking and what she considered to be sharpness.

"You're a sife", said Lars, "You siphon," looking towards Chantal.

"I can see it in your eyes. Show me your wrist." Chantal extended it and Lars noticed the plexi was almost red.

"That's how we feel," he said to Chantal," No tropus and a much clearer was of thinking, acting and living."

Chantal replied, "But that's because you are immune to the Nero virus. I'm taking a risk."

"Not really," replied Carolin, "I'm not immune but have lived in the Tract all my life. My parents had the immunity, but it didn't pass on to me. The whole Nero thing is a big conspiracy now, but no-one would ever listen to the Tract on this. The conspiracy works well because everyone is frightened and therefore won't let the Tract people into Ellipse, except under stringent control."

Chantal was thinking about this. She knew that she'd been siphoning for at least a couple of years. She'd started with an eighth, but now it was half that she took out. When it dropped below a quarter, her blood started to change colour, and now it was almost the raw red colour that they had been taught was dangerous.

Janie looked at Chantal and then back towards Lars. "I'm trusting you on this far more than I should do. I need some way for you to prove you are on the level."

The taxi had just crossed Westminster Bridge and the rail it was following ended by Parliament Square.

"We're here," said Janie. "Chantal and I are leaving you now. I'll help Scrive Mallinson if he shows up at the office, but I don't know what I'm supposed to do."

"Neither do we at the moment," replied Carolin," Scrive is a tracker. He'll be looking out for signs of what has been happening. He'll need access to some things in your area, but we just don't know what."

"I'll be in contact again," said Lars. He opened the cab door on the pavement side and pushed it wide. Carolin climbed out, and he followed.

Shred

Makatomi cursed at the news that one of the trackers he'd hired was creating a problem.

He'd have to get it fixed. In a low-key way so that no-one suspected or followed any link back to Biotree. It couldn't be a disappearance, the hiring of Scrive was too recent, and the limited people who did know would think Makatomi was implicated.

He decided to call the fixer that had just helped him out to locate the heavies in Los Angeles. He needed something done quietly, and without fuss.

Makatomi reached into his pocket. The disposable cell-phone was still there. He smiled; a low-tech solution and the same phone he'd used to set up the earlier appointment. He should have already destroyed the phone. On this occasion, it worked to his benefit, because he still had the link. He hit the redial, and the phone indicated two minutes left.

"Arusen, please terminate Scrive Mallinson. I will message you his address. Use a cartridge. I want it to look clean."

In turn, Arusen sent another fixer the information about Scrive's land address as well as the IoH address of Scrive's current Plexi unit. With the

Internet of Health address, the fixer could arrange for Scrive's old tropus cartridge to be decommissioned, necessitating a replacement.

4 Scarlett's ride

"This isn't a ride you can take again,
but one, I'm guessing,
that is simply impossible to get off."

— Scarlett Thomas, The End of Mr. Y

Head-Up, Head Down.

Scrive was playing with a Head-Up Display. He wanted to try the virtual support before he took it outside. It was a new unit, and he wanted it to run very fast. He'd pressed the button-shaped device into a small carbon fibre frame attached to his left ear. It was one neat device. As he ramped up the speed, he heard a beep. It was his cartridge. It was giving out the depleted tone. He looked at it and could feel the little ripple on his wrist where it sent a small alert requesting a replacement.

Scrive thought about it. He had only put a new one in that morning. They were supposed to last for four weeks. He'd never had a defective one before. He looked for his supply. The package he'd left out this morning had been tidied away. It must have been Charlie. She'd have put it away, as part of a tidiness campaign. He flipped open the cupboard, and there was the pack. He opened it and reached for another cartridge. He was supposed to keep the old one if it was defective, but in truth, he couldn't be bothered and pinged it into the waste disposal.

He flipped the new one in and instinctively looked at his plexi. The tropus hit caught him unexpectedly, the one he'd been using before had given him a rush, but once he was back to an

335

average tropus level, the replacements didn't usually have the same effect.

He wondered briefly whether the previous cartridge had been faulty from the start.

Then he saw the impact in the plexi. Instead of staying yellow, his blood colour was running towards red — the danger colour.

He gasped.

It was turning past red to a bluer colour. He struggled for breath. The blue colour could have two meanings. He was de-oxygenating. But he could also feel a kind of pain increasing all over his body. Too fast to be Nero, but something was attacking him.

He slid towards the floor, and as he did so, a couple of the neatly arranged cooking utensils also crashed down alongside a large cooking pot.

He couldn't speak.

He couldn't breathe.

He couldn't see.

There was another crash, and he felt a sharp pain in his chest.

Pulse

Useful nanofibre pressure suits

Denny had booked some tickets on the Hypersonic out of LAX. It was a regular route, if somewhat pricey - what with enviroset taxes added to the ticket price. It didn't have the comfort of the V-Blade that Makatomi had arrived on, but as the Mach 6 speed meant the journey only took around 30 minutes it was worth the annoying pressure suit.

Suze and Denny were both frequent travellers, and both had nanofibre pressure suits of their own. They could go through the business check-in wearing the suits which would be tested on the diagnostic before they boarded. It was a lot less irritating than having to use one of the suits provided by the airline.

They decided to go hand luggage only. Makatomi had given them a decent amount of upfront money, so even if there were no separate expenses, it was still worth it for the convenience.

Denny packed a small box with some gadgets and dropped them at Fed-Ex on the way to the airport. They'd receive the items at an address in London later.

The flight was being called as they arrived and after the bios scans, checks of their cartridges, declarations that they hadn't been in contact with the Tract for the last 30 days and the pre-diagnostics for their suits, they boarded the flight.

A head-up display forced its way onto their vision. It was a feature of the suit and forced them to watch the pre-flight safety instruction. Danny referred to the suit as 'boil in a bag' on the basis that if anything were to happen there'd not be much left of anyone, except whatever it was would be entirely contained in the armour proofed pressure suits.

A few more minutes and the HUD flipped to a news channel. It could be adjusted by eye movement, but Denny just watched the programme, which was giving updates on a few political situations, a couple of scandals involving a movie star who'd been in the porn industry and some football results. Next up was a pop video, but as it started, the screen cleared. They'd landed and were in London.

The combination of the HUD, the pressure suit and the cartridge modification during flight meant that the entire sensation of take-off, flight and landing had been replaced with a somewhat intrusive video show.

Denny felt the suit pressure release, having not noticed it tighten at all during the flight. The haptics were designed to bring passengers down to reality again at the end, and he could feel circulation and pulse for a few seconds as the suit ran its post-flight diagnostics of his blood pressure and pulse.

He's joked that suits had a mind of their own, and in a way they did. They'd stay inflated and rigid if a passenger was suffering adversely from the flight, or indeed if there was any security concern when they'd been run through the passport and customs processes.

Thankfully for both Denny and Suze, the suits depressurised and returned to a skin-layer level. Denny pulled a tee-shirt over his head and Suze wrapped herself in a jacket.

"Let's go," said Denny, "We've some tracking to do".

Suze nodded. They had been awake since 4 am LA time, and after the flight against the clock it had turned back into the London evening.

Charlie's titanium SIG

Charlie heard the crash from the kitchen. First a small one, then a larger metallic one and finally a scraping and a thud. She was in the bedroom and reached into Scrive's bedside cabinet to bring out a SIG pistol. Titanium, lightweight and with minute rounds, it could fire as fast as many machine guns.

She flexed and moved fast and silently towards the kitchen. She could see Scrive inert on the ground. Scrive was on his back, arms upturned, but still breathing. She could see the plexi and the blue, almost black, blood display.

"Scrive - stay with me - this is Charlie," she called, "this is a nano crime". She glanced around and slid back to the bedroom, returning a few seconds later with a small handheld wired to a little T shaped connector.

She lifted it above her head and smashed it down onto Scrive's arm, smashing the plexi and the cartridge. She held it there as black blood oozed onto the floor. She was holding in two buttons on the device and kneeling in the slow ooze that was dripping from Scrive.

The T connector glowed blue, and she could see it was processing Scrive's blood.

Ten minutes passed. Charlie looked down at Scrive. The thin slick of black slime on the floor by her knees was changing colour. Scrive's face was also evolving from a grey colour to one that was notably redder. She could hear him breathing very quietly. She gently removed the unit that she had slammed into his arm. She stood and hurried to the freezer, selecting a polypack of frozen vegetables. She ripped the packaging with both hands, allowing the small, chopped vegetables to fall to the floor. She took the packaging and wrapped it tightly around Scrive's wrist. She could see his eyes moving rapidly under his closed lids.

Then he shook violently, and with a sudden and sharp gasp he sat up.

His eyes opened.

"What's happened?" he looked at Charlie, "Something has gone wrong."

"Yes," replied Charlie quietly, "I think someone has just tried to kill you."

Scrive looked at his arm. "I was adding a cartridge; the last one was defective."

"Maybe...", replied Charlie," But I think the new one was contaminated. Deliberately".

She looked again at the liquid on the floor. It was now a blood colour, having been black a few minutes earlier.

"Ew," she said, "Kind of messy."

"Someone had added a payload to your cartridge", said Charlie, "I've seen this effect when I was working with the nanotech. Someone has added nanobots to the cartridge. I think they were intended to mash up your inside."

Scrive looked around and spotted the handheld device on the floor. "You used your experiment on me?" he looked at Charlie.

"This time, bad science has a good outcome. I saw your black blood, typical of a nanobot incursion, I reckoned that my device would finish off any normal nanobots and it seems to have worked. I inserted a huge block of nanoreductives into your bloodstream - it worked because even the blood spill has been processed. The zappers half-life is one hour, so we'd better keep an eye on you. What's interesting is that the zappers have been so thorough that your blood has reverted to red. Not yellow or orange. There's something else weird happening here, normally nanoreductives restore balance but are not powerful enough to completely neutralise tropus."

Scrive pulled himself up using the edge of the kitchen counter. He did it slowly, aware that his blood level was a little low.

"Lay down, you've lost some blood" said Charlie, "and where do you keep the first aid kit? I don't think you've lost that much. It looks worse than it is. I'm going to guess less than half a litre. That's less than a blood donor session."

"Weak tea with sugar and a cookie then?" asked Scrive. He was aware that his reaction times had slowed. He seemed to be running at normal rates. He wouldn't push it at the moment until his cardio had stabilised.

"I'm a little bit worried that whoever has done this might want to come back and take a look," said Charlie. "I think we may still need to fake your demise".

She slid across the floor and retrieved the SIG pistol. There was a quiet click as she replaced the safety catch.

Non-linear cubism

Denny had gone non-linear. He was strafing various global systems to try to find any patterns that would help trace what had been happening in London, LA, and wherever Makatomi seemed to be taking his presence.

He'd brought a Cube with him; it was a small offboard processor that he could use with a regular handheld to boost the power and number of connections. He'd brought them both in his carry-on luggage as well as a somewhat cumbersome looking mains adapter so that he could run his system in the UK. It wasn't the voltages; it was just the over-engineered plugs and sockets that the Brits seemed to prefer.

Suze was also browsing online, but at a more leisurely pace. They'd checked into a central London hotel and needed to run their operation without attracting attention. She was gently reprogramming the network around the hotel so that Denny's activities wouldn't draw undue attention. He was hitting the system so hard that any schoolkid could probably spot the activity. Suze was creating an electronic cordon to stop it being detectable.

Denny glanced up.

"Makatomi is a real spinner," he muttered, "One minute he's in London, then LA, then Tokyo. He's carrying Arusen around with him on that V-Blade, but there's another electronic presence that seems to skitter around even faster."

Suze nodded," Yeah, I saw that too. Holden. He seems to turn up everywhere. I can't tag him properly; it is like he's somehow in the system. He's also been to Norway and Arizona."

Denny nodded. It must be evident if Suze had picked up on it too. Their work division was unbalanced, and he'd been doing most of the heavy tracing. Suze was quietly folding some of the quaint but expensive grey hotel stationery into the shape of a swan, with her spare hand. She was already wearing the colourful courtesy gown and had now pushed two of the chopsticks into her hair, making an instantly more Eastern look.

"Was that the influence of the room service?" he quipped. They'd ordered Japanese as a sort of homage to Makatomi and been enjoying maguru tuna sushi with nori seaweed. Suze had spotted a pineapple dessert, but neither of them had expected the laser cut slices laminated with microlayers of a ginger flavoured wasabi.

"Yes, it's auto suggestive, I think," replied Suze as she flipped another firewall. "The ginger and pineapple must be talking to me."

Chinese wake up

Guangdong Province, China and Shenzhen Ruby were due to play an important match. Den Xiapau had tickets to the Shenzhen Stadium for what promised to be the Super League match of the season, as far as he was concerned.

But.

Now he'd alerted everyone about that hacker attack, which seemed like two different sets of people on the same mission, he was notionally chained to his desk.

The station commander had called for a lockdown. No-one in or out. Sure, he could get a bowl of noodles from the canteen or spend half an hour playing some table tennis, but he was fundamentally locked in until they'd got to the bottom of whatever was happening.

There had been an initial burst of system scans apparently from all over the world. Den Xiapau knew it was typical of an advanced hacker - probably a tracker - who was trying to find out something important. It looked like a coincidence that the second series of probes had appeared but the second set (which seemed easier to trace back to London), had also been shooting probes into Biotree.

He wasn't sure what to make of this. Was China being challenged, or was this some corporate protestor trying to dig into Biotree's vaults?

The main reason for all of the escalation was because of the sophistication. Two sets of almost simultaneous probes with Chinese corporations and governments as a target.

After the first alert, they'd decided to inform the US and European Governments. If this was someone playing silly games, it was better to let them know that they had been discovered. The longer they left it, the more the West would get cocky about such things.

They spent a little longer deciding whether to inform Biotree. Still, because they also had a sizeable Biotree manufacturing plant right within the province, it seemed like the right thing to do.

The challenge now was that both the Americans and Europeans had escalated it within their jurisdictions and Biotree's Chairman had been on the red phone to the Chinese President.

It was all causing something of a meltdown.

Then, after the initial attacks, everything had gone quiet for a couple of hours. What was interesting now was that a new series of high-intensity probes seemed to have started. From London, although someone very sophisticated was running a cloaking operation on them. It was fortunate for Den Xiapau that he'd spotted the source and knew where to be looking. Otherwise, the cloaking would have worked before he'd been able to get a fix.

5 the enemy

And all I need now is
intellectual intercourse
A soul to dig the hole
much deeper
And I have no concept of time
other than it is flying
If only I could
kill the killer

All I really want – Alanis Morissette

Bodø

Sheri was originally Canadian, although she had studied in the USA, as well as a short spell in Switzerland and was now into her second year at Biotree's facility in Norway.

The Bodø environment was surprisingly familiar, a mix of her childhood's Vancouver waters and the nearby ski areas, where she had spent winters skiing as well as getting something of a reputation around Whistler for her freestyle snowboarding.

The cold end of the Pacific had first raised her love of nature. She would still think of times spent with her Grandfather out to look for whales with their tail splash, fishy snorts and the rippling radiation of the water as they would dive near to the boat.

The Pacific had also stimulated her study of marine biology and the organisms that maintained the ecology. Then her time at Harvard where the study of very small things had eventually led her to Biotree. Harvard had taught her how the organisms worked and then CERN in Switzerland had taught her how to build them, ironically by first showing how to smash things apart.

Now she was working with mechanosynthesis, construction an atom at a time. It was beyond a watchmaker's precision, to know how to bolt the atoms together to make the tiny machines that formed the basis of the Biotree business model.

She'd learned how to build these tiny structures, how to make them operate, which parts would simply refuse to work together because of the still only partly understood and apparently tiny forces between them. Forces she knew were big enough to destroy the machines to which they were attached if they were not coupled properly.

She sometimes thought of it as being inside God's head. If a God existed, the God would need to know this stuff really well.

It was her work in the USA had lifted her profile considerably. At Harvard she had gained extra letters after her name. She had also met some of the super-scientists that worked in her field. They had told her about the opportunities at Biotree, but initially, she had remained sceptical.

Then she'd worked on ultra-transformables, which were a branch of the science that could help significantly in healthcare, the machines having a squishiness which meant they travelled well inside humans.

The spell at CERN had been mainly using the accelerators to smash a few things apart and see the effects. There was something mysterious about the power needed when humans tried to do these things, compared with the weak forces that were apparent in the nano-machines and which could do exceptionally more powerful things if assembled incorrectly.

Back to God's head, it was like His way of saying, "No, No, don't do that."

Inevitably she'd also run into nanotoxins as part of the research. Still, to people in her community, there were some basic rules about what to attempt and mix, and most of the 'No-no's" were very obvious. It was more the effect of a constructed machine and its erstwhile operating system that became the challenge and the thing that led Sheri to Bodø, Norway.

She'd been working on nanoparticles to improve foodstuffs and found that the addition of inert machines as a way to deliver medicinal payload didn't work very well.

The human body (or any other living organism for that matter) detected and destroyed the nanobots on the way in. If they couldn't be destroyed, they were at least neutralised although this could leave residuals in the body.

What was fascinating was that the residuals were stored in almost homeopathic quantities.

Sheri had worked out that an average human adult could eat nano-processed food every day of their life. Unless the human system changed the form of handling, the effect of the residual "neutralised and stored" would still only amount to something which in homeopathy was called the 60X formula. This wasn't one sixtieth, it was ten to the power of minus sixty. Something like the equivalent of a single pinch of salt into the Pacific Ocean. This level was so far below the 24X considered to be the limit of any homeopathic remedy, that the little broken machines couldn't pose any threat at all.

Of course, that assumed that the body had done its 'repel all invaders' thing and broken the machines down and expelled them.

It was the very power of these tiny machines that fascinated Sheri. They were already being packaged and consumerised by Biotree, and she knew there were many more practical and positive uses.

When she'd left for Bodø, it was with the idea of spending a couple of years, to make enough to live well and then move back to the West Coast. What had been seductive about her time was both the work and her discovery of Nathan, a fellow Canadian worker, with whom she now lived on the extended Bodø complex.

There was also the feeling in the Biotree Bodø R&D facility of being in the core of the core. It wasn't just in the Ellipse, Bodø was the inner sanctum of how everything worked. And there she sat, in the Advanced Technology Area, right at the centre of the centre.

Sheri knew that both she and Nathan had found something worthwhile and challenging. They'd both talked recently about settling longer in Bodø, and with her approaching birthday, she wondered whether Nathan was getting ready to ask her a big question. She smiled to herself as she knew the answer already. And it would give an excuse for a trip back to Canada.

V-Blade departure times

From their London hotel room, Denny was convinced that there would be a way to track the other suspicious person that might be following this assignment. He was sure that something had happened around the time he had run the last trace.

He was still hitting the Chinese sites and the Biotree locations but was now interested in tracing back Makatomi's steps.

"Suze let's see what happened inside Biotree this morning". He looked across at Suze, and they both laughed. Was it really the same day?

Suze was testing a few links around Canary Wharf. She found a tourist weather cam monitoring the general view. She jacked into its archive and ran back to the morning period. In reverse, a purple flash appeared over one of the buildings. She stopped the rewind and hit a button.

"That's the 'out' point", she said and continued to run the video backwards. Another purple flash, "and that's the 'In'," she added," Now we can see when Makatomi arrived and left."

Denny nodded. Suze had found the arrival and departure times of the V-Blade. There wouldn't be many of these crafts around, and a direct docking with the building was still relatively rare. They both knew this would be Makatomi's ride at the headquarters.

"Now we have a time range, we need to work out where he's meeting his visitor", said Suze," I've got the building directory here. It is the one used for visitors to the headquarters. Here we are, floors 60-70 seem to be the ones for the special meetings."

Danny looked at the list. There were a couple of boardroom floors, some training areas and three floors with individual meeting rooms, including two levels with ultra-secure screening.

"It will be one of these," he said, pointing to the secure facilities. Makatomi won't want to take any chances."

Suze nodded. "Let's play with the elevators". She flicked into a few more screens and then a few more.

"This isn't working, there's no access", she said.

Denny picked up his handheld and tapped a number. He was calling the building's foyer. "Hello, I'm a guest in your building. I seem to be stuck somewhere between floors 58 and the top." He hung up immediately.

The foyer receptionist punched another number.

"Gotcha," said Suze, "watching the number ringing through and tracing it to Westinghouse, who had a facility in the Canary Wharf area. They have people on hand to manage the elevators in case of problems.

"Now I know where you are, I might just need to borrow a couple of your documents," she said as she dragged a couple of diagrams onto her own machine's desktop. They were high-level plans of the elevators, with the network number for the various individual shafts.

"Okay, now to find their camera feeds". Suze tapped a few numbers and soon had three TV pictures displayed on the screen of her computer.

"Now I need to run the archives". Suze tapped some numbers and three pictures simultaneously wound back to the time of the V-Blade landing. Then she ran them forward at ten times normal speed. They watched as one floor became busy, but the other two had no visitors.

Then suddenly a lone individual appeared on floor 63. He walked to the reception and was met by a third person. It wasn't Makatomi.

"It's him," said Suze," I know it is."

Contamination

Scrive looked towards Charlie after hearing the safety click from her weapon. "You found my pistol, then?"

"Old habits," replied Charlie.

"What happened?"

"My handheld, the one I told you about. Complete with my failed blood reprocessor - the one that destroys nanobots. That's what I used on you. I could see your blood. It was turning black. Someone has contaminated you with nanobots.

"It looked like they were attacking your blood. It was clotting, instead of flowing. I managed to smash into your circulation through the cartridge and to release my little App. It is the first time its failure has been useful. It caused a signalling failure amongst the 'bots. Whatever they were doing stopped.

"It looks to me as if they were attacking via your cartridge. Someone has contaminated the tropus. "

Scrive looked around. The kitchen was a mess. He was a mess.

"Thank you," he said to Charlie, "I think that could have finished me."

"Yes, and I'm not sure whether whoever did this will be back?"

"And they don't know about me at the moment", said Charlie.

"That's a good point," replied Scrive. "We should keep it that way, no sense in showing all of the hand."

"And maybe we can throw them a scrap too? See if we can draw them out?" Charlie grinned.

"Okay," said Scrive, "Maybe I need to disappear for a while, make them think that I've gone."

"And not show them that I'm around either?"

"...Or we do something more reckless?" Scrive pulled himself up, "Maybe I need to carry on for a few days as if nothing has happened? They won't know when I'm exchanging the cartridge; they may even know that I'm a little lax on such matters. I'm going to carry on normally for the next few days. The difference will be my traceability, and you can help with that."

"Sure," said Charlie, "But I think I'd better move away. It's better that they don't know about me at all, and it will be a lot easier if I'm somewhere else.

And from one tracker to another, don't even think about tracing me."

She was already walking to the bedroom. Scrive could hear the sound of a zipper bag and realized that Charlie was already planning to move out.

"You'd better take this," he said and handed her the high fire rate SIG, "I guess you only found the one," they both smiled.

"And I'll be listening out for you", said Charlie, "Let's just agree on some signals." Scrive tapped a series of numbers into his handheld and then pressed a few keys. The numbers transferred to Charlie's handheld. They now had a private key between them, which could be used for tracking and tracing.

"And we'll use the transit station as a drop," if we need it. By the left-hand side of the entrance. Use a marker. I'll find it."

"Daily," said Scrive, "But don't tell me where you'll be".

Scrive knew that he and Charlie could operate a hidden analogue protocol between them. Charlie could track him, but he wouldn't know where she was. If she needed to exchange information, they would use a physical location, which Scrive would check every day. Charlie would not go there herself but would use an intermediary if there was to be anything placed there. If Scrive needed to pass information to Charlie, he would use the tracker to signal a location, leave the material and then continue. Scrive knew that he was now on an electronic leash, which was managed by Charlie.

Camtran mission

Suze and Denny continued to watch the replay. Suze was intently watching for the return of the visitor to the elevator. Some 20 minutes later, he arrived again, still accompanied by the same person. It was difficult to make out what they were saying, the sound channel from the elevator monitoring was faulty, and the wide angle meant they were both shown tiny.

"That was a short meeting," said Denny. Suze nodded. "Maybe he's not getting paid well as us - that can't have been much of a negotiation?"

Suze was referring to the negotiation they'd run with Makatomi, once they had realised that money wasn't a problem to resolve this case.

"Okay," said Suze, "Now let's see you leave the building", they flicked to another camera view from the ground level and saw Scrive exiting towards the transit station.

"Go in, go in," said Denny, willing the image of Scrive to enter the Tube. He did, and Suze and Denny smiled. We'll get him at the gate. Scrive passed through it as it automatically opened. Suze tapped a few buttons and moments later retrieved the cartridge RFID for Scrive.

"Radio Frequency Identification - Thank you for some old technologies," said Suze.

"We have you now, Mr Mallinson," she said, as the RFID sequence yielded its information. And you seem to have good health records too, Mr Mallinson of Chelsea Bridge Wharf."

Denny calculated that Scrive Mallinson's home address appeared to be less than three miles from their hotel in Central London.

A visit would be necessary. There were two ways to do this. Overt or Stealth. Denny ran some further scans to try to find the back story for Mallinson, but it wasn't straightforward to trace more than the most perfunctory information.

Denny thought this added up. He knew that a similar scan for his own ident wouldn't yield much, so the sequence of events seemed to form the right type of trail. Mallinson visits the office of Makatomi. He has as a meeting, leaves and then within a couple of hours there's a series of scans being run on China and also on Biotree. Checking on Mallinson yields little. Denny was pretty sure it had the hallmark of another tracker.

"Suze, I think Mallinson has the other half of our contract." said Denny.

Suze nodded. "Yes", there's a pattern that looks strangely familiar. And now we know where he lives. I don't think we should try to deep scan though, if he's as good as us he'll notice it in seconds.

"Do you fancy a ride in a London taxi?", asked Denny to Suze. "I think I know someone we can visit."

"Now about tomorrow morning?" answered Suze. "We'll need to be sharp." She pulled the chopsticks from her hair.

"You stay here", said Denny, "I'm going on a little mission with a camtran. It would be good to keep tabs on Mr Mallinson." He flipped open his hand luggage and retrieved a couple of small camera transmitter devices.

"I'll see you in about an hour."

Denny slipped from the hotel and took a taxi across town. It was one of the black cabs that could go anywhere, although their route seemed to be along the major streets. He was dropped outside a large apartment block, next to a hotel complex.

Denny looked around and then reached into his bag. He identified the entrance to the apartments and set up a small camtran monitor across the way. It showed the main entrance and part of the thoroughfare either side.

He looked across to the adjacent hotel to the apartment buildings, then crossed the street and made for the reception. Ten minutes later, he returned with two cardkeys.

He and Suze were about to become Mallinson's neighbours.

Norway

Sheri was at home with Nathan. She looked out of the window, across snow towards some distant hills. Beyond them were mountains. She had grown to love the Norwegian scenery. The water, the sparkling ice and snow. It was like every cliché she'd heard about the place, an extraordinary land where bad vistas were not permitted. It made the already enjoyable work even better. She had a good location, good job, good money and most of all she was now sharing it all with Nathan.

Although she and Nathan both worked for Biotree, their roles were very different, and they didn't see each the during the working day. The campus was vast, and they would usually go to work separately, and because of the distance and security inside the facility, it was best to stay out of contact throughout the day. There were ways to communicate, but it was much the same as if they worked in different towns.

At the moment she was reading a report on her handheld. It was related to engineering advances in India and the possibility of some breakthroughs in molecular design. Sheri could sense immediately that this wasn't a plausible article. There were holes in the logic, and the approach was one she already knew to be flawed.

Nathan was preparing some food in the kitchen and called through from time to time to inform of progress. She could generally tell by the variety of hissing sounds, aromas of onions, garlic and the tell-tale sound of a bottle cork being extracted. The evening meal was almost ready.

Sheri knew she had the more demanding and specialised job of the two of them. Her work was at the (almost literally) cutting edge of the design of the nanomachines. Originally, there had been a set of standard assemblies that worked well together. Most of the incremental designs were based around these pieces.

When she explained it to others, she likened it to a car. Four wheels, one on each corner, some seats an, engine, steering and brakes. The nanobots had a similar basic construction kit.

At the small sizes of the machines, it was the use of protein as a fuel and the same effect that makes creme in coffee eventually spread through the liquid that created for a bot a power source. It also provided a significant physical limitation to the small machines and their deployment. The 'coffee-cup' Brownian motion made everything shake dramatically at this small scale. Nothing was every static and the trick with the scientists was to harness this motion as a power source.

The building blocks that Sheri used were similar to the components of the car. There were components for movement, components for sensing, components to join things together - the so-called fixtures - like the chassis of a car and elements to provide grip and contact between the machines - the end effectors.

Then, as Nathan referred, "JASMOP" - or "Just a small matter of programming" to create the operating systems for these small devices. Just a small matter was an interesting point. The technology of Scanning Probe Microscopes used to view the assembly work had never really scaled itself and relied upon clean, secure environments and mega-voltages. It was not surprising when a speck of dust would be like throwing a planet at some of these machines.

Sheri's work was within the so-called 'exotics' division. As expected, there were pictures of palm trees and Pina-Coladas stuck to the walls inside, but fundamentally this was the area where Sheri tried to outdo her maker.

It was the place where new elements were designed. Original elements to create the missing shapes of matter needed to extend the constructor kit of parts for the nanomachines. The pieces that God forgot. There were practical physical limitations to how they could be used. Apart from atomic forces that would blast structures apart, the continued jittering from Brownian motion and the protein fuel consumption of the tiny devices, there were still some basic components that were proving impossible to construct. It was like the car but with only a few degrees of steering and no gears.

Sheri and the team were attempting to build the new shapes. The missing piece parts that would extend the nano constructor kit.

Nathan entered the room, triumphant. "Dinner is served!", he quipped and gestured towards their dining table. They would still eat together whenever possible because the nature of the work often meant irregular hours, and this would give a chance to spend some time chatting. There was an inverse luxury to 'dining in'. Most of the time, workers in the facility would avail themselves of the vast eco-system of restaurants and cafes that had established around the complex.

Pretty much all cuisines were catered for, from fast-food Americana to the fanciest French or Japanese food. Most evenings they would eat out, sometimes alone but often in company with others from the facility.

Being alone in their living quarters was an excellent time for decompression, even if Sheri had started the evening with a scientific journal article.

"I can't believe it's nearly two years that we've been here." started Nathan. "I know it will be after your birthday that its officially two years for you and about three weeks later for me."

Sheri alerted herself as Nathan started this line of conversation. A meal at home, talk about how long...would this be leading to a discussion of 'them'? She decided to see where it was going, but Nathan moved in another direction.

"I hope tonight's 'dish of the day' is okay?" he inquired, "I had to scratch around for some of the ingredients".

Sheri relaxed. She was keen enough for a talk about their future, but tonight it didn't somehow seem to be the time. She was just too strung out on the current work. A few tonal changes were creating some new upsets. Makatomi's business plans were at odds with Sheri's personal beliefs. Instead of the Biotree being about healthcare and the future, it seemed to be moving towards more ominous goals. They had recently brought some contractors into Sheri's department who seemed rather more lackadaisical about their approach to safety systems.

"I've been working on some new secure perimeter systems this week," called Nathan," It looks as if Biotree are getting even more paranoid based upon the recent share prices and business news."

Nathan worked in another area of high technology, but rather than being progressive and forward-facing, like the pure R&D that Sheri conducted, this was more related to the protection and security of the Biotree complex. Beyond obvious physical defences, there were rings within rings of security measures that could both give an impression of a relaxed environment but at the same time could become extremely strict in moments if something inappropriate was detected.

Nathan worked on the improvements to this world. A guardian role that also meant he spent more time around the whole complex that Sheri.

They had initially met just after Nathan had joined. Sheri had been out for a weekend skiing in the adjoining mountains, with a couple of new-found girlfriends when they had run into an 'induction team' part of which included Nathan.

Sheri was snowboarding at the time and noticed that Nathan also seemed pretty accomplished, and they'd broken away from the group to try a particularly exciting route. At least that was what Nathan had said and - on reflection - Sheri had also thought the course unexpectedly delightful.

They'd been together ever since that first encounter. After a short time, they had decided to move into what was considered one of the better apartment areas. Their facility looked out to the sea on one side and hills and distant mountains on the other side.

Tonight, Nathan had placed candles outdoors in the Norwegian tradition, and Sheri could see a distant twinkle from boats on the sea and stars in the sky.

But no, tonight wasn't the one to have deep conversations about the future.

Backtrace

Captain Taylor had a problem. He'd almost singlehandedly scrambled the U.S military to a state of high alert. All based on what started as a routine intercept in China.

He had seen some routine tracing by a hacker; it had tripped his high-intensity alert. He'd selected the trace, and watched it hop about between Biotree and SuzGene, a Chinese biochemical producer. Then he'd seen a second trace. It looked as if it was doing something similar. Taylor worked out that the traces looked like they were coming from professionals. London and, more worryingly, Los Angeles. The LA one was a problem because if it was left undeclared, someone might challenge it as a state cyber probe. The kind of probe invoked before a cyber-attack.

Taylor knew there was no such thing on the horizon and thought he'd better call it in. That's when everything disappeared and seemed to go back to normal. Except, it looked as if a Chinese listening station had intercepted the activity and was now sending a few quiet probes to try to backtrace to the source of interference. He could see that the Chinese were running their monitoring from Beijing, and he thought it highly likely that it originated in the Chinese Ministry of State Security (MSS). The event was troublesome because it could easily pop up as a diplomatic incident.

He worked out that the president would already have been alerted to the situation. Taylor was old-school, from before the Great Leap, when most comms was still stuff with dials and buttons, rather than something one gestured to or voice activated. He'd paid extra to have his automobile fitted with a facsimile retro radio, one that had a rotary volume and push buttons for the channel selection. He knew that behind the covers, it was all solid-state, but it somehow felt better to dial things up.

Taylor had a secret habit. He was one of the people who ran low dosage from the tropus cartridges. Taylor didn't sell on his excess tropus, which meant that hardly anyone knew about his habit. He'd discovered the phenomenon by accident after a boating injury, that depletion of tropus made him feel somehow sharper. He could also remember more. He doubted, now, that he would remember about the analogue stations if he was on full dosage.

He'd still had to deal with the shadowy worlds of sifes and Tractwalkers though. He knew that the standard cartridges produced logged usage reports which could be monitored. He'd arranged through dealer connections for the fitting of a cloned cartridge holder, which somehow reported normally, despite his unconventional usage. He'd even paid over the odds to get it fitted because he didn't want to be linked to a dealer to whom he would supply his surplus tropus. It was another link in the chain that he'd prefer to avoid.

Taylor knew he could still fire up the old listening station and with the improved access from an array of satellites plus the old links created when Camp LeJeune, NC and Naples, Italy were still front-line monitoring stations, he might be able to uncover some further information. A few of the old links were worth re-activating

Outside was heavy rain. A squall from the Pacific had whipped into the harbour area, and there was a grey blanket across everything. Taylor was wearing a weatherproof cape, and a walked at an angle of forty-five degrees towards the four-wheel drive. The usual transit rail system was available in the main town here, but he'd be trekking into the forest and paying a visit to a couple of smaller buildings seldom used nowadays.

He booted up the four-wheel drive and reversed onto the road. Then along to a lane and then to the end of a road with a gate and wired fence.

He pressed an access pad in the four wheel's cab. Two large metal bolts slid open in the gates. Perimeter lights lit and a series of green spots briefly twinkled. He had disarmed the perimeter field, which would otherwise prevent his progress. He drove through the gates and into the area, which was buried in the forest. Then further along the track and sharp left. Two buildings, overgrown with what he'd have called kudzu or foot-a-night, back home in Tennessee, but here it was some other kind of vine. It did the trick though, and the two buildings were almost absorbed into the forest and undergrowth.

He tapped a few more access codes and hopped from the vehicle. A splash from underfoot, despite the now more distant sound of the wind and rain. The trees provided a groaning shelter in this part of the forest, although he knew that the weather further afield was still savage.

Now he entered the first building. He reached to the side of the door for a still familiar monitor light. Flipped it on and heard a countdown sequence. He had 60 seconds to disarm the building and punched in the access codes. 3-2-7-0. He still remembered it, despite the years of disuse. This facility had been a staple in the older days when picowave interception and traditional listening posts were still the order of the day. Now the remnant of the facility might give an unsophisticated way to access what was happening in Beijing. He really was going 'under the radar'.

6 rust promises

What you see is what you get
You've made your bed, you better lie in it
You choose your leaders and place your trust
As their lies wash you down and their promises
rust
You'll see kidney machines replaced by rockets
and guns

Paul Weller – Going Underground – The Clash

Follow me

Scrive was walking towards the transit station on his way back to his apartment. It was time to check for any update from Charlie. They were using a purely analogue protocol so that they could not be tracked by listening in through phone links or similar. Scrive knew that if anyone wanted to, they would probably now have access to his RFID and could be tracking his movements. The main thing was to make it all look normal as a way to reduce suspicion.

He'd patched his arm from Charlie's 'smash and grab', but it meant he couldn't use tropus again until he got the cartridge's plexi fixed. That would require medical attention which would signal that he'd already taken one of the doctored cartridges.

He'd be using the cocktail of his blood's residual tropus, any remnants of the hostile nanobots and the hacked concoction that Charlie had forced into his bloodstream. His blood was already running red from after Charlie's intervention, and now he simply didn't know because of the lack of a tropus fix.

Yesterday he'd used a pin to prick the end of one of his fingers, to see what colour emerged. It had been red. He'd been taught this meant danger, and it had made him feel unexpectedly queasy at the thought. Then he'd rubbed the finger, the blood had stopped, and since that point, he'd tried not to think about it.

Now he certainly felt okay. He felt better than he'd expected. He felt sharper with a kind of clarity and freshness. Today, it was beyond his ability to rev his thoughts to a higher speed. This feeling was offset by the sense that part of his mind was awakening that was usually dormant.

He took his routine walk to the transit and already had in mind the point he would approach to check for any marking from Charlie. It was only a day, and he didn't expect there to be anything yet. As he approached, for the fourth time he saw the slim girl he had noticed previously. She was walking across in front of him. He saw the tattoo. She was singing quietly to herself as usual. 'Follow me', she was singing. "Follow me, follow me, follow me Scrive. Follow me Mr Mallinson."

He looked startled.

Then he decided to be as deadpan as possible.
Was this something from Charlie? He doubted it.
It was too much of a coincidence to have seen this person four times consecutively. But there had to be a link. He thought quickly. He'd seen the girl before he'd been to Makatomi. Before the threat to his life. Before he'd made arrangements with Charlie, but this had to be related.

He decided to follow.

Around the street corner was a small pavement cafe. Busy with local well-dressed people sipping small coffees and tapping into various handhelds. His guide showed him to a table in the corner. An athletic-looking and tanned man was sitting there.

"Mr Mallinson?" he inquired, "My name is Lars, and this is Carolin. We think we can help you."

Scrive looked at them both. He noticed Lars looked weathered in a way that wasn't very common. Then he realised, Lars was from the Tract. He looked back at Carolin. So was she. He hadn't picked it up when he first saw Carolin; maybe he was looking at, well, other aspects.

"You're both from the Tract, what are you doing here?" he asked.

"The same thing as you, "Mr Mallinson," we are looking at what Biotree is trying to conceal. We know it is not about a Chinese plot, and that the story has been put out to help explain the share price drop and the strange activities within the company."

Scrive was tallying this new information. He'd only been involved with Makatomi for a day, and already someone had tried to kill him, and now he was being intercepted by a couple of Tractwalkers.

"Okay, say I go with this," replied Scrive, "what else can you tell me that would make me want to work with you?"

"The main thing we have to offer is immediate access to someone working inside Biotree, close to where Makatomi operates. But before that, I should tell you some other things."

Lars began to explain his background, that he was from Norway, that he lived close to Bodø, near to the Biotree R&D complex. He stated that he was one of the first people discovering their immunity to the Nero toxin. He had been forced into the Tract, along with others that were also immune.

He lived close to Bodø and seen the growth of the R&D facility and the influx of scientists.

"But the thing I need to tell you about is that Biotree has been working on its second-generation business. Nanobots. Before that it made its money from the tropus and cartridges."

"Now it is working on a third-generation business, and that's the thing we need prevent."

Scrive knew that the original business model for Biotree, along with several other large pharmaceutical based companies, had been the creation of the tropus that had formed the basis of the inoculation against Nero.

He also knew that the pricing for this had started high because of patents, but that global governments had forced a rapid change in order that the tropus could be manufactured on the industrial scale needed to provide it for large parts of the world population.

Everyone in the Ellipse took the regular medication of tropus through the cartridge system, and it had grown to be a basis for other services, based upon the addition of the radio links and secure identity capabilities.

Of course, Scrive also knew that this had effectively given everyone an electronic tag, based upon the need to use cartridges permanently and that the cartridge support clip was mainly where the tracking, wallet, comms, telemetry, health and other functions were held.

Scrive had also been noticing that his own system didn't seem to be affected negatively despite his lack of tropus. He had never been designated as immune and wondered how long Nero took to have an effect.

"You're both immune to the N3Ro?" he asked, "I can see that you don't have the plexus, so I assume no cartridges?

"Correct," said Lars, "but that's one of the things you should know. The Nero story has been convenient misinformation for the last several years. The cartridges that are in use now are for another purpose."

"...and what might that be?" asked Scrive.

"To stop you from seeing", replied Lars. "To hide what is happening".

Scrive looked quizzical, "To hide what?" he asked, "to stop us from seeing what?"

"You'll have to take what I say on trust," continued Lars. "We've no way of proving this here, although we do have other things we can show you later."

Scrive said, "It's already been a helluva week. I'm prepared for just about anything at the moment." He absentmindedly scratched at the place on his arm where a bandage covered his broken plexi.

"This is what we think has been happening," continued Lars,

"The Biotree corporation are being employed on several missions. They are one of the manufacturers of the tropus that all of you use to protect against the N3Ro virus. That has been a great continuous source of revenue for them and allowed them to create the facilities to build their other businesses."

"The main source of new business now is the Nanotechnology, although they seem to have a hit a wall with their R&D and at the same time there's a rumour that the Chinese have cloned the technology."

"We think that's why Biotree has employed you, to track down if there's a leak to the Chinese. We don't think you'll find one though, we think that the Chinese will be as surprised about this as anyone."

"How can you say that?" asked Scrive," You don't appear to have the resources, especially being based in the Tract."

"It is because we are from the Tract that we think we have an idea about what is happening.

"We believe that the Tract is the source of the new nanotech and that there is someone building products to compete with Biotree. In the Tract, we hear of things by the old-fashioned ways. Some of it isn't as reliable, but there are too many stories of people in an area deep in the Tract where this work is being done.

"If that's the case, how is it we haven't spotted it from the Ellipse?" asked Scrive, "I can't believe that anyone would be able to do this without someone from the Ellipse knowing what was happening.

"Normally I'd agree, said Lars, "But I think there is a reason, part of the way that the Ellipse has been engineered means that there is a huge blind spot for what happens in the Tract. It is as if we don't exist."

Scrive thought for a moment. It's true; he knew the Tract. Aside from the warnings to stay away and that they may be carriers of the Nero virus, he had little idea of what else went on.

"What? Is it like a return to the 'flat earth'? When people used to think of the earth as flat, before they discovered it was round, then the corners of the map would sometimes be printed with 'Here be Dragons'?" asked Scrive.

"It is a little like we've returned to that concept." continued Lars.

"Except we do all know that the world is round, indeed that's where we get the Ellipse term. There are still some areas that don't ever get mentioned by Ellipse dwellers. It is mainly an area south of Japan and China, an area we still call Australia."

Scrive replied, "But there isn't anything south of Japan, it is the exclusion zone, where the Nero toxins wiped out everything and where there is still a huge risk to anyone that attempts to go there. Part of the process of creating the Ellipse was to block off the remaining toxic zones. That's really where the terms of the Ellipse."

Lars nodded, "Even the Tract dwellers knew there were still some areas off-limits. The Ellipse dwellers had such advanced technology that they could fly just about anywhere, but the safety systems in the craft would ensure that they did not stray into dangerous territory. The Tract dwellers didn't have this technology, which limited their ability to move to all but the edges of the Ellipse. Without technology, Tract dwellers would not have access to the Exclusion zone, even if they wanted it."

Lars continued, "The last 30 years have seen a formidable rate of increase in technological progress compared with any period before. The nanotech, the transit, the handhelds, the tropus are all examples of the changes. I know it's called 'The Great Leap', but it does sometimes seem kind of far-fetched that so many changes happened so quickly."

Scrive thought, there was superb sophistication of technology and rapid advances in many things, all within his lifetime. The eradication of Nero by use of tropus cartridges created what media referred to as 'The Great Leap Forward'. A rate of unparalleled technological advancement, creating newer and faster transits, better media, a plethora of new devices including the ubiquitous handheld and the majority of what had become the secure networks and RFID based access systems.

A modern citizen could walk around safely and securely within their designated zones. By a simple application, they could visit other areas, but the system was well regulated to also keep crowds and supply and demand for goods and services under control. People with higher status (he included himself in this category) had greater freedom and the ability to travel more widely. And it had all been created without the politics and controls of a Big Brother state. The modern Ellipse citizen considered themselves pretty well off.

Carolin added, "We think there's a code word for where many of the secrets are held. It is called

'Australia' and we believe it is an area somewhere within the exclusion zone."

Scrive shook his head, he didn't recollect hearing of Australia.

"What or where is Australia?" he asked.

"That's one of the things we hope you'll be able to help us find out", answered Lars.

Analogue Tracking

Taylor found the inside of the tracking station remarkably familiar. There was a clinical look at complete odds to the run-down, camouflaged appearance of the outside. He had walked through the double entrance chamber, and by the time he had closed the access doors, there was a cool charcoal filtered air-con smell permeating the building. He knew he'd soon get used to it, with the slightly boosted oxygen levels to keep anyone in there sharp.

He flicked on a few of the displays and waited for the technology to settle. Sure, it was from before the 'Great Leap', and so some of it was a little dated looking, but he twisted a couple of the satisfyingly analogue controls, knurled surfaces with a real tactility missing from the modern devices with their flat screens, HUDs and haptic feedback.

He started with a probe to Beijing, mainly to see whether anything still worked. Sure enough, he was able to access a clean picture from a conference room inside an important building in Beijing. There was no-one in the room though, so he thought he'd try jacking into some of the other video links. He soon had access to some camera phones and a further and larger conference facility. It was all quite easy, but he knew that was because he was using the modern-day equivalent of listening through a wall with a glass beaker. It may not be possible to scan for individual devices, but, still, a basic cup could be a useful spy aid in the right hands.

The lack of any activity to watch made him wonder if the Chinese had been smart and that he was now watching a doctored video feed. Maybe they had tied these links down, and now there wasn't anything left to view.

He flipped a few dials to see whether there were any other live feeds still running. He found the continents menu and switched to the USA. It would be interesting to see if he could get back into the Maddox meeting undetected. In this case, there was nothing, so it looked as if someone decommissioned the link. He flicked back to the menu and noticed an extra entry at the top.

Australia.

He had never heard of Australia before. Yet there it was on the menu.

He flicked to enter the menu structure. A little series of towns appeared. Was this a test environment, maybe.
He clicked in.

Adelaide. Nothing
Canberra. Nothing
Darwin. A signal, but a blank screen.
Melbourne. Nothing.

Perth. A scratchy signal. A bad picture. There was something on the screen. It looked like a storm, although there were some large cylindrical structures running across the view. It wasn't easy to make anything out clearly.

Sydney. Nothing

Two more options. Woomera and Yulara. He wondered whether to bother, or whether to flick back to Perth to try to make out more.

Instead he clicked Woomera.

Sixteen squares appeared in his view. 16 cameras, 12 clear ones and four that had blank or static.

He opened the first one. Sound as well. He had a full monitor signal from this place. If it was a test site, then it was pretty impressive.

He studied the picture. A blue/green coloured scene, with what looked like some almost molecular structure in front of him. There were small red flashes in the image as well, which he initially thought were interference, but were tracking the shapes of the molecular structure.

It was difficult to scale the picture though, and he couldn't tell whether he was looking at something substantial or if it was a scale model of some kind.

He flicked to another screen, and then another. Whatever it was, the people who had set up the monitors had wanted to get a good view of it.

He bookmarked the page and hit the recording system in the control booth. He'd run this onto disk so that he could relay it later. He wasn't sure whether the links would stay reliable and whether he'd just struck lucky with this viewpoint.

He selected a few more of the screens.

They showed different external perspectives on the same structure.

Apart from the molecule shaped item, there were tubes running in various directions. They looked like pipes - uniform and machined. It was like a very large oil installation or something similar, but he could not see any fractionating towers or flames or any of the other tell-tale signs. This was a very clean installation and looked as it if it didn't need human intervention.

He kept each feed running for several minutes, partly to look for any signs of activity, but also to ensure he had something committed to his recording.

Then he decided to flip to the last spot.

Yulara

Yulara

Captain Taylor had never heard of any of the places in this area of the system, and he began to wonder if they were tests, a secure facility somewhere or even 'off-world'.

As he flicked to the Yulara system he was again greeted by sixteen small screens, although on this occasion only three of them had a signal. He immediately saw that there was movement.

He tapped the screen to zoom into one of the pictures and recoiled as he saw the detail of what was in the feed.

Looking like similar pipe and molecule structures to the earlier Woomera feed, here was a system that seemed to be handling a viscous liquid.

He let this run and he could see that the process was probably something that ran for several hours, or perhaps even continuously.
There was clearly audio on the link, but no signs of human voice just a background ticking and clicking sound.

He eventually flicked away from the sight, after dutifully recording around 20 minutes of what appeared to be a section of a compound being reduced to its core constituents.

He decided to flick to the next screen. It showed a similar scene, but this time he could see that there were more of the same processes. It was like a processing plant. He didn't know how large.

He ran the recording for ten minutes and then flicked to the final active feed. A different scene. Red flashes mainly, some kind of glass structure. He didn't know what he was looking at or whether it was real or some kind of interference.

Again, he recorded it.

This camera had an extra set of numbers along the bottom of the feed. He recognised them as GPS co-ordinates. -25° 20' 51.90", +129° 51' 12.93" .

He had a fix for the camera. He already knew it was deep inside the Tract. Off the radar. In terms of the Nero toxin, this was a deadly area. He started to think it was deadly for other reasons.

The monitoring station he was in didn't have the level of communications needed to contact the Pentagon securely. He realised he would need to leave and make his way back to his main station to be able to explain in detail what he had discovered. Nonetheless, he thought he'd better send an update so that by the time he reached the secure facility they would already be prepared for the discussion.

He decided to send a short message:

"No further activity detected on Sparrowhawk. Another development. Complex called 'Australia' / 'Yulara' in the Exclusion Zone has relevance. More when back inside Brookings Main Complex."

He read the message back. It was suitably basic for a forward alert. He pressed the button. It was gone.

He carefully made a copy of the disk onto which he had recorded the video, and then made a further copy onto a removable unit, which he put into his pocket. He'd use that to transmit his findings from Brookings Main.

He left the complex switched on. The monitoring for Yulara would also stay running. For Beijing, he would need to make a further visit to see whether any of the screens became active. He somehow doubted it. It looked as if the Chinese had blocked the view and were sending back dummy pictures.

Outside it still rained heavily. Taylor moved back to the four-wheel drive, revved the engine and started the short journey back to Brookings Main. He would upload his videos to show the content of Yulara.

Ailartsua

Scrive looked across to Lars. "I'll try a few search terms now," he said, "I want to see what I get on a basic search."

He flipped open a small computer.

"This is a clean environment", he explained, "I keep the image on here to look like a factory specification machine. It is usually better to look like an amateur that's just trying something for the first time, rather than a calculated tracker."

Lars nodded.

Scrive selected a search engine and typed "Australia". There was a long pause while the search spawned other searches cross a wide range of systems.

He tried all capital letters. Alternate capitals and small letters.

Nothing.

Scrive tried another search 'Australia'. And another 'Aust*'

Now he had 'Austria' which immediately returned thousands of results.

He reversed the term 'ailartsua'.

Nothing.

He would need to check the code word. Something was wrong. Somehow, he recognised the name. He was fairly certain that "Australia" was the key to something.

He repeated this process with several other search tools and on several systems.

Still nothing.

Lars looked at his watch.

"I guess I'm boring you", commented Scrive.

"Not at all", came Lars reply, "Carolin and I had arranged for another meeting here today. They are due to be with us in a few minutes."

"How did you know I'd go along with this?" asked Scrive.

As he did this, a single word came up on his screen. Woomera. He blinked. He tried to run himself at hyperspeed. It didn't work. He decided it was the lag from lack of tropus.

Woomera.

It reminded him of something that had once been in his mind. Australia. A country. He couldn't remember how he knew about it and why it had disappeared from his thoughts so completely.

It seemed to be something to do with the sharpness he'd found since the tropus dose had worn off.

He could even remember that Woomera was famous for something. Travel, planes, space travel. That was it. Australia was a big country on the lower part of the planet. He thought it was probably outside of the Ellipse now.

That's why he couldn't' remember anything about it. The television maps of the planet didn't show it, and there was no reference in books or online.

He began to wonder if it was a made-up place, but somehow, he was sure that it was real. Canberra. Melbourne. Sydney with a harbour. Woomera, where they tested rockets.

He thought of Tajikistan. Plenty of people wouldn't name that as a country. Or know its capital — Dushanbe, which used to be the main market town.

He flicked to blank the screen on his handheld. He flipped a bookmark to Charlie. The first time, the only time, he'd broken protocol since they had separated.

"We didn't", answered Carolin, "We didn't know you'd go along with this." Those in the cafe were entirely unaware of the search result that Scrive had seen, his thoughts and that he'd now effectively concealed his finding and signalled it to Charlie.

Carolin continued, "We had to make some assumptions though. We thought that you'd be impressed if we told you what we already knew. We also realised that even if you were the best tracker, you'd probably have difficulty finding anything without some further help."

"What do you mean?" asked Scrive, he looked concerned.

"It is okay; it is not another tracker - you won't be sharing trade secrets," answered Lars.

"We've done better than that," added Carolin," We've found someone with access to what happens inside Biotree."

Scrive looked questioningly," You are not just expecting me to trust you - you're now expecting me to trust a Biotree employee as well? All of this seems wrong."

"Okay, at least meet her," replied Lars, "she, and her friend will be here in a few minutes."

Scrive considered, "Okay, but I'll want to talk to them first."

"Not a problem," answered Lars, "...and by the way, they have seen a picture of you from the ones that Carolin took over the last few days."

As he said this, the cafe went quiet for a moment. The door had opened, and Scrive saw two good-looking women enter.

One was dressed conservatively as if for work, the other had a gold metallic micro-skirt and a white tutu. She appeared to be carrying an umbrella which looked suspiciously like a wand.

"This is Janie and Chantal", introduced Carolin, looking up and down Chantal's outfit. There were several others in the cafe doing the same. Chantal wiggled onto a chair, and Janie sat beside her.

"Hello again, Lars," started Janie, "...and hello also to you, Mr Mallinson, or should I call you Scrive?"

"Scrive is fine," said Scrive, "You both seem to believe in making quite an entrance."

Scrive couldn't help admiring the remarkable impression that Chantal was having on the others in the cafe. She had stopped conversations. Scrive noted that they were both very attractive women.

"Let's just say that everyone will remember we have been in here this evening," said Chantal. "We want our movements to be very noticeable at the moment."

Carolin nodded, "This is the situation. We've told Scrive about what we think is happening; we are sure that the investigation he is on has more to it and that it affects the Tract.

"We believe that there's something to do with the Tract hidden somewhere with a code word and have said the word to Scrive. He has already run searches but not found any links. We believe that he will need to go into the Biotree systems to get further. That's where Janie comes in. She can help Scrive get access because of the area that she works in and her access to codes and security idents."

"What were you looking for?" asked Janie, "Maybe I already have some ideas."

"Okay to tell?" asked Scrive, and Lars and Carolin nodded.

"The codeword is 'Australia' or some variation of that spelling."

Janie looked blank. Chantal looked up. Scrive stayed straight-faced.

"I know Australia," Chantal said, "Or rather, I know someone who says they were once from Australia".

She looked across to Janie. "It's one of my acquaintances from the - er- fundraising."

Janie looked back. She knew that Chantal was referring to the illegal trade she did with the tropus. Chantal's little habit meant that she got to know some rather unconventional people.

Janie said, "Look, we are getting into this quite deeply. I only brought Chantal along to give me some backup when I was meeting Lars. I didn't want to get Chantal involved further."

Scrive was interested in the Australia comment. It could be more of a breakthrough than rummaging around inside Biotree. He'd prefer to take both options, and he certainly wouldn't mind working with these two women for a while.

"Okay," ventured Scrive. "I'd appreciate your help. Both of you actually; Janie to help with the Biotree systems and Chantal to introduce me to a friend from the Australia Project".

"Actually, Australia isn't a Project," answered Chantal, "I'm pretty sure it's a place. A town in the Tract somewhere, I think."

Chantal continued, "You know what, I'll - we will both - help, BUT I would expect Mr Mallinson to find ways to compensate for my other possible loss of income."

Scrive smiled. He was dealing with an unexpectedly ad-hoc group of people. He was used to working with hard-nosed professionals who would set a task, agree some parameters and then leave him alone.

This situation was different.

The people he was working with seemed to be making it up as they went along. He guessed that Chantal would have no idea how much was in play, although he suspected that Lars would have a more realistic idea.

He surmised that Chantal was a tropus dealer. He noticed all of the bangles on her wrist around where the cartridge would typically be. She was probably a Sife. Cutting her tropus doses and selling on the residuals.

It was good money but dangerous to do this. Scrive had experimented with this, not for money, because he wanted to get a sense of effects of tropus deprivation. He was getting that experience big-time at the moment because of the damage to his last cartridge by Charlie when she saved him. If anything, he felt the lack of tropus was surprisingly good, and seemed to be clearing areas of his mind and thought. He had not expected that.

He spoke to Chantal, "Okay, I'll pay you for your help. There is one thing though; you'll be responsible for your safety as we move this along."

"Okay," answered Chantal. "Let's see; my fees will be the equivalent of a two thousand sales of tropus, with half up front...and the same for Janie. That's

20000 tropes now and 20k at the end," Chantal dropped into the street slang which treated the cost of tropus as if it was a currency.

Chantal was trying to get enough cash to mean that she didn't need to sell tropus anymore. This would be the equivalent of several years' worth of sales.

Scrive blinked. He didn't know how much the tropus sold for, but it couldn't be that much compared with his usual fees. He'd accept the deal but haggle slightly to ensure that Chantal thought she was pushing him.

"Okay, but I'll do half upfront, then a quarter when we've done the work and the last piece two weeks later."

Chantal looked at Janie. Janie hadn't expected this to turn into a commercial haggle in any case. Janie nodded to Chantal.

Chantal said, "Okay, we'll accept, but the first money is due immediately." Chantal was delighted. The immediate payment of 10,000Ts was 4 or 5 years work in one hit. And more to follow.

Scrive nodded. "Okay but remember I'm a tracker. Any funny business from this and I'll be on to you." He tried to make the threat sound both realistic but also friendly. He would prefer the simplicity of them being on-side with him rather than having another conflict to handle.

Janie nodded and looked to Chantal. Janie briefly held Chantal's hand, as if talking to a child. "You heard what Scrive said, Chantal. We need to play this straight."

Scrive assumed that Chantal could be a random element in many situations and that Janie was the best chance to tame her. In reality, he'd consider the payments to them would be small change. A lead that helped him move forward would be great, but he'd not worry too much if they both legged it after the initial interactions.

Scrive and Chantal both clicked their handhelds and placed them on the table.

Scrive said, "You request it, and I'll make the payment."

Chantal pressed a few buttons. Scrive noticed an amount displayed upon his machine. It wasn't much at all. He tried to keep a serious face.

Chantal looked across, "We're ready," she said.

"Okay," said Scrive and pressed a couple of buttons. The money transferred.

Chantal smiled, briefly jumped to her feet and clapped her hands together very lightly. Then, as if remembering she was just finishing a negotiation, she looked at Janie and said, "Great, Janie, now let's see how we can help Mr Mallinson.".

Tokyo

Makatomi was back in his office in Tokyo. He'd found the little trip to the UK and LA somewhat irritating because its primary purpose had been to find out about the leak affecting his business.

It was all so negative. But also, a chain reaction.

Someone had started a theft which he was now being asked to clean up. He'd had to hire two sets of people who were on the edge of legal. He'd then found one of them creating ripples which had upset his boss. He'd been forced to use an extreme measure to remove that person from the investigation.

Then he'd had a call from Holden.

The person running the task with Mallinson had told Holden that Mallinson had somehow survived. The indirect approach using a tampered version of the tropus hadn't worked.

If it had, they could have used the nanobots to clean up any evidence. Instead, somehow the 'bots had been eliminated. Mallinson was not to be messed with.

"Use a Trigax," Holden had said, "Make it fast and remove Mr Mallinson. Use one of the Biotree units.

They are at our test facility," added Holden. This was a whole further dimension of escalation.

Makatomi knew the Trigax was illegal for aggressive use within the Ellipse and that it was only used for peacekeeping in the Tract.

He also knew that the Trigax would be effective and leave no trace. Makatomi knew the Trigax as a finalising device although the technology was still not fully understood.

He'd assumed that his 'no questions asked' assassin would have ready access to such technology, which he considered to be almost alien. Instead, he was now having to provide the unit as well. This was really getting out of hand.

The Trigax units had a global range and were individually tuned to a specific target. They could be fired from anywhere and the beam that they asserted was efficiently scrambled until the point at which it reached its target. It used quantum principles and a focusing of the wave using discrete photon focus.

Most people had said the technology was impossible and poked a cat's paw at great physicists like Bohr, Bohm and von Neumann, but the results had spoken for themselves.

The devices were now licensed and maintained for Tract management and scientific experimentation. If used, there was no evidence that they had been fired, because of the wave dispersal of the energy.

It was like a perfect weapon.

"You'll have to use it from Bodø," said Holden, "You will have one chance".

Makatomi nodded. He knew that Holden would remove him if he didn't remove the mess he'd inadvertently created. On the other hand, if he fixed it, then he stood to gain exceptionally from the expected change in fortunes of Biotree.

Chantal

Chantal's penchant for the low dosage form of tropus had led her into some fascinating communities. They were people who regularly ran the risk of redding their blood and to some extent ran an alternate if somewhat privileged lifestyle.

It also meant they had access to some things that were not at all in the mainstream. It was as if the reduced tropus gave them abilities to see things that others could not. They could also remember things from before the time of the Nero toxins.

Today's outfit had been Japanese manga and Chantal had included combat boots. It worked to her advantage because there were no inconvenient heels. Useful around the stairs and cobblestones of this old part of London.

Chantal took Scrive to the door of the club. It was an old railway arch from the 19th century. There was a small door, she pushed it open, and it led into a sort of cave.

"There's miles of this around here", she explained. "They were originally built under the old railway systems in London and used to be a kind of retail space. Then as the Transit was introduced, they became bypassed with the new generation of retail environments. Now they are vestiges of an older London."

She continued, "Look - when we meet Crispin, don't tell him that you know he is from Australia. He might not co-operate. Ask him about Australia instead as a concept. Let him decide how he wants to talk about it."

Scrive replied," I'm ahead of you on this. I kind of know when I'm dealing with secrets that it is easier to have them fall out than to try to push them. Like that Chinese finger trap. The more you pull, the tighter it gets."

"Just don't say that around my friends," said Chantal, suppressing a smile.

She showed the way through the labyrinth to a dimly lit area.

Candles flickered in between several people sitting together.

Despite the candles they had power as was evident from the powerful workstation that a couple of them were using. "Hi Chantal," said one, looking up, "You've brought some more tropus?" he asked.

"Yes," replied Chantal, "plus a friend who'd like to ask you something."

"Is he cool?" asked the same person.

"Crispin - He's fine. As a matter of fact, I think he can help us, but we will need to help him as well. You know the Vaults? I think they would like to take a look for something."

"Hi," said Scrive, "I've tried the reddening way too." He held up his arm from the smashed cartridge. He'd had time to tidy up the fragments from the attack, and it now looked as if he was mainlining on red blood instead of the usual tropus mix.

Crispin looked at his arm.

"Jeez – That's messy – and unusual," he grimaced, "you look as if you are completely red? I've never seen a plexi manipulated like that – it looks awful. It looks as if the telemetry is still working, but I can't see how you'd ever get another cartridge to be accepted?"

Scrive looked at the mechanism. He'd play it hard. "Yes, I've broken the tropus injector. If I don't get surgery, then I'll be completely red in a couple of days. It doesn't seem to affect me though, if anything, it makes me see things more clearly."

Crispin looked at Chantal, "Whoa hardcore - you do hang out with some crazy people, Chantal. So, your name is Scrive. How can we help you?"

"Other way around, "said Scrive," I want to help you- but I have a few questions first."

"We are told you have preserved a lot of the old ways," said Scrive, "That you have ways to review old files and information from before the time of the Nero toxin."

"That's correct", said Crispin, "but you'll have to ask us to find what you need. Ask us, and then leave us to trace it."

Scrive wrote down a few words on ordinary paper. Australia. Woomera. Where and What?

Crispin looked at the paper," I can do that." he replied, "And I think I may know something about it too. Please leave now. Wait outside. We will find you."

He turned to another person working the workstation. "Look Lucas," he said, pointing to the paper.

The second guy turned back to Scrive and Chantal. "We can get you this, but Chantal, your friend looks as if he might have some fees for us?"

Chantal looked at Scrive, "I haven't discussed this at all; this is all freestyle," she commented.

Scrive replied, "Look, I won't kid you both, I need this help, and I will pay for it. I can also be dangerous, but I'm a friend of Chantal, so I want to do this the right way. You both understand me. I'll pay, but please, I need to know about this project 'Australia'."

Chantal answered for Scrive," He'll give you the equivalent of 100 units of tropus. I know he will. I think that's his limit."

Scrive nodded. "Yes, Chantal knows my limit. I can pay you right now, handheld to handheld, but I want to see the information first."

"He's good for it." said Chantal, "And he won't trick you. I will vouch for him."

Scrive looked across to Chantal. She was doing more than her fair share of handling this. He decided that she'd got emotionally committed to the situation.

"Okay, we'll do it", said Crispin. "Lucas, help me with this." They pulled a couple of boxes together, and all sat around.

RFID

Suze had been pleased with herself since she'd found the RFID for Scrive. His Radio Frequency Identity. She had his address, his transit chip identity and could easily follow his movements around. She'd seen him take a walk from the apartment to the transit and had been about to follow him when he changed course to a cafe.

She'd seen him spend time and assumed it was a meeting that would lead further towards the answers to the various questions.

Who was trying to kill him?

Why?

Was Makatomi or Biotree involved?

What about the Chinese?

Suze was also suspicious about whether Scrive was operating alone. It was quite normal for trackers to work by themselves, but they'd often have some kind of safety system, particularly if they were on a bigger quest. If Scrive's pay was anything like theirs for this, then he'd have an accomplice somewhere.

Suze discussed this with Denny.

"I think we may want to pay Scrive's place a visit, when he is out on a long journey, said Suze.

"I agree," said Denny, "We'll probably need half an hour just so that we have time to check for any safety precautions he may have put up around the place."

The little cameras they had installed gave an easy indication of Scrive's departures and return and he didn't seem to be taking any particular precautions.

There had been two other visitors they had seen accessing his premises since they started monitoring. Both had a key. One had been wearing a hooded jacket and had only stayed for around five minutes. They didn't have any idea who this was, and the other was an attractive woman who had been in and out of the apartment a few times, but since her last departure had not returned.

They had no idea who the people were and had not had time to set up any form of monitoring.

This time they decided that after Scrive's next departure, they would attempt to break into the apartment, but using the fine equipment in Denny's holdall, which should not leave a trace. Of course, a tracker could probably work out what had happened, but they would be careful to leave no mark of their entry.

The hotel they had moved to was just across the road from Scrive's. It was a modern location and overlooked the River Thames. The chimneys of Battersea Power station were also close by, as an embedded part of a south of the river luxury shopping and entertainment complex.

This helped because there were plenty of people around the area, so the movements of Denny and Suze would easily blend in.

Denny had left the tracking systems running as Scrive had left and then taken a transit across to the area around London Bridge. Even if he took a taxi, it would take him twenty minutes to return — enough time for them to make their move and break into Scrive's apartment.

"Okay, we'll be looking for memory blocks, idents, Hitech and signs of who the other visitors are," said Denny, "We'll need to work fast and not take anything away."

Suze nodded, she knew this well, they had done other 'information gathering' sessions together and speed and a very light touch were what was required.

Denny carried a couple of small devices. One was an e-burster. This created a small electronic pulse that sensed all of the electronics in a room. It would inventory them and provide an exact location. The devices were available domestically for locating remote controls and missing keys, but this version was a military grade device that could pinpoint any form of technology. Its second function, not available on the domestic ones, was to be able to read the content of a device and store it. The third function, chillingly, was that it could destroy the device, both silently, and destructively. Denny was only planning to use the first two modes, find and copy. And to do this selectively.

The images copied would not necessarily be readable straight away, because of possible encryption, but that was something they could worry about later, back in the comfort of the hotel.

The second device was simply a form of basic self-defence. A needler was device that could generate high volume sound that would disorientate an attacker, and which could also fire small electronic probes which could then deliver variable voltages. The technology was based upon an older wired technology called taser, but in this version the darts had the power charges attached and could be separately triggered at a selectable threshold.

Suze and Denny both carried these devices, which were also silent in operation but very painful to any recipient. They had never been in firefight situations with them but were both aware of the need for some self-protection in their line of business.

They both inserted ear systems which included personal communications as well as filters for the sound wave defence. Denny made sure they had tuned it to the device he was carrying so that the cancellation effect would work if they needed it.

"Okay, let's go", he said. Scrive was over in London Bridge, in the tunnels and vaults to the south of the river.

PART TWO

7 long song

As the verses unfold
and your soul suffers the long day

And the twelve o'clock gloom
Spins the room, you struggle on your way
Well, don't you sigh, don't you cry
Lick the dust from your eye

Life's a long song
Life's a long song
Life's a long song

We will meet in the sweet light of dawn

Ian Anderson

The long song

Crispin started to speak, "Sometimes, people make fun of the way I speak; I pronounce a few words differently from most people. I have some extra words in my vocabulary. It's because I'm originally from a real place called 'Australia'. Australia isn't a Project, Scrive. It isn't even a town. It's a whole country. Or rather, it was.

"I'm from the Tract originally, and my parents had me when the Nero virus was at its height. They were both immune and so am I. We lived in the country called Australia, in a town called Darwin. When the virus was at its height, Australia was the original centre for it and the whole country was quarantined. We'd had the Flames a few years earlier, and so much of Australia burnt to the ground.

"The Flames was through climate effects, we'd seen the wildlife progressively eradicated over a few years, because the huge bush fires kept coming back. The fireys just couldn't keep up with it. It was relentless, year after year, like the fires in Southern California now.

"It was a modern-day tragedy, a side-effect of global heating. Most of the livestock and about a third of the population were killed.

"It made Straya a very dangerous place to live. The old joke was that The Northern Territories (which is where I'm from) was filled with nasty critters. They were all out to kill and eat one another or passing humans. Freshies, Salties, jellyfish, sharks, spiders. You name it, they'd kill and eat one another.

"Now add to that the Flames and then the viral attacks which led to the introduction of tropus. I guess the decimation from the Flames left many places with unsafe water, which created some of the contagion. The evolution of the Australian virus ran away from the engineering of its vaccine. They couldn't keep up with the variants. Some said it was because there was a clone vaccine introduced from China. I don't know.

"What I remember is that people went a sort of black colour when they caught the virus. There was no way to stop it. Their blood didn't run yellow or red, it went through a blue colour and then to black.

"And despite the death of so-much wildlife, there were still the buzzards, vultures and the rats which seemed to thrive.

Lucas chipped in, "What I remember is that the whole landmass was horrific, like something out of a disaster movie. But it is strange. Only people that were actually on the landmass seem to remember it. It's like it never happened to everyone else."

"Did people try to leave?" asked Scrive.

"You couldn't go in or out of it. I'm not talking about a small landmass here. I'm talking about something the same size as North America.

"Biotree had been trialling their newest versions of tropus and nanobots in the territory. Speculation was that they were trying to cover up for something. It was hard to get accurate news because most of the comms infrastructure was down too.

Then we heard that quarantine restrictions were to be boosted. We were told that Biotree was helping the effort to instigate the new processes. It turned out they were implementing a series of geostationary satellites, which were nicknamed 'The Bracelet'. The Bracelet applied an electronic border around the landmass. It seems ironic that they had to shoot things into space to do this, after all the fuss about global warming.

"How come none of this was in the news?" asked Scrive.

"This was like the End of Days," said Crispin, "The politicians and world leaders opted for discretion to avoid a world panic. It was thought better to contain it than to let panic set in globally."

"I'm not so sure though," said Lucas, "The Chinese were open about that Wuhan virus, and it probably saved lives because it gave the rest of the planet a chance to prepare."

"Yeah, I agree, and it showed how some of the plans were quite piecemeal. They ran out of protective clothes; the facemasks were not to the right filtration; Planeloads of possible contaminants shipped offshore," added Crispin.

"Pressure was put onto Biotree to come up with a resolution. I'm guessing it was desperation which led to the Bracelet. The Low Earth Orbit monitoring system for the boundaries of the Australian zone. Then they added the so-called 'charms' to the Bracelet, which provided the enforcement.

"I've not heard of either of these things," said Scrive.

Crispin continued, "You wouldn't, out here in the Ellipse; it is only Tract dwellers and anyone still in Australia that needs to know about these things."

Lucas added, "The charms operated with a railgun. It was a sneaky way around space-wars legislation because the railgun is not technically classed as a weapon. There's no explosive warhead or anything."

"I know about railguns' said Scrive, "they can be pretty lethal. They fire a massively high-speed projectile which can cut through just about anything."

Yes, that's right- the exit velocity is about 3km per second. Enforcement of the territorial edges of Australia was via the Bracelet and charms. Typically, a breach would be spotted, triangulated and then three railgun cannons would be deployed to stop the escape. There was even a Biotree branding for the technology: Trigax."

"Ah, I've heard of Trigax," said Scrive, "Biotree put out some seemingly jolly marketing videos about this."

"That's right," said Crispin," And also about their final clean-up systems."

"They didn't come up with a pretty-sounding name for this piece. I think the Brits invented it and called it something like APKWS laser-guided rocket."

"APKWS?" Asked Scrive.

"Yeah, Advanced Precision Kill Weapon System, "answered Lucas.

"These were standardised laser-guided rockets, which fired from anything from a hypersonic pursuit vehicle down to a regular drone. They were handed out to the cops to put next to their tasers. They would have the handset, lock it on with a bluedot and ka-boom."

Crispin continued, "I was eight years old when the quarantine restrictions came down, and we had no chance to leave. The country was also in the middle of a vast area of sea and so the idea to leave by, say, a small ship was impossible. And the exclusion zone around that was built up around the country was intense.

"When a few people tried to leave by ship or plane, they'd be detected and destroyed. The argument from authorities said was safer to keep the virus in one place on the planet than to have it spread. It's another reason why there was such a news blackout from the region.

"What about wildlife, birds, insects, migration?", asked Chantal, who Scrive noticed was taking this story in for the first time as well.

"We used to call it the Glimmer," said Crispin. "There was a sort of sparkle that you could see from the seashore and sometime in the night sky. We used to think it was somehow magical, but it was the perimeter systems destroying anything that was flying or swimming out of range."

"If there were so many defences, then how come you are now on the outside?" asked Scrive. "The way you describe it doesn't make it seem possible to leave."

"Correct," answered Crispin. "There wasn't any way to go. Basically, those of us that were not affected by the Nero were effectively prisoners. We had a small amount of sea edge, then a cold zone and after that was the exclusion area that effectively killed anything that entered it."

"That's some pretty big weaponry," said Scrive.

"I know," said Crispin, "I used to wonder how we'd created the technology for such a thing, but it all happened at around the time of the Great Leap when the planet's technologies also accelerated. Of course, I didn't know that at the time, because the Australian communications systems were destroyed as well. We didn't have television, radio, computer communication or phone. It was like one of those EMF pulses that you hear about that destroy electronics, except in this case it was communications but not other forms of technology. Cars still worked, for example.

Crispin looked across to Scrive, "We then saw a period where the technologies in Australia accelerated almost as fast as they have done in the Ellipse. It went on for a while and for the survivors wasn't so bad. The weird thing was that the people affected by the virus didn't hang around like in zombie B-movies. They seemed to disappear almost within minutes of dying. Faster than the vultures or rats could handle."

He paused to think about this.

Scrive noticed that Lucas had been watching as well and nodded a few times during the description. Lucas seemed to know as much as Crispin while

Chantal looked on, fascinated at what they were hearing.

Crispin continued.

"Those of us that survived couldn't get to the bottom of what was happening. Most stories were word of mouth, but it sounded as if there'd been some kind of riots in another town quite a long way to the North of us. You'll need to remember that Australia is - was - a vast country with not so many people. There was also quite a lot of desert land, and the reduced infrastructure meant most people stayed in the areas they knew to be safe.

"We heard that this area to the South near to a wild area called Uluru was where there had been the riots and a lot of people killed. They were not killed by the virus, but by the fighting that took place. We never had a chance to find out though because that's when Lucas and I managed to find a gap in the "Glimmer".

"As kids, we'd play dare games with one another. We used to take the small boat out to the edge of the safe zone. It was to where the air started to get cold. We could see the area where you couldn't go further and sometimes, we'd throw things into it to see them spark.

"One day we'd been onshore, and there'd been an extra wave of new bodies die and disappear from the virus. I was around twelve by this point. There had been some big rumbling sounds which we

427

thought were some kind of hurricane or earthquake or something.

"We cycled as fast as we could to some high ground in case it was a big sea or something that could harm us. Our thinking was that if the virus wouldn't get us then we didn't want to get struck down by a wind or a flood.

"It turned out that it was an earthquake. The buildings shook, and we could feel the world moving underneath us. It was a pretty scary feeling.

"Then we looked around and could see some geese flying away from what we assumed was the source of the 'quake. We never actually saw the 'quake ourselves.

"The geese flew towards the Edge (where the force-field starts). As kids we were waiting for the sparks as a whole flock of birds were vaporised.

"But they flew on. They flew past the Edge. We looked at one another. We both jumped onto our bikes and shot down the slope towards the town. There was almost no-one around. The official warning system was sounding, which meant people had gone to shelters.

"We went down to the harbour and picked up one of the larger fishing boats. This was a deep-sea boat for catching sharks. Metal hulled, big engines and very fast. Even at our young ages we knew how to skipper those boats.

"We gunned it out to the Edge. No coldness. We took the harpoon guns from the front of the ship and fired one into the Edge. We expected it to vaporise. It just kept flying through the air.

"We took another one and did the same. Same thing happened.

"We looked at one another. Crispin looked at Lucas now. We asked each other whether to risk it. We were going to take the boat through the Edge.

"We trickled the engine and headed for what we knew would typically be the point of vaporisation. We threw everything overboard in front of us as a test. But we just kept going. We headed north west for another 100 miles. Then the boat's radar suddenly started working again. We'd never seen the radio communications working so this was a novelty. It was a pretty cool system too, with flat panel displays and maps of the sea and land. We could see ahead of us a large belt of island, behind us was some sea and then - nothing. The place we had come from had blanked from the system.

"We worked out that the Edge must have started working again. We'd managed to get through a gap, probably caused by the earthquake. It had taken us about three hours to get to our current position, and it was only when the radar came on that we could tell that the Edge was back surrounding Australia.

"The radar showed us various islands, some of which were huge, but we could see that our best chance was to use the fuel we had to get as far as possible. We charted a course that threaded through a belt of islands and eventually wound up in Indonesia. This was around 800 miles from where we'd started!

"What we didn't realise was that we'd navigated from the Edge through the Tract and landed at a place that was within the Ellipse.

"It was mainly luck, but by doing this, we'd avoided the areas where if we'd landed, we'd have been stuck there for good. The lack of infrastructure would have prevented us from going any further. It was simply the fact that we'd had a boat from inside Australia, where there was still fuel and that we'd then passed the desolated areas and arrived on the Ellipse with its super high technology infrastructure.

"We were both used to living by our wits, so we brought the boat in by night and let it drift the last part of the way. We snuck ashore and effectively became two more of the shadowy people on the edge of the systems. We've lived via the economics of trading tropus and other street skills ever since.

"The strange thing is that no-one here has ever heard of Australia. It is as if it has been wiped from people's minds. We don't think there are many people who did what we've done either and managed to escape."

Lucas added, "Although scattered around the city and in other parts of the country are others like us but from other places usually at the edges of the Tract.

"They are the ones that have managed to get inside the Ellipse but are not citizens.

"None of us has the cartridge but selling the tropus creates an economy to keep us alive.

"How did you get from Indonesia to London?" asked Scrive; he was still trying to assimilate this story.

Crispin picked up the story, "As we said, most big cities have a few people like us. We live in the underbelly of the city and so eventually we find one another. There are a few routes to move us around, mainly between the big metropolitan centres. We usually have to stow-away on the big planes between centres. Strangely enough, since the security became so tight it is easier for us, because the main security relies upon the transit tags that everyone has in their tropus cartridge. It's very easy to find anyone in the wrong place if they have been tagged.

"It's a lot less easy if you don't have any identity. Stowing away usually requires someone to create diversion when we go through the airport scanners. It can take days to get through an airport, but trust me, it's possible. The main thing is to not be too impatient. And the other thing is to look for flights that don't seem to be too busy.

Chantal interrupted, "So you've come from Australia (which none of us have heard of), through the Tract in a boat and then snuck into the Ellipse from where you've stowed away in planes to get to London?

That's about the sense of it, replied Crispin.

"And now you make your living dealing in tropus?"

"Yes - you'd be surprised how many city types pay for boosts. They say it takes away the pain. For us, we say 'no pain, no gain'. " Crispin and Lucas looked at one another and smiled.

"We can jack you into the Vault system to see if there is much more information. There is some, because we've looked at it previously. It is part of how we pieced together our route from Darwin to London."

Alert

Holden had been alerted. There was more network traffic than usual checking the sources of the domes. There had been radar and lidar pings to the Bodø location as well as additional fly-bys of the location.

There was unconventional analogue activity on a series of locations in Australia, including the largest one in Yulara. The US Desert locations had also been probed.

Holden understood the ways that the domes were connected. That one could interconnect with another and it was this powerful set of linkages that exerted some of the power which was often referred to as the Great Leap.

Holden was uncertain what was causing this level of activity. Unless it was the trackers that Makatomi had invoked. They were having the opposite effect to that intended. They were drawing more attention to the domes and creating new forms of activity.

Holden decided to track down Makatomi. Perhaps Makatomi was losing his grip on the situation?

Makatomi was back in London. He had put some miles on the V-Blade running around the planet. Maybe he should try Holden's approach? So much simpler.

Holden called Makatomi.

"This is getting out of control. They are probing some of the domes. That includes the Biotree complex in Bodø. You are letting this slip away from you."

Makatomi answered, "No, I have a couple of trackers working on this. They are finding the source of the problem. I had to use Scrive as bait to force a sighting."

Holden's voice shifted to a softer tone, "Your explanations are wearing now. We cannot afford any more mistakes. Consider this a final warning. The next time you will go like Scrive. Get this fixed, as we agreed."

Makatomi bowed his head slightly towards the monitor. He hated to bow to these passive aggressive screens like a Nam June Paik installation.

He felt something like the fingers of a pressing across the top of his head. He knew it was from Holden.

Punching out

Lucas flipped the workstation screen so that they could all see it. He tapped a few access codes and was soon browsing secure sites. They were not those inside Biotree but places that were part of an American defense network. It showed some basic Information about Australia through something called the CIA Factbook, dated in the early part of the 21st Century.

"This will give you a sense about Australia," said Lucas, "and you'll see there's way too much for it to have been something that we prepared earlier today."

He found the map and showed them the whole of Australia. "Here's Darwin," he added pointing to the top. "Here's the ring of islands and here's Singapore. You can draw the Edge of the Ellipse by Singapore, then the Tract to just off of Australia and then Australia itself in what is now the missing part."

"I can understand that from a geographical perspective," said Scrive, "But I can't understand it from a memory point of view. Why don't people know about Australia? It is not as if it disappeared thousands of years ago. Maybe 20 years which is well within living memory."

"This is something we don't understand either," said Crispin. "But when we test people on this, no-one shows any recollection. It's as if we dreamed the whole thing."

"But we didn't," said Lucas. "I am Australian."

He emphasized "am" not "was", "am".

Scrive pondered his next steps. He asked Lucas to copy some of the data onto a memory block that he could review later. Lucas flipped a small card into the side of the machine. "I'm using a non-rewritable stick," he said. "It will burn this image on but once it's there it can't be erased. I'm going to copy this whole site for you," he said, and his fingers flipped a few select software constructs as he piped the content of the secure vault across to the memory block.

He pressed a mechanical catch, and the block slid out. He held out his hand. A small block the size of a gaming dice. "It's all on here," he said. "You can read this on most systems. I've encrypted it. The password is 'Australia'," he smiled, "not very original, but you'd have to know it to find it. Heck, it's not even a forbidden word; the password scanner didn't even flag it as weak."

He passed the block to Scrive. Scrive nodded his appreciation. Chantal smiled. "Scrive, I think Crispin and Lucas have been incredibly helpful." You should give them their money, and when we walk away, you should promise not to see them again."

"I'll do that," said Scrive.

They exchanged electronic money. Scrive added a huge extra sum to that which they had requested. Lucas and Crispin smiled at the thought of their improved economic status.

Scrive and Chantal walked back outside into the street, with Scrive still holding the cube.

"You know what", said Scrive, "I'm going to trust you some more. Don't try to scam me on this, but I'm going to hand you the cube. It's better that it's separate from me. We both know the story, and we both know what's on it. Keep it somewhere safe."

He handed her the cube. They walked along towards the nearby transit. "Thank you, Scrive," said Chantal.

Scrive started to reply; as he did so he looked down.

Three holes appeared in his chest.
Perfectly circular, each the size of a fist.

Chantal stepped back; there was a crackling sound, like sparks, she watched as Scrive's body appeared to be sucked into the three holes.

Then she ran. As fast as she could manage.

The combat boots helped her speed. She still had the cube in her hand. She'd need to get inside. Call Janie. Figure out what was happening.

A *quiet stopover*

Charlie had been tracking Scrive ever since she'd left his apartment. She thought that this trip to London wasn't at all how she'd expected it to be. A quiet trip to Geneva, with the London stopover to see Scrive had turned into something altogether different. Still, the money was good. Scrive had already paid her a third of the fees, and with that, alone, she was already substantially better off. She'd worked out she now owned her apartment in New York and had enough to buy another one in London, if she wanted.

Her new hotel was close enough to Scrive; she'd decided to hide in plain sight, rather than to move a long way away. That way if she needed to intervene, it would be easy. Scrive wouldn't expect her to be almost next door, and the Portuguese themed hotel was enjoyable.

She'd dialled to get a penthouse room and could see the river, the power station entertainment complex and even an area where people seemed to be exercising dogs. She'd asked the concierge about this and been told it was a home for stray dogs in London. Something she'd never thought about in New York, but that the Brits seemed to like doing.

Charlie had made sure Scrive was wired for sound as well as tracing when she'd left. It was pretty easy to do, based upon the accidental suffering she'd imposed after smashing his tropus cartridge when she was rescuing him from the nano toxins.

She'd inserted a small transmitter and had sound, but no video, as well as the RFID tracking. It was enough to have a pretty good idea of what she was hearing. She was sure that Scrive would know she'd done this, although they'd not talked about it and had followed the silent process of splitting up without agreeing to any specifics that would allow one of them to betray the other if they got caught. It was a one-way situation though because Scrive would be the one caught, and he didn't have a clue where Charlie was.

Charlie also recorded the movements of Scrive onto a Geosystem and bursts of conversation; buying a baguette at the local store, the meeting with Carolin at the transit, the cafe session with Lars and the introductions to Janie and Chantal.

Charlie had also seen Scrive meet Chantal for the second time and visit London Bridge.

The sound reception from the arches was poor. They were underground and in a damp area where radio frequencies were finding it hard to penetrate. Charlie had a broad idea of what the meeting was about but although the tracking device worked well enough, the sound was awful.

After around twenty minutes Scrive had obviously reappeared at street level and he could hear him talking to Chantal.

"You know what", said Scrive, "I'm going to trust you some more. Don't try to scam me on this, but I'm going to hand you the cube. It's better that it's separate from me. We both know the story, and we both know what's on it. Keep it somewhere safe."

Charlie could hear them walking, "Thank you, Scrive," Chantal was saying, as the sound disappeared, and then the signal from Scrive.

Charlie looked at the systems she was using. They were still working.

She reset the communications; She rescanned for Scrive; she tried a transit link as if polling to book a journey. Nothing worked. Scrive had gone from the system.

She knew what this would mean. Scrive had been terminated.

Charlie felt nausea overcome her. She wasn't used to this feeling, even in combat situations where things could get pretty robust. But she'd been with Scrive. Right in his room across the way. They'd planned together for the mission. Scrive should be invincible.

She snapped herself together. Think - Either Chantal had murdered Scrive, or possibly they had both been killed together?

Charlie knew she'd need to erase Scrive's presence from his apartment and that it needed to be fast. Mainly to remove the electronics and to check them at her own pace. If he'd been killed there would likely be people showing up at his apartment.

Charlie mentally considered her presence of mind to be staying so close. She could be into and out of Scrive's apartment in ten minutes. She knew which devices to take. Scrive had plenty of technology, but there were only a couple of critical small units that contained the information that mattered. It would take anyone else an hour to locate and then remove the technology, and even then, they would have a long task to sift to find the important stuff.

She picked up the pistol and pocketed several silver devices about the size of a small fuel cell. They fitted easily into her hand and had a flip-top safety catch. Each was a stun grenade that used a combination of sound, smoke, EMF and potentially ballistics to create a defence cordon. There was a large arrow embossed on the front. It pointed away from Charlie. Charlie knew that at the time it was deployed, she needed the arrow to point towards the hostiles because the device was radial 270 degrees. The person using it should be in the 90-degree shadow.

She'd decided they would be a last resort, but one was ready to prime in her hand.

She crossed to the apartment and let herself in.

Immediately she could hear a sound. There was someone already inside. She felt for the grenade and could feel the raised arrow pointing away from her body. She looked around the corner towards the main room. Two people were frozen. One was holding a metatazer. Then she noticed the other one. Also holding a metatazer. They could see she had the stun. It was stalemate. If they hit her, she'd fire the stun, and they'd be blasted, assuming she had it set to ballistics. She didn't.

"You know Scrive?" asked the woman.

"Do you?" came her reply.

"We all do." said the man.

"We're trying to help him." said the woman.

"How did you get in?" said Charlie, "I live here, it's my apartment." she lied.

"We've seen you come in here. You haven't been back for two days," said the woman.

"You know why?" asked Charlie. Her arm was beginning to ache, holding the stun grenade forward.

"Do you?" asked the man.

"Someone tried to kill Scrive," said Charlie. She'd worked out that if they were professional assassins, she'd be dead by now. They were probably clients of Scrive. Not friends routinely armed like that.

She noticed the technology that the man had scattered in front of him.

"You're trackers." Charlie realised. "You're tracking Scrive."

She decided that they didn't know he'd been killed. But their appearance here couldn't be coincidence.

"Look," she said, "Let's put down the armaments. I think we are doing the same thing."

She moved her arm down slightly but didn't take her thumb off of the top arrow on the grenade. It made her look more peaceful rather than fundamentally changing anything.

She realised that the pistol would take too long to reach and fire, so the grenade was her only chance if they got nasty.

The woman started to lower the metatazer and looked at the man.

"Okay," said the man, "let's talk."

"How can I trust you?" asked Charlie.

"You can't," said the man, "Look, my names Denny and this is Suze, we're working for Biotree, and we think Scrive has been doing the same thing."

"Okay," said Charlie. She pressed the safety catch back onto the stun grenade.

"One more thing? Tea or coffee?"

Charlie had to decide how much to say to Denny and Suze. they didn't seem to know that Scrive was dead. They presumably didn't know about the other people he'd been meeting, but they were pretty good because they'd been able to find Scrive.

Charlie wouldn't give too much away about being a tracker herself, but the small arms she was carrying were a pretty big hint that she wasn't quite like other local people.

"Okay," she started, "If I'm going to tell you anything, you'll have to tell me what you know first. Scrive will be back soon, I can alert him, and you'll never see him again, or you can tell me enough that I reassure him that all is cool."

"Here's the deal," said Denny, "We think Scrive and maybe you have been given the same task as us. To find out who has leaked information about Biotree to the Chinese. We are supposed to trace the leak and report it back to someone in Biotree."

"Who?" asked Charlie. This was a good test of how much they knew.

"Makatomi, and a guy called Arusen," answered Suze. "They met us in L.A yesterday and briefed us and then offered us a deal."

"Did they say anyone else was involved?" asked Charlie.

"No," answered Denny, "But we suspected that Makatomi would want an insurance policy in case something went wrong. Then we noticed someone else on the grid searching for Chinese secrets and around the Biotree site. We are almost certain it is Scrive. We tracked him back to the Biotree building a few hours before Makatomi flew to meet us in L.A."

"Neat," said Charlie, noting that these must be pretty good players and wondering why she didn't know about them. She didn't want to ask too many specialised questions, though, or they'd realise her level of involvement.

"I'm Scrive's girlfriend", she lied, "We've been together for a couple of years".

"There's not much sign of it here," said Suze, looking around.

"This isn't Scrive's main place, and I live in New York," embroidered Charlie, "Scrive was using this place for the current project."

"Okay," asked Denny, "So what do you know about what has been happening?"

There was a ringing sound. It was the entry phone to the apartment.

"This is getting crazy." said Suze. She looked over to Charlie.

"Is this anything to do with you?"

"Nothing, I swear - er - it's probably Scrive," she lied, "He'll probably let himself in but is letting me know he is back."

Suze walked to the entry-phone.

She put her finger over the camera.

She lifted the receiver.

It was another woman. She looked like a Japanese cartoon character. Somewhere between stylish and absurd. She also seemed quite distressed.

"Hello, can I help you?" asked `Suze.

"I'm a friend of Scrive," answered the person. "Can I come in, please?"

"Ask her what her name is," said Charlie, "If it's Chantal, let her in."

Suze relayed the question, "Just a second, what's your name?" she asked politely.

"I'm Chantal, a friend of Scrive," she answered.

Suze looked at Denny, who nodded, "Let her in".

There was a buzz and then a few moments delay as Chantal made her way through two sets of entry system and reached the front door of the apartment. The regular front doorbell rang.

Denny opened the door, while Suze and Charlie remained seated. Charlie had a hand reaching towards her pistol.

As Chantal walked in, she burst into tears. She was shaking. Suze stood and walked over.

"There," she said and lightly touched Chantal's arm, "Sit down, take your time."

Suze looked at Denny, who had been ultra-vigilant as Chantal entered. He relaxed slightly. Charlie was next to speak.

"It looks as if we all know Scrive one way or another. I guess Chantal was the last person to see him."

"That's right," said Chantal. Her appearance was still of bright colours, in an outfit that looked as if it had been outlined with a neon pink pen. But her expression was anything but bright. She looked decidedly ashen, and her skin complexion reminded the others of a Tract person. Denny had already looked at her arm and seen that she did have a tropus cartridge, but he wondered if there was some illness affecting her.

Chantal spoke, "I was with Scrive when he was killed. About half an hour ago. We had just left a meeting and he kind of turned into a vapour. I've never seen anything like it before.

"One minute we were talking and the next he had holes in him and then sort of vaporised."

The others looked at one another and then Charlie asked something.

"Did you see who did it?" she asked, "who killed Scrive?" She was angry that someone she cared for had been removed without trace. In her line of business, she had heard of this kind of thing happening, but never had any first-hand experience. It didn't look as if Denny or Suze had seen this either.

"It was so fast," said Chantal, "and very selective. I was almost next to him when it happened. I heard a crack, like a spark, saw three holes appear and the next thing I remember was that he had gone. I just ran away as fast as I could."

Charlie remembered the conversation before the signal had dropped. She had to decide whether to ask deeper questions, with the risk that she'd show more of her hand. It looked as if Suze and Denny were trackers, but Chantal seemed to be a regular person that had been pulled into the situation.

Either that or she was very good and throwing them all off of the trail.

"Did you find out anything from Scrive, or did he give you anything?" asked Charlie.

"Yes," replied Chantal. "That's why I've come over to Scrive's place. I have a memory cube that may have some answers, but it needs to be played in one of Scrive's devices. Otherwise it doesn't work."

Cube root

Chantal fumbled into a side pocket in her outfit. It was zippered but somehow made to look like a lightning flash.

"Here, I know I've got to trust you. Scrive gave me this after we left the meeting with the people in the Vaults"

Charlie knew what Scrive had done. The cube would have two information partitions; a public one and a private one. Anyone picking up the cube to casually browse its contents would get the public version. The secret content would only work in a machine devised by Scrive. Charlie was also pretty sure that Denny and Suze would know how this worked.

"Okay, we'll handle this together," said Denny, "I guess that you two - Charlie and Chantal - are now both in danger from association with Scrive. At the moment Suze and I have covered our tracks pretty well, although Biotree and Makatomi know about us. It doesn't make any sense that Makatomi would set us on a mission and then send someone straight after us to kill us. I can't think why he'd do that with Scrive though, yet he appears to have done so.

"I can't think that anyone would be onto Suze and myself yet - we have moved so fast to get from L.A. to London so that probably Scrive's place would be the last one on the planet anyone would be looking for us."

Chantal said, " Look - I'm not supposed to be involved with this. It's only through a friend of mine - Janie. She asked me to help and to meet someone who, it turns out, wanted to give Scrive some help at getting into Biotree."

Chantal explained how she'd accompanied Janie to meet Lars. How Lars had led Scrive to them, via Carolin. How Lars wanted to help Scrive get into Biotree. That Lars knew Janie's work colleague Karin and she had also disappeared. But mostly how Chantal was only involved because she'd known about Australia.

"...And Scrive paid us too, it was 20,000T at the start and another 20,000T at the end," she knew she was taking a chance.

Charlie Suze and Denny looked at one another.

"3x7s?" asked Charlie.

"Okay," answered Denny.

"We'll cover Scrive's money; we'll give you 21,000 tropes," answered Charlie, "If you stick with us."

Chantal nodded vigorously. Eight years' pay, in total.

Charlie had heard all of the conversations about Australia, but Suze and Denny were hearing for the first time.

Chantal explained how Australia was outside the Ellipse and even outside of the Tract and that it had somehow been erased from memory.

"I don't know how they could do that," continued Chantal, "Two of my friends said they escaped from Australia in a boat. Australia had an exclusion zone around it. My friends are both red. They don't use tropus, although they are both dealers. The reason I know them is because I have a bit of a habit myself. She looked at her arm and showed them the depleted cartridge.

"You cut your tropus doses", said Denny. He realised that it would also account for Chantal's different complexion from most people in the Ellipse.

"Yes, I deal", said Chantal," I'm trying to make some money so that I can move away from London. And you know something, I feel better and sharper when I don't use the tropus."

Charlie interrupted, "That's what Scrive said too. He'd lost his tropus delivery system after someone tried to kill him a day ago. His blood was turning black until I destroyed whatever was attacking him. It was a nano culture of some kind. Because of the smashed cartridge, he ended up with blood that was turning red, but he also said he felt sharper with the red blood. He was a little concerned about the loss of Nero immunity, but otherwise felt fine."

"That's what Lars said," added Chantal, "When Janie and I met him, he told us that the Nero toxin was nowadays a hoax. Misinformation put out to keep us scared and needing the tropus."

Get your coat, you've pulled

Taylor had arrived back in the main Brookings complex.

He'd saved the information from the old monitoring station in a way that meant he could access it from his central systems. He wanted to find out if there was anyone else that knew about Australia and Woomera. The nature of his monitoring meant he could easily check for this as a correlation and his own systems could dig deep into secret files.

Taylor was thinking how useful it had been to get some additional information from the ancient monitoring systems.

He was bypassing the gatekeepered modern ways to obtain information by simply referring to older systems 'frozen in time' in the way they looked at things.
Taylor remembered the old stories about concreting the gun emplacements, but that it meant you had to know which way to expect the enemy.

On this occasion the systems had been pointing the right way.

Back onto the modern connections he now had a variety of information to use to help him find more depth. Within ten minutes he'd found a link and the name of someone who seemed to have a connection.

Bruce Henderson. He was a retired ex-marine Captain who nowadays lived on Cape Cod. Taylor smiled at the cliché of this. He'd seen those old movies where a black multi-role vertical lift arrives to extract a one-time military hero for just that one special mission. He envisaged that would soon be happening to Bruce. Heck, Bruce even had the right first name.

Taylor realised he needed to pass the information to the Sparrowhawk gathering at the Pentagon. If he went any further himself, it would create problems.

He decided the fastest way to alert them was by sending a short message to Colonel Maddox.

"Sparrowhawk, relates to earlier Yulara files sent. We have now located assistance. Captain Bruce Henderson of Cape Cod. He attached an electronic ident for Henderson to the message."

The uploaded files covered an area of land that was not normally visible to any of the US surveillance systems.

Maddox had a team of National Security Agency experts review the material as a high priority. They described the location that Taylor had recorded as within the land formally described as Australia.

Technically this was now part of the exclusion zone and regarded as off grid.

They had also checked for experts that would be able to assist with this and following Captain Taylor's lead they also identified Captain Henderson, who had been on black ops missions to the southern hemisphere.

A military Levitor was sent across to the airport at Otis Base on Cape Cod. At the same time a call was placed to the National Guard based there and a small h-Rover was dispatched to pick up Henderson. The h-Rover had its full complement of weapons on board and the riders looked alarming enough to scare most people, although Henderson hardly blinked when he saw them.

"Guys," he said, "Welcome, but I'm guessing you've come to get me for something."

Henderson could still handle himself but realised the arrival of fully armed National Guard in a high-speed hover transport could only mean that he was to be accompanied somewhere.

"Let me get a coat," he said, at the same time priming a few switches on his desk console.

He then left quietly, wondering what was creating the fuss. Henderson had been through mind conditioning as part of his Marine role so most of what was happening now was for him, relatively routine, even if there had been a gap since he'd left the service.

His preconditioning also meant he was able to withstand severe torture and questioning. He wasn't expecting anything bad though, he'd kept a straightforward and low-key existence since he'd left the Marines.

Like many, he'd suffered from jitters once he'd left the service although he'd found, against prevailing advice, that he was able to steady it better by not using the tropus in the way that most people did.

He self-administered a reduced dose, realising he could be adding a different danger, but at the same time knowing that he felt better on the lower dosages. He knew other people that did this and that they were generally considered part of an alternative lifestyle and that they often sold the spare dosages to those that wanted a different kind of buzz. He didn't do this, preferring to wash away any surplus and at least stay the right side of part of the law.

The ride in the h-Rover was strangely soothing. It was a new model and didn't look as if it had seen any active service. The seats were relatively plush, still with the metal grid that was uncomfortable to sit on for long journeys, but overall the speed and comfort were slightly better than he remembered.

The same applied when they arrived at the airport. Even in the military he'd often spend ages waiting to get through various controls as well as all kinds of weapon checks to be performed.

This time it was through the gates and across the airstrip to the waiting Levitor to take him quickly to Washington.

The Levitor had both passenger and freight capabilities and the h-Rover drove straight into the back of the unit.

They disembarked and walk through to a stairway and up about ten stairs to what looked like a medium class airline seating area.

"Buckle in," said the steward who was waiting for them. You'll get some refreshments when we are airborne. This will be a short flight and we are not expecting any turbulence.

They all clipped in and the plane rose vertically before firing forward on what was a levitation rail. The predefined parts of the route gave the craft speed and accuracy.

Henderson felt the handoffs from the Lev-routes as the plane switched on what was the equivalent of invisible rails in the sky, while it made its way rapidly towards Washington.

Around thirty minutes later they were positioning for a landing, and again they used the stairs, back into the h-Rover and away to one of the complexes on the edge of the Pentagon.

Henderson was again surprised at the lack of security as these processes took place and as they drove through the final gate into a building, he realised they were entering the tunnel system, to the north-east of the Pentagon. Henderson knew about this area and also that it was a highly secure part of the base.

He looked at his watch. It had been around 90 minutes since he'd first spotted the incoming h-Rover back at his home. For Maddox, it was still less than two hours since he'd heard from Taylor about Henderson.

8 metaphysical doughnut holes

"Whether you take the doughnut hole as a blank space or as an entity unto itself is a purely metaphysical question and does not affect the taste of the doughnut one bit."

— *Haruki Murakami, A Wild Sheep Chase*

Holden doughnut

Holden knew he had a reputation that he lived in the wires. That he was reclusive. It was a convenient story.

Holden knew just about everything driving his mission and its link with the domes. He secretly knew more than anyone about the domes and how they linked together by micro-hysteresis.

A slight change in one dome would send tiny magnetic ripples to the other domes. The tiny granularity of the field changes meant that enormous amounts of data could be exchanged imperceptibly.

Holden thought of it like gently washing the world.

And that was how Holden preferred to work. Stealthily, imperceptibly.

And now, through links to the multiple sites where research was conducted, he could keep track of all developments.

Holden knew about other so-called secret developments around the planet. Some were laughably public.

A prior meteor incursion had landed in the Groom Lake salt flats in Nevada. The Americans had gone for the old trick of hiding it in plain sight. They named it R-4808N, as a flight zone close to Coyote Alpha. They could not stop there, though, and went on to call it Area 51, which attracted plenty of sci-fi nerds who thought there must be something else happening there.

Holden knew the mundane truth, that the Americans wanted somewhere to test captured Russian aircraft and for years flew dogfights of Russian MIG-17Fs and MIG-21Fs across the area. Have Drill, Have Doughnut; the missions were code-named. The foreign planes were not claimed in a superspy sting. No, they were simply residuals from defecting Israeli Airforce pilots.

And then there was the recent super bolide which hit Chelyabinsk. It was a small piece flaked from the larger 2012 DA14 asteroid. Like an accidentally detached car wheel on a freeway, once it was off the vehicle it could accelerate away and surprise earth 16 hours ahead of schedule.

That was in the times before the Great Leap. Such events were waxed by what came later. Curiously, no one had made any connections of the behaviours of the historical meteors compared with those forming the Ellipse.

Holden held the data and the process capacity to out-think most of the occupiers of earth.

He could see attempts to use Great Leap thinking, but only in the most basic of ways.

For example, he knew that Makatomi was using the quantum effects of the Trigax as the way to eliminate Mallinson. The Trigax. A great Leap, but now being reduced by humans to being little more than a murder weapon.

That second attempt on Scrive was still much better than Makatomi's first botched attempt to use a sabotaged cartridge to kill Mallinson.

Holden had monitored Makatomi while he took a V-Blade all the way to Norway from Tokyo to set it up. Holden considered it blind squandering of resources.

Then Makatomi used Arusen to acquire a fixer. Double blind handling of the request. Arusen had handled the fixer for the Trigax too.

For Holden it was just a flash along the wire. But it was easier to let the real people perform these acts against one another.

Holden worried that Makatomi's actions were getting less dependable.

He dialled into Makatomi's biosystems and ran a trace. Makatomi was stressed, but still functioning well. There was no need to intervene.

Ed Adams

Apartment 123

In Scrive's apartment, Denny, Suze, Chantal and Charlie sat around a table.

Denny spoke next, "There is a pretty high chance that both of you are known to whoever has killed Scrive. We're not known at the moment and I want to keep it that way. We can help you get new idents which will help keep you out of view, but in return I'll want to know if there's anything else that you know."

Charlie wasn't particularly interested in this type of deal. She'd spent enough time on the road to know how to look after herself. She'd also acquired quite a lot of new money from the deal with Scrive, where he'd already passed her part of his share at the start of their working together.

"I'm okay," she said, "Don't get me wrong, but I can look out for myself." She looked down to where she had clipped the small stun grenade to her belt. Denny and Suze knew immediately what she meant.

They looked at the data cube that Chantal had been given by Scrive.

"Nice," said Denny, "This is a very high-tech device. If we access it wrongly, it will still give us information, but it will also destroy the part where the valuable stuff has been stored."

"I can help you," said Charlie. "I've worked with Scrive and I'll know how to find what is on it."

"Back to our early questions, how do I know I can trust you?" asked Denny.

"Let's just say I'm also motivated to see this thing through. Two reasons, firstly for Scrive and secondly for money."

Charlie didn't want to mention the machines she had stored in the hotel.

Denny asked, "so you know Scrive well - do you think you know how to access this cube?" He held up the data cube that Chantal had produced.

"Sure," said Charlie, "Let me have it for a few minutes. I'll find the right system, and access."

"Remember if you get it wrong you will wipe it," said Suze.

"Not a chance," said Charlie confidently and clicked it into one of Scrive's decks. It was the machine he used the most. She knew she'd need to use a code sequence to read the cube and pressed a sequence of keys. It was a sequence they'd used together in the past. She knew it would work and immediately a small blue light appeared inside the data cube.

Seconds later the cube had offloaded its contents into Scrive's deck. It was going through a decryption process and arranging itself into a series of hyper-walls. Each wall was a set of archive materials. The ones that had originally been downloaded from the Vault by Crispin and Lucas.

"It's going to take a time for us to search this," said Suze.

Not if we type in "Australia", said Charlie and started a search based upon what Scrive had sent to her.

The cube was starting to give up its secrets.

Cube

Denny and Suze watched as Charlie manipulated the Cube using Scrive's technology. Charlie had worked with Scrive on enough jobs that they could both work through each other's patterns, and they had frequently stored safe codes to help one another break into each other's technology.

"Scrive must have really trusted you?" said Denny, watching Charlie at work.

"The trust was total and the same from me," replied Charlie, "I'm doing this for Scrive."

"...and the money." thought Chantal, but she didn't say anything.

"Okay... Here's what we have," continued Charlie. "Australia looks for real, and it seems to be outside the Tract. We all know about the Ellipse and the Tract. The Tract seems to be a barrier layer protecting what is a third layer. The main part of this Third layer seems to be in the southern part of the planet. It's as if we've all somehow stopped noticing a whole piece of the world. I can't work out why none of us even remember it?"

"You know something," said Chantal, "I have this very faint recollection. It is like something from childhood. It's so strange. We all grow up, and despite living intensely through our childhoods, there are large swathes of it that none of us remembers. It's a bit like that with Australia. It is like a childhood party I attended that wasn't very good. Not bad enough to remember because someone fell in a pool or we got chased by a clown, but on that edge where if I think really hard I can just about remember something... Kelly," she said, "Somebody Kelly, a folklore robber."

The others looked at one another. Charlie said, "What is even stranger is that even while we are doing this, I'm thinking that I won't be remembering this tomorrow. It is as if there is something erasing part of my mind even while we are having this conversation."

Denny and Suze looked up. They were having a similar thought. They could remember Scrive clearly enough, but already the name of the place was starting to fade again. Denny had written it down. It was in his handheld. But what had he filed it against?

"There's something extraordinary about all of this," said Chantal." I'm not getting that same feeling. I remembered that two of my friends claimed they were from Australia. They told me that months ago. I haven't forgotten it."

"We'll still need to move fast on this," said Denny, "I suggest we start the process with Lars to get inside Biotree. If we can contact your friend Janie, we could try to obtain some improved access codes, and that may increase the effectiveness of our search."

Chantal looked up, she agreed this was the best thing to try next. She just wasn't sure about bringing Janie into the middle of all of these new people.

"Let me contact Janie alone. I'll need to explain to her what has been happening. Then I can see whether she is still prepared to help. She didn't meet Scrive, but she knows that Karin disappeared as a result of helping Lars and Carolin. I'm getting worried that this could start to happen to all of us."

Chantal stood as if to leave. "You know what," she said, "I'm a little concerned that you are all going to forget some parts of this."

Charlie nodded. She also felt that the aspects associated with Australia were already becoming less clear.

"You're all going to have to trust me on this," Chantal said decisively.

"I'm going to take Scrive's system and the data cube. I'll also upload it to Wolkerech - you know that German cloud system. Purely for safekeeping. I've got money riding on the outcome of this. I don't want you all to forget where you've placed the system or something."

They looked at one another in the room. Denny gestured to the others. "Okay, I can see why you are saying that. I'm also concerned about the way this information is slipping away. But we won't forget who you are, Chantal, nor Janie, so please remember that. If there's an attempt to be too smart, we'll have a way to find you. "

He put on a menacing look. Chantal's own expression overrode it. She was treating Denny like a comic playground hero when she'd already discovered smoking and boys. Kind of "yeah, right."

"Like I said, I want this to work." She moved to the door. Her manga cartoon outfit somehow seemed stronger to the others.

Search

Chantal left Scrive's apartment and made for the street level. She would get a taxi back to her place and then tell Janie what had been happening. The death of Scrive was huge news, plus the disappearance of Janie's friend Karin were both big alarm signals that all was not well.

Chantal was also worried that so many people were now getting involved. It had been a whirlwind couple of days.

She called ahead to check with Janie that she would be indoors and sure enough, Janie picked up the phone.

"There will be a lot to explain" began Chantal.

Thirty minutes later she was back at home and started to tell Janie what had happened. She also felt that they needed to decide who they could trust because it still wasn't clear whose side people were on.

Janie was also cautious, "We've said to Lars that we would help him get information for Scrive, does that mean the mission will also automatically transfer to these other people? It almost seems too convenient?"

"Well Charlie, who I met, seemed to have a genuine affinity for Scrive," said Chantal, "I think I trust her based upon the conversations today. It also looks to me as if Denny and Suze have been asked to do the same task as Scrive. I wonder why Makatomi asked two teams to look for the information?"

"Maybe for exactly the reasons we see now, where Scrive has been killed. He might even have been a decoy for all we know," speculated Janie.

"Tomorrow, I will go to work as if it's normal, but I will also try to find the codes or searches for what people are trying to access."

"Great," said Chantal, "Although I think we need to tell Lars some of what is happening. He did ask us to let him know as things move along."

"Do you think Lars is on the level?" asked Janie, "I am less certain of whom we can trust by the minute."

Chantal responded," I think we should assume that Karin was already working with Lars. She then disappeared. That all seems consistent.

Lars already knew about Scrive and was trying to approach him to get Karin and now you to provide access to some of Biotree's secrets. The other team of Denny and Suze seem to be an insurance policy by Makatomi - or if as you say, Scrive was a decoy. I somehow doubt it though, he was too good to be expendable in that way."

Chantal didn't mention Charlie. She was upholding her part of the deal to keep Charlie a secret from as many people as possible. It had been the deal with Scrive, and she thought that she owed him, - them - that much at least.

Janie contacted Lars, and they arranged to meet again that evening, in a different location and this time without Carolin.

Chantal was interested that like Scrive and Charlie, Lars was putting distance between himself and Carolin. She decided it was part of the way these types of teams operated - maybe with the exception of Denny and Suze.

Chantal similarly realised that such an approach wouldn't work with herself and Janie because they were already sharing an apartment and it wouldn't take anyone more than about five minutes to make the connection. But on the other hand, neither Chantal nor Janie had any previous connection with this type of activity.

The cafe they had arranged to meet in was in the west end of London, in a busy tourist location close to Trafalgar Square.

Lars arrived at the cafe just after them. They were seated on the pavement, at round metal tables, closely spaced and on the edge of the theatre district. The swaying area was crowded, mainly from early evening theatre goers.

Janie explained to Lars that she would help but needed to be told what she was to look for.

Lars had explained that it would be information firstly about the manufacture of tropus. Then it would be information related to the nanomachines and their design.

There was also a couple of projects; code names seemed to be "woomera" and "australia."

Chantal had already told Janie about Australia 'the place', and they all now assumed it might be something to do with the location rather than a specific project.

The way that Janie would need to try to discover anything was to type these searches into a highly secure system and to see what happened.

If the system itself could be located, then Denny and Suze would be able to drill into it securely from afar. The challenge was to know what they were looking for. It was far easier to discover this from inside rather than to try targeting everything from outside.

Lars said Janie might strike lucky and get to some actual secrets, but it was far safer to find the right system and then to let Denny and Suze take over.

Janie was to use another data cube which would provide trace recording of what she did. It would give the system addresses that Denny and Suze would need.

Lars asked Chantal if she believed Denny and Suze were working for the same side.

"I don't even understand what the sides are," replied Chantal.

Chantal added, "Denny and Suze seem to be on the level as much as anything could be over the last few days."

Chantal decided not to mention the little matter of the fees she would be accepting from Suze and Denny, although she would do right by Janie once this was over.

"I think tomorrow we should link whatever we find through to Suze and Denny," said Chantal, "that way we bring them along and also have capabilities similar to those from Scrive."

Lars nodded in agreement.

"We should split up tomorrow," he said, "Janie will go to work, you, Chantal, will go to Suze and Denny and I will coordinate."

Like they had all agreed, Chantal had kept Charlie out of the story. Scrive and Charlie had decided to keep Charlie involved but separate, and it was an excellent ongoing plan.

Chantal took the Transit system across to Scrive's apartment where Charlie, Denny and Suze waited.

Janie's next day was a regular working day, and she would be back at her main offices. Suze and Denny decided to accompany Janie on her route and to wait in a nearby WorkSmart location, which was a managed office facility, with high bandwidth communications and coffee.

Janie worked in an area close to the facilities that Makatomi used when he was in London. Makatomi travelled the globe and so was more usually a virtual presence.

That was nothing compared with Makatomi's boss Holden, who was never seen in this office in person. Holden had a large suite on one of the top floors, but even when people attended for meetings it was usually supported by telepresence and through the agency of another person who would stand in for Holden during the session.

None of this phased Janie, who was used to the way of the modern and heavily virtualised workplace.

She arrived at her workspace and began what looked like a typical day. She would create a situation where she needed to go to the area that Makatomi usually worked and needed to create an issue that would warrant a visit rather than just a call or email.

She decided to use some further information about the Chinese and adapted some numbers in a report which implied even faster erosion of Biotree's financial position.

It was a clumsy adaptation, but enough to mean she needed to visit Makatomi's floor directly. It was the type of sensitive information that would usually be treated with the highest confidentiality because of its potential impact upon share price.

She made her way up to Makatomi's floor. As she arrived, she was aware that the V-Blade deck was occupied. There was a V-Blade on the building. She hadn't noticed it when she arrived, but it probably meant that Makatomi or Holden were actually in the building.

Janie thought this slightly unusual. Makatomi had been around half the planet in the last couple of days, and it would need to be something special that that had caused his return so quickly to London.

She carried on along the corridor with the special report she had manufactured and was able to get through to the area where Makatomi worked. It wasn't an exceptional feat; the reason Lars had selected her was that she had this access.

The area was deserted. If Makatomi was in the building, then he was elsewhere.

She could see a workstation in the corner off the room. If Makatomi had left it switched on, she would be able to bypass the security screen and get into Makatomi's workspace.

As she walked towards the terminal, she caught sight of a reflection in the nearby glass. It was the glow from the terminal screen. It was switched on. It would be more straightforward than she had hoped.

She used her access code to bypass the screen security and then flipped into Makatomi's area of the system. As one of the trusted people in the team, she had generic access to many of the privileged areas.

The difference was accessing the systems from Makatomi's physical system. It automatically gave her a better status of access. She clipped the small data cube into a port on the workstation and watched a short blue light pulse. The cube that Denny had provided for her had found its way onto the network. Now she could record everything that she entered along with the responses from the system

She brought up the search screen.

Australia.

Nothing

Woomera.

A response came back immediately.

"Access to this area is restricted. Physical presence in Biotree's Research and Development facility required. Access from Bodø Only."

9 – *Jenny*

I am not free,
because I can be exploded at any time

Jenny Holzer, Tate Modern, London 23 July 2018

Jog-shuttle

The memory cube quietly captured the response and Janie could tell that it was running further analysis based upon the code from Denny.

She assumed that her typing of a few words was somehow filling up the cube with various types of useful information.

Janie reinstated the workstation and removed the cube. She was toying whether to leave the faked report when she heard the door to the office slide open.

"Hello" she called. "Is that Mr Makatomi, I have a report for you. It's rather confidential."

Makatomi looked startled. He crossed the floor towards Janie.

"Hello," he said graciously, "I'm slightly surprised to see you in my office alone. Did someone show you into here?"

"I'm Jane Southern from Advanced Analytics, and I thought I'd better bring you this analysis without alerting too many people." She handed him the eSlate with the information. He flicked his way through the first couple of pages.

"If this were true," said Makatomi, "Then I think we'd have another serious issue to deal with. You say you are from the Advanced team? There seem to be some basic flaws in this analysis. I'd have expected you to have spotted them very easily?"

Makatomi pressed a circular control. Janie looked at it and thought that it was an ancient way of accessing some kind of playback system.

"You know what," said Makatomi, "I decided not to have the old systems replaced when I move into this suite. This jog-shuttle control is a quick way to wind back through the video recording that is built into this area. Let's take a look, shall we?"

He twisted the control anti-clockwise. Sure enough, there was a video projection which now ran along the back wall of the room. It was displaying four images, each a metre across. In three of them, Janie could be seen walking in reverse, and then a few moments later she could be seen at Makatomi's screen intently typing.

"This is interesting," said Makatomi, "Ms Southern, you seem to be using my system? I don't recollect giving you that authority."

Janie was holding the small data cube in her hand. She felt it get hot. Almost hot enough to drop. Makatomi looked more intently at the screen.

"I think you'd better hand over the data cube you installed," he said as he aimed a microcordon at her, "I'm going to need to hold you here."

The cordon snaked from Makatomi's small pistol-like device. Janie felt the electrics pulse as it surrounded her. She was captive.

Charlie could be a stunner

Denny had been monitoring the cube's system all morning from the WorkSmart location, since before Janie had arrived at the office. He was waiting for any sign from the cube that Janie had accessed the system.

Sure enough, at around mid-morning, there had been an alert, and the channel to Makatomi's office had opened up. Denny hadn't told Janie, but the cube had several functions. It was a recorder, but it would also drop a small payload into the Biotree systems. It would give him some direct privileged access so that he could work fast without waiting for Janie's return.

He kept a monitor screen running with Janie's image as she worked while he dropped a few more small packages of code into the Biotree environment.

Then he noticed the entrance of Makatomi. He flicked to the surprisingly dated monitoring system for the suite and noted that Makatomi was challenging Janie. Makatomi had found the cube. This was not good.

Suze was also working and had followed the link back to the Bodø system, based upon the original search created by Janie.

"It's a tough one," said Suze, "there is only one way to access the Bodø environment, and that is from Bodø. It's completely cut off from all of the other access routes. They looked at one another.

"So how will we get to Bodø?" asked Denny.

"Makatomi's roof has a V-Blade on it", said Suze. "But we need someone who can operate it."

"That would be me, then," said Charlie. "I worked with these with Scrive, when we were in the middle east. I have a licence to fly the military versions. Makatomi's is a luxury model by comparison."

"All we need to do is break into Biotree, rescue Janie, steal the V-Blade and fly it to Bodø." said Denny.

"We might as well kidnap Makatomi while we are at it," added Charlie, who was counting out some stun grenades.

Sky fire

Captain Henderson took the short flight to Washington. He was led from the h-Rover and escorted through another couple of corridors. He expected to be taken into something resembling a cell or a white room. Instead, he entered a room which looked comfortable. Leather sofas, military pictures on the wall, a coffee pot and some fine porcelain cups. Not a standard intel room. It looked like the Pentagon had money.

"Captain Henderson?" asked a serious-looking uniformed man, who turned from a seated position across the room. He had been in discussion with two other people and also with a video wall.

"Yes, I am Captain Henderson. You've brought me here on some sort of military matter, I assume?", he decided to stay unruffled, as much as he was capable.

"Correct. My name Colonel Maddox, US Marines. I know you recently retired from the Marine Corp, and now it is information we are requesting, and it's the sort that I think you will freely give. We are trying to understand something that is happening and think you may be able to take us further. We'll re-commission you for active service and provide significant years served pension increase if you'll help us."

"I was just about done with fishing, anyway," said Henderson.

Maddox proceeded to explain what he had heard from Taylor, and how Henderson's name had occurred during the various searches.

Henderson nodded while he was being told the story and decided it was better to come clean on everything.

"I worked on an assignment in a country which used to be called Australia." he began. "We were there as part of international peacekeeping and running quiet exchanges of information about rocket propulsion technology. I was a part of an active monitoring unit asked to validate specific claims made by one of our allies. The technology they were using didn't seem to be possible, according to the scientists and so they wanted to check a few secure areas to see what was behind particular closed doors. My team were part of a black operations mission to find out what was happening.

"Frankly, that was routine, and we were just waiting for our chance. Compared with what else happened there, this was an almost irrelevant diversion.

Maddox listened suspiciously to what he was being told.

"We were at the site during what turned out to be small meteor showers. They had been forecast and were vectored towards our location. The scientists said they'd burn out in the upper atmosphere, so I wasn't' too worried about it. We were all thinking about how to get the doors open on our mission."

"This was maybe 20 years ago. There was a kind of sky-fire as the first meteors streak across the sky. It looked a little like the trails from a modern-day V-Blade actually. There were several more, but they all seemed to burn out.

"We assumed it was the tail of a vast comet somewhere much further out in the solar system. The early reports made the news and the science community was quite excited. At the time, the individual showers didn't make it through the earth's atmosphere. Like most things thrown at earth, they were burned up on entry."

Then a larger item was spotted, on a different trajectory and a course for earth. This one was different though because it seemed to have steering. No -one could work out how it changed course and speed as it approached the earth and it then took a path that allowed it to glide in, to where it landed, which was in the middle of the Woomera rocket testing ranges in Australia.

"So how have I not heard of it?" asked Maddox.

"It is part of what happened," continued Bruce Henderson, "And it is something that only a very few people are aware of now. Australia is unknown, the Woomera ranges are unknown, and the landing is also unknown"

"The meteor, or whatever it was, landed in the middle of the test range. The Australians took this kind of thing very seriously. There was a significant base of the former Joint Defence Facility at a place called Nurrungar, which was about 15 kilometres south of Woomera. The planes from there were in the air long before the 'meteor' hit.

"I should explain that the test area around Woomera is huge. About the same size as the state of Alabama, or the whole of England, as a matter of fact.

"This facility, and another one, the one where I was based, was administered by the Department of Defence. I was based at Maralinga to the far west of the Woomera Testing Range, on the edge of the facility. We had planes too, but we were also on special instructions to evacuate if anything untoward were to happen. To be honest, we'd all worked here for years, and it was mainly quite a sleepy place.

"But on this occasion, we had the Aerospace Operational Support Group of the Royal Australian Air Force appear over the horizon like a swarm of bees.

"They escorted us into the air in fast fighter planes and told us we would be permitted a non-intrusive fly-over of what was happening. We took an SR72 with all the spy gear fitted. We were to use this to brief about follow up action.

"Several crew flew the Lockheed and the rest of us took an Australian chauffeured ride in fast planes back to the site of the impact.

"The strange thing was that the site of the incursion was almost antiseptically tidy. Instead of a huge hole in the ground or a long scar, there was an elegant teardrop shaped glassy looking structure on the ground. It had a slight movement from within, like light, but that was the only sign that there was anything active.

"My immediate reaction was that it was like an egg or spawn or something and that we'd better stay clear of it and create a cordon. By its landing in the middle of a rocket testing range we already had a head start. My plane was the one on the outside of the formation. We flew at a half kilometre wing separation but did a vertical bank over the structure. We were also to look at whether there was anything else unusual in the desert.

"We returned to the base. It was a typical scorcher of a day with temperatures up around 42C at the middle of the day. I remember stepping out of the plane and being hit by the heat.

"Then we regrouped inside the facility and started to compare notes. There were a few things we'd noticed, ahead of getting the telemetry from the SR72. The flights lower over the structure had sighted some red trails moving along its surface. Thin lines that wove around. I didn't see this. There was no sound and the area around the structure looked undamaged. It looked more as if the structure had been built there rather than had somehow crashed or landed there.

"We also tried a replay of the structure's incoming flight path. There were a couple of adjustments on the way in that looked more like flight corrections than something that could happen by chance. It was also a very fast entry; beyond anything we'd expect to land without a massive impact zone.

"Of course, we were on the communications link about this, and it had also been picked up by plenty of other satellite tracking systems. The Americans, NATO and most of the super-powers were on our case about what had occurred. The Australian Prime Minister had also been alerted and had said he would allow the situation to be treated as a global one, rather than through just the resources of Australia, but that it would stay under Australian command.

"The opportunity to get anyone to us quickly was low, except for a couple of American stealth planes which were on the ground in Nurrungar within a half an hour. It is one of the times when everyone realised that Australia is quite large and quite a long way from everywhere else.

"I can still remember that one of the American-marked stealths was a Chinese design, J22-Dragon or something

"We had various cameras and radar on the planes, but we are talking about using the technology from before the Great Leap."

Henderson looked at Maddox, "Of course, now we know that this was one of the reasons we could have the Great Leap."

Maddox was looking at Henderson with an uncharacteristically softened expression. "I'm not sure about any of this. You seem to be telling a true story, but frankly, it is all so far-fetched I'm wondering if your mind is altogether stable?"

Henderson looked back, "That's the irony, that is exactly the problem that the landings started."

"You said landings", said Maddox," Are you telling me there are more than one?"

"Yes," replied Henderson, "The one we found was one of several outliers. It turned out that the main landing was in Yulara. The way it worked was a kind of teardrop shaped scattering of six separate impacts. It took us several days to realise this, though, because of how the structures worked.

"Yulara, how far from Woomera?"

"About a thousand kilometres, north west, 12 hours by land transport. Australia is vast. The Gibson desert is to the West of Yulara, it's about 60,000 square miles. But it's even bigger really because it is sandwiched between two other deserts - The Great

Sandy and The Great Victoria. The Great Sandy is over 100,000 square miles and the Great Victoria is more like 160,000 square miles. We are talking about a deserted spot in a desert within a desert.

And that's within a continent that no-one has ever heard of?" interrupted Maddox. "Like the lost city of Atlantis, or Eldorado? - only as big as the United States? I don't think so."

"Let me continue," said Henderson. "It will be difficult to prove any of this directly, but you will need to know if there is any chance for us to change anything."

"The various authorities decided it would be better to clamp down news about the structure until we knew more of what was happening. This was standard protocol for us in any case, so we had already been running everything encrypted and secured."

"The story to the world was that we'd had another meteor shower that had burned up in the atmosphere. We'd had all the fires earlier too, which only added to the mystique around everything.

"A few people had seen the flashes, but we said it was the burn up. Some of the civilian astronomers asked more questions, but we created an explanation which also covered the apparent changes of direction. Sunspots, I remember, came into it."

"Because we were based in 'deepest Australia' it worked to our advantage when we were explaining any of this. I think the conspiracists thought we'd let off a rocket or missile that had somehow crashed in any case."

Bruce sipped at the coffee. He ran his finger around the pressed-bamboo lid and fidgeted with the small hole through which he could drink. With a fingernail, he was absentmindedly counting the corrugations in the holder.

"Of course, that was before we realised that scale to which the structures would develop. We couldn't approach them at that stage and Woomera and the Gibson Desert made automatic cordons to prevent people from getting nosey.

"By the next day, we had our own monitoring systems in place. Secure cameras, telemetry and a side presence of serious military strength."

"That's also about the point when everything changed..."

Illicit Trigax

The Bodø facility where Sheri worked had some of the most cutting-edge technology. The rest of the site referred to it as the ATA, which was supposed to stand for Advanced Technology Area, but most people called it the "alien technology area".

The Trigax was one of the most secretive devices available. It was intended to be for use in atom separation as part of building new nano structures. It could operate with extreme power and could select individual atoms for isolation. Such was the nature of the device that Sheri used to think of this as a 'god device'. Although having a local range which they used for experiments and building, the boosted calibration could deliver the same power and capability anywhere on the planet and also probably as far as the moon into space. It was simply a matter of getting the coordinates set.

There were various safeguards included that limited its range to a minimal area designated within the R&D facility and the levels of failsafe were such that it effectively had a huge electronics and software guard to prevent it being aimed anywhere else. The technology was classed as munitions, and the current peaceful use came with stringent conditions.

Sheri wasn't sure how the original design had been created. It was something to do with Holden, who was now in a top company position. The science within the device was still beyond her, despite her ability to use the machine with high precision. But, she rationalised, she didn't know the details of how Transit system engines worked either, but she still used the Transit.

This time she had arrived for work and gone directly to the lab. She was surprised when she noticed that the system for Trigax had its coordinates adjusted. Apart from her, there were only three other people with routine access to this device, and they wouldn't interfere in the middle of one of her experimental protocols.

Not only were the coordinates changed, but they were also outside of the guard rails. She looked again; the access had somehow broken the deadlocking system that prevented the Trigax being re-calibrated to other locations.

She pulled up a second screen image, this time a planet model. She punched in the coordinates that the Trigax had as its focus. It wasn't even in Norway.

They were pointing to London, England.

Then she noticed that the Trigax had been deployed, and on a high-power setting.

It looked to her as if someone had used the Trigax as a weapon. It also looked as if they had done so on her watch and with her access codes.

Sheri tried to think who could do this. Other than the three other users, she could only think to ask Nathan. He didn't work in this area or have access. He would need to get through several sets of security, although that was his role. Managing the security of the site.

There was something highly irregular here, and she needed to take great care to find out what had been happening.

Sheri decided to call Nathan. She knew he would be on-site, but probably a long way away.

"Nathan, its Sheri, something's happened. To do with work. I need to talk to you. It is important"

They arranged to meet, but Sheri decided it would be best if they went back to their apartment. She told her department chief that she was unwell and that she needed to get out for a few hours. She made a point of saying that Nathan would be joining her.

Sheri arrived back at the apartment. Nathan's four-wheel-drive transport was already there. She walked inside, and Nathan was sipping tea.

"What's going on?" he asked," this seems pretty unusual."

"Nathan, I'll need you to keep this secret," started Sheri. "It's to do with work, but I can't let this go elsewhere."

"Hey, Sher, what is it?" asked Nathan, "This isn't like you." He looked concerned.

"I think someone has managed to steal some of my codes, my ident," said Sheri," They've used it to gain access to a critical piece of equipment." She knew Nathan wouldn't automatically know about the Trigax, and it wasn't something she'd routinely discuss.

"A special device for handling atoms has been tampered with. We use it for science, but it can be used for other purposes too."

"As a weapon?" asked Nathan.

Sheri knew that Nathan would be aware of the nature of some of the specialised equipment in her section. The whole of Biotree had many esoteric and potentially lethal devices, mainly because of the power that they were using.

"Yes, it is something that can be used as a weapon. There's only a few of them on the planet, and the Pentagon holds one. As well as its experimental potential, it can target an area and then send a huge pulse with an accuracy of a few centimetres. It could take out whatever is at those coordinates, which are measured in three planes. Effectively it can address any point on the planet.

"Whoa," responded Nathan, "So it is like a 'death-ray'?" he asked, "That does sound a bit far-fetched."

"It would have been before the Leap," answered Sheri, "To be honest even me and the people I work with don't understand what makes it operate. The Science of it seems to be almost too clever."

"Another piece of your alien technology?" joked Nathan. Sheri knew they had conversations sometimes about some of the things she worked with, which just seemed too good to be true. Even Nathan had found technology in his security line which seemed to be beyond anything he could have imagined, yet it all seemed to work and as if it had always been there.

"The machine is called a Trigax. It uses three types of energy as an output. When they combine, they create an exceptional power surge which we have been able to use to extract unusual atomic structures. Effectively it is building us atoms that shouldn't exist. The three pulses have to focus at a single point - the coordinates are set, and then the pulse travels like a wave through the air, but when it converges, it will take the source at which it points and de-materialise it. The resultant effect is like something out of Einstein - kind of E=MC Squared, but we are re-building the Energy from the matter.

We have been using the side effect of it to isolate the atoms we've needed. The energy seems to disappear somewhere, but we've still not worked out how that part works.

So, you are creating massive energy but then losing it? asked Nathan

"Yes," said Sheri, "it's as if it's all transferring somewhere very quickly, in a way that we can't trace."

"Another part of what Einstein described?" asked Nathan "- something to do with the speed of light?"

"Maybe," said Sheri, "we just don't know."

"Anyway. The Trigax has been primed and fired at a location outside of our controlled range. I thought it was impossible to do that because of the way the Trigax has been locked down."

"But I can see the Trigax has been fired, and it was pointed to somewhere in London, England."

"Now you want me to help you figure out what has happened?" asked Nathan.

"Please," said Sheri. "Everyone in my area will know that there are only four scientists with the access codes for the device. The triggering is showing my ident, and my codes have been used, but I know I haven't done it, nor have I told anyone- not even you, Nathan, about the codes."

"It sounds like you need me to take a look at this in a professional capacity." said Nathan, smiling, "This'll be a first!"

"You're sure you couldn't have done something to affect this?" asked Sheri. "Please tell me if you think there's a way that you'd give information to someone."

"No," said Nathan," I didn't know about the Trigax until a few minutes ago, and I've been offsite for the last two days when this is supposed to have been happening."

Sheri nodded, " Let's think of a more basic security leak that would give you a reason to be on site. I know. I'll say we've had a hack into one of the lower-tier systems. It would be exactly the sort of thing we'd call you over for."

"Okay," said Nathan, "I'll be nearby to ensure that I get the call."

Nathan left in his security patrol vehicle and parked across from Sheri's block.

Sure enough, Sheri called through the request for assistance and Nathan got the call because of his proximity.

Nathan worked his way into the area where Sheri worked following normal procedures. He didn't want it to look unusual, yet he did need to be in the secure area if he was to be able to trace back what had happened.

He had brought some of his security checking systems with him. Compared to the technology used by the trackers, he knew his own technologies were still rather primitive, but he did know his way around the Biotree environment and with his access codes had a head start on most people.

"First of all, we should check where the Trigax was pointed, he said, that will give us some extra information."

He looked at the coordinates and traced them back to London. Sheri assisted, and they found a particular location. It was by an old part of London called London Bridge. They zoomed tight and found the target.

Jumpy RFID

Nathan then zoomed using his own technology onto the same area. He was able to see the RFID of the person targeted. He looked it up in his directories and noticed that the code jumped around before settling.

"Unusual," he muttered, "...the way it is jumping suggests it is someone that didn't want to be found. A tracker probably, they all do this but the toolkit here at Biotree can get past that type of camouflage."

"A Mr Mallinson, Scrive Mallinson. But his trace has stopped after this point. It looks as if he has been wiped. I don't know how, except its exactly where your Trigax was pointed."

"Yes," said Sheri, "That would be the Trigax. An accurate tracking and then elimination. I have never seen it used at that range or in that way before. Its way outside of our safe tolerances."

He froze the grid reference and looked at neighbouring proximity. There is another RFID close by he said. It's got a signal; Someone else was very close by.

He dialled the codes. "Chantal," he said "'Chantal le Strang'. So, who are you Mademoiselle le Strang?"

He continued to work and noticed that Chantal's other known movements were around London. The tracing he was using was only at thirty-minute intervals and he could see that she had stayed in London for the last two weeks prior to the firing of the Trigax. He couldn't easily check her off against all of the other people she had met, it would take separate time and analysis for that. He decided to flip forward to now.

"My god," he said, "the trace for le Strang is showing from this facility here in Bodø. I'm certain that Chantal le Strang knows something about what has been happening."

"So how can we track her down?" asked Sheri, "If she was in London when the other person was targeted then she can't have been here also. But maybe she is part of a gang, and she's come to contact or collect the other members?"

"Perhaps," said Nathan, "or she was a friend of the person targeted. But in that case, I don't think she'd be able to find this place so quickly, let alone the Trigax."

"And another thing...the speed of her movement from London to here is phenomenal. She'd have needed to use a V-Blade to get here that fast. I'm going to check for flights from London in the last couple of hours."

He looked at the system, "It says here that Makatomi San has taken a flight from Biotree

London to here in the last hour. The V-Blade is over at Dock Three. Do you have comcam in here?"

"Sure," said Sheri and flipped on a display. "Where do you want to look? I'm going to add my control layer onto it so that I can control the video," he said. He flashed some codes and the security system become operable from the comcam.

He flipped to Dock three, via a couple of preview screens. Then he flipped through five or six displays.

"There's no-one here, but that V-Blade is in cool-down. It's been on a fast flight in the last hour. I'm guessing it is in from London and that Chantal le Strang was on it."

He flicked to another area of the base. "This is the dock area. There's a route back to the main levels and a holding area for security clearance. I'm guessing that Ms le Strang won't have the right idents to get inside. I expect she is in the arrivals area somewhere."

There was a large and comfortable set of lounges for arrivals and departures and he ran a quick scan through the area. "It will be easy enough to check her ident again," he said and commenced a scan.

Within a few seconds, he had located her spot in the area. She was with several other people. He snapshotted their idents too so that he could start to trace who had arrived.

"This is getting very interesting," He said and started making for the exit. "I'm going over to find out what is happening. I can contain this group in a secure area whilst we decide upon next steps."

He was speaking into a handheld as he left, and Sheri realized he was getting a secure unit to take him to the arrivals lounge. She saw him clip on another level of insignia to his uniform as he walked out.

Bounce

Chantal couldn't believe the last couple of hours. They'd been in London, and then they'd been to Janie's office - she'd never seen inside it before. Somehow Denny and Suze had kitted the three of them plus Charlie with special idents that let them get to Janie's floor.

They'd taken an elevator to a high floor and as they exited Chantal had noticed the V-Blade docked to the side of the building. It looked pretty cool up close. She'd never travelled in one before.

Then Charlie had walked away from the group, with a rather serious expression. Chantal had heard a couple of loud thuds from inside a room and saw it filling with a greenish tinged cloud. There had been a crackling sound and Charlie had re-emerged with Janie, who looked totally horror -struck.

"Chantal," Janie called, as they emerged, "These people are lethal."

Charlie said to Denny, "We won't be able to take Makatomi with us – something crackled out of a wall and took him down. I think he was getting a little carried away with what Janie has been doing. Shew as secured in an e-cordon until Makatomi was fried"

They flipped a few controls and walked into the V-Blade area.

Glass panels opened; they entered the holding area. "Not a typical departure zone," said Charlie.

"This is just for individual execs. I'm pretty sure where we are going there will be a full-blown arrivals reception area."

They entered the V-Blade. It had ten seats in two rows, plus a frontal cockpit zone with two places next to one another.

"Strap in," said Charlie, "I'm going to fly this thing."

"Are you sure you know what you are doing?" asked Janie.

"I've flown these plenty of times, military grade, with Scrive. This is quite a luxury vehicle." responded Charlie. She was plugging in a headset and adjusting some of the controls.

She flipped a few buttons, asked Denny to take co-pilot and started a short countdown sequence.

"I'm going to fly this the military way," she said, "no niceties, so stay buckled in."

She hit the airwave, and the cabin pressure kicked in. Chantal felt as if she was in a bubble and found it difficult to move her arms. There was a juddering, and the lights went blurry. A slight sensation of movement and then a loud suction noise as the pressure re-balanced.

"Something wrong? asked Chantal, "What, more than everything else?" asked Charlie, "No - we've arrived. I said it would be a short flight."

"Where are we?" said Chantal, "I thought we were going to Norway?"

"We just did," replied Charlie, "These things are fast."

Charlie could again see out of the windows, which had some sort of polaglas, which had just resumed vision. It was a different scene. Snow, Hills, Mountains, a huge airport-like lounge ahead.

"Welcome to Bodø Arrivals," said Charlie. "Thanks for riding wingman, Denny."

"Now we've got to figure out how to get through their security. This may take a little time."

Chantal sat looking at the coffee table in front of her, wondering what had happened in the last few minutes, when a man in a uniform approached, smiling.

"Chantal le Strang?" he smiled, "We have been expecting you."

He held out his hand to shake hers and nodded towards the others seated in the area.

"My name is Nathan, please come with me, and I'll escort you through the system."

Chantal looked confused. How could anyone possibly know she was here? Even she didn't know a few moments before.

"I'm a friend of Scrive Mallinson," he added, "I thought you'd be along around now."

The team looked at one another. This was either very good or very, very bad.

Charlie fingered one of the stun grenades in her pocket. Denny looked around for signs of other security people in the background. To his surprise, Nathan appeared to be alone.

"Look," said Nathan, "I'm going to help you, but you will need to trust me. I am taking you all to a secure area and then I will take Chantal to a further area. We know about Scrive and think we know what happened to him. I can't talk about it here, so we will be creating a locked-down space where we can all talk."

Chantal looked around towards the others. "I'll only go if I can take Charlie with me," she said. "Otherwise, we won't cooperate."

Nathan looked across. He looked at Charlie and then back to Chantal.

"Sure," he said, "I'm asking the rest of you to stay in one of our VIP lounges for the next hour. Then I'll be back - in person - and make sure it is me - to accompany you through security. Let's go," said Nathan, "I've someone else I would like Chantal to meet."

Sheri had been working on the Trigax whilst Nathan was away meeting Chantal. She'd noticed the firing sequence and the long-range setting which was something they were forbidden to use and also which she didn't even know how to override. They had always been told that the Trigax had been locked down. The thing that interested her was whether the longer range would give better signals to help identify the way that the device worked. Their routine use, which they deployed for stripping out atoms was a slow process, partly because the power range of the device was always set to minimum. It was like trying to listen to great music on a very low volume.

She analysed the outputs from the Trigax. She was interested to see where it was dumping the energy it had created as a result of the terrible thing it had done. If it had vaporised Scrive, then there would be a trace in some form, somewhere.

The amplitude of the deployment gave her a chance to check this. Sure enough, she found a signal from the Trigax. It seemed to have streamed the energy to a further co-ordinate. She tried to locate the position globally. It had a type of arc, like a pebble skittering across a lake, except the bounces seemed to defy normal logic. After the weapon had been fired there was a small arc from Norway to London and then a much larger arc from London to the other side of the world. The energy had bounced to the edge of the Ellipse, to a point almost opposite London on the earth at the extremity of the Ellipse.

Sheri looked more closely. It hadn't stopped at that point. It had bounced again. The next bounce was smaller than the one before, but it seemed to arc over the Tract to an area that shouldn't really exist.

As Sheri was thinking about this, she could feel the knowledge that she was gaining starting to go away again. It was as if she was forgetting something whilst she was still discovering it.

She wrote 'bounce, London, Japan, over Tract, further' on a piece of paper. She found herself picking it up and throwing it away.

She wondered why she was forgetting what she had just discovered.

At that moment Nathan re-entered the apartment, with Chantal and Charlie.

Charlie caught Sheri's movements as they walked into the room. It looked as if Sheri was acting oddly, but she wasn't sure why. Almost as if Sheri was trying to hide something.

"This is Chantal, and her friend Charlie. They have said they will help us. We'll need to tell them what has been happening."

Sheri started to explain what had happened, but it also seemed as if she was forgetting part of the story. Charlie and Chantal listened. Sheri tried harder to explain what had been happening, but part way through said," You know something, I think I'm forgetting some important parts of this. It's as if my memory is being erased while I'm telling you. It has something to do with the Trigax and something I found, but I can't remember what. This is stupid, I only discovered it a few minutes ago. I guess it wasn't important.

"Is this room monitored?" asked Charlie.

"Yes, regular cameras only in here," said Nathan.

"Can we replay them?" asked Charlie, "Just the last hour should be enough."

Nathan picked another small handheld from his bag and tapped some codes. Now he had a wall display showing their current room. They could see they were all standing together, until Nathan started to wind the display backwards.

He sped it up and ran it like a video shuttle, until they saw the points where Sheri had been analysing something and then wrote something on paper. They had been watching the video backwards and so they could see that Sheri had written on paper and then thrown it away. Nathan reached for the paper disposal. He opened the unit and the single sheet was still inside.

"I'm not sure why I did that," said Sheri, "to write something and then immediately throw it away."

Nathan read back the words - "bounce, London, Japan, over Tract, further"

"What does this mean?" he asked.

Sheri shrugged. "It's my writing, but I don't understand it"

"Bounce. Something bounced from London to Japan? and then bounced over the Tract?" suggested Chantal.

"There's some things I discovered with Scrive, and I think we need to talk about them. It will need to be here somewhere where we can't be recorded," said Chantal.

"Okay," said Nathan, "Let's move to the diplomat lounge. We can fine-tune security in there."

PART THREE

10 larger beings

Norway…Pretty didn't do it justice.
I felt like we'd sailed into a world meant for
much larger beings,
a place where gods and monsters roamed
freely."

— Rick Riordan, The Ship of the Dead

Diplomat

Nathan showed Charlie and Chantal the way back to the Diplomat Lounge near to where the V-Blade had docked.

Nathan said, "Sheri, I think you'd better introduce yourself more fully."

"I work with the nanotechnology," she started, "We are building the new constructor kits here. - that's the components that the nanobots are built from."

Charlie intervened, "Before Scrive was killed with that ray, or whatever it was, something else happened to him, you know. Someone doctored one of his cartridges, and the tropus contained nanobots which sought to terminate him. There was something strange though, I injected his bloodstream with a big blast of nanoreductives, and it stopped the nanobots in their track. But here's the thing, the nanobot and nanoreductive also spread across to Scrive's spilt blood. It had turned to a blue-black colour but then went back to red after I'd applied the nanoreductives.

"Nanoreductives?" asked Sheri, "That's some trixy stuff you've been handling. Where did you get them?" She looked at Charlie in a whole different way now.

"It's a long story, but I've been places, you know,"

"C-beams glittering in the dark?" joked Chantal.

"Yes, kinda," said Charlie, "I had some time tinkering with simple 'bots, to use as trackers.'

"Okay, let's see if we can work out what happened with the nanocrime first," said Sheri, "I don't mean yours, Charlie, I mean the original cartridge swap."

"I assume Scrive had a standard plexi-cartridge?" asked Sheri, "Not some sort of booster clone?"

"Yes," said Charlie, it was a standard A-port, with all the digital engineering included as well. At least until I smashed it to get my reductive injector into Scrive's system."

"And your nanoreductives were relatively simple devices? I'm not being funny, but I assume you'd not been into heavy-duty nano-engineering?"

"No, that's right. I used basic nanotech using small molecules and proteins. It's what I call my failed experiment."

For the first time Sheri smiled," Yes, we call them nubots - they are a reasonably basic machine type. I guess you ran into problems with them not working and destroying one another?"

"Yes - that was the outcome. I thought of it as 'let the weakness become a strength', and that's how I learned the term "nano-reductive". The ones I built reduced the number of nanobots wherever they were let loose."

Sheri looked at Charlie, "The interesting thing is that they travelled so far and kept working - for example, into the blood spill. That is usually a feature of a more advanced biohybrid. You'd need to be making them in the kind of lab we have here in Biotree."

"Trust me, I only had a cooking lab, I was simply trying to place unique tracers into the cartridges," said Charlie.

"For tracking? That's mega-illegal, you know? Hacking a cartridge and then adding something to it."

"I know," said Charlie, "but I guess that's what has got me here, so you know…"

"It is the same with the boosters," interrupted Chantal," People will swap out their secure and safe A-port plexis for a generic, and then shoot boosted cartridges into their arms."

Sheri nodded, "I guess I get kept in bubble wrap out here at the ATA."

Nathan nodded too," Yes, everything that Charlie and Chantal describe happens around the unsecured edges of the site. Biotree is the stable end of a very ragged set of processes."

"Something I don't understand," said Charlie," Is that Scrive reckoned his thinking was improved after he came off the tropus."

"Yes," said Chantal, "We notice that too when we are selling it. There are two types of boost. Speeder and Kalm. Speeder, like its name implies, jacks the metabolism and certainly the thought processes of the recipient. Kalm is an altogether more mellow experience."

"Nowadays you can usually tell which kind it is by the cartridge markings. Chinese writing is for Kalm and English for Speeder. It didn't used to be the case. They both worked the same originally, more like Kalm."

Sheri nodded, "That would make sense. Biotree changes the formulation every so often. One of the changes was to optimise deployment. Biotree brought in some hackers, and they changed one of the mechanisms inside the nanobots, when we needed the new variant to try to solve the widespread virus in Australia.

"I remember when the packaging changed on the Biotrees. Something about "New, faster acting," said Chantal.

"Originally, Biotree had always built them fail-safe, with a kind of small valve inside. The difference is that the newer bots can replicate; the original design had a so-called Brownian ratchet inside which was like a little cog inside a clock. It made sure the machines could only run up to a certain speed. It was an elegant fail-safe which stopped people's systems becoming overrun with self-replicating nanobots. They ran slower than a body's metabolic speed, which meant the body could handle them without getting overloaded.

That's what seemed to happen with Scrive until I stopped it," said Charlie. His blood was clotting, and I reckon it was black from the incursion of nanobots."

"Yes, the second exploit, which was really to weaponise the nanobots and to force even greater speed. It was an exploitation of the Laplace-Beltrami theorem for narrow escape. Think of it like air escaping from a balloon. It meant the bots could speed excessively and the effect would be like that which happened to Scrive. In battlefield the bots could run riot for a short time.

"And you know something, Makatomi's last business plan? Weaponisation of the nanobots. His idea of a way to save the company."

"That would explain there being two types out there now," said Chantal.

"Yes, the unconstrained variant from Biotree and the safely managed clone from the Chinese. The old economic model of the cartridges was to ensure that the tropus had a half-life measured in weeks and then you'd need to buy a new one. It lost most of its potency by around week four."

"So," said Charlie, "we've got a desperate Biotree that is building weapons and the Chinese trying to steal a copy?"

"If you want to put it like that," said Sheri.

Meteor disturbances

At the secret Mil-base, Maddox had been busy. He'd been following up on the meteor disturbances that Henderson had described. It meant looking into quite ancient archives, from around 20 years ago. Sure enough, there were a few reports' although none in the mainstream press. He'd called Henderson back to his room at the Pentagon.

"It's as if this thing you describe never happened," he said, "There's more news of B-List celebrities than of a major earth strike by meteors. But there's some information which seems to concur with your account."

"I had the guys track a couple of astronomy sites, including the crowdsourced eurekalert, which did feature the meteors. Yale scientist Denison Olmsted was referenced in an article, from 1833, which talked about a prior meteor shower and the need for tracking. Then much later there's the establishment of the KELT follow-up network with two low resolution small telescopes. These enthusiasts tracked the meteor shower you describe to its collision with earth. One of the amateur reports described them as jellyfish-shaped structures – a dome-shaped body and elaborate trailing tentacles. They are outside of all the military and government networks, which is, I'm thinking, why the records are still available."

"Here's the thing," said Maddox, "According to KELT, the meteors seemed to be flying under guidance, which bears out your story. Not only that, there's another couple of hits in the northern hemisphere, both in areas with low populations. One near Fort Resolution in Canada, the other near Bodø, in Norway."

Henderson smiled, "Fort Resolution is in the North West Territories. It's right in the middle of nowhere; I'm guessing that Bodø is the same?"

Maddox nodded, "Not exactly, Bodø is also the R&D centre for Biotree Industries."

"They're the ones in trouble in the city pages at the moment, aren't they?" asked Henderson.

"Yes, and they seem to be under copyright attack from the Chinese clone manufacturers."

"No smoke without fire. Do you want me to take a look?" asked Henderson.

"I've already cleared your path to take a small detail to Norway," said Maddox.

"Okay," said Henderson, "but I don't want to put heavy Navy boots all over this, the fewer people that know, the better."

"But you won't turn down a V-Blade and some 'accessories'?" asked Maddox.

"Thank you, but I'll still travel light," said Henderson.

Bodø

In Bodø, Nathan was leading Chantal, Charlie, Denny, Suze and Janie back through the corridors from the Diplomatic Lounge. He was arranging for each of them admittance to the facility under his supervision.

Of them all, Nathan noticed that Janie seemed the most shaken up by whatever had happened to them. He took her one side and asked, "Are you okay? - Let's get you checked over."

Nathan spotted the tell-tale signs that Janie had been caught in an e-grid or some other form of containment, by her occasional little tremors. Nathan hated the devices and wondered what kind of idiot would use one on Janie.

"It's been a helluva day," said Janie, she looked over to Chantal, who nodded back.

"I'm going to put Janie through some base screening," said Nathan, "Just for peace of mind. Chantal, would you like to stay with her?"

"No, go, Chantal, please go," said Janie, "I'll be fine."

Chantal realised Janie was keen to be left alone. Chantal could only guess what had happened to her in Makatomi's office, and then what had happened when Charlie arrived, no doubt guns a-blazing.

"It's a massive complex you have here," said Denny,

"Yes, it has its own public transit system around it," explained Nathan.

"The transit runs pretty much around the entire perimeter and crosses over the centre in a couple of places."

"Do you mind if we take a look?" asked Suze.

"I can do better than the transit system for you," said Nathan.

"I'll get one of the jetters, and we can do a quick lap of the facility."

"I expect you'll all want to come along for the experience, so I'll get a ten-seater"

He spoke into a communicator and arranged for a pick-up.

An almost silent craft appeared, and they all climbed aboard.

"This isn't going to be as fast as Charlie's piloting, is it?" asked Chantal.

"No, we'll keep to a low speed and make a pass over the site," said Nathan, "We'll start by heading back to the dock where you landed the V-Blade. Then we'll do a circuit.

The craft crawled forward, and they approached the landing docks. A shimmer on the horizon denoted the approach of another fast craft. It was a military specification V-Blade. United States.

"Looks as if we have visitors," said Nathan. He called a central control number.

"It looks as if they are here for similar reasons to you," he said, half-listening to his communicator.

"Does anyone here know a Bruce Henderson?"

Everyone looked blank.

"Okay, here we go," said Nathan as the jetter made for the runway.

"This is a bit more interesting than flying in a V-Blade," commented Chantal.

"We can see things as they go past."

The pilot banked the jetter over the site. It was immense, with a series of modern blocks and a few scattered outbuildings. In the distance was another long teardrop-shaped structure, surrounded by further fencing.

"What's that place?" asked Suze.

"It's the old NATO stores," replied Nathan. It is almost dormant nowadays and we use Robocarts to gain access.

"Why is that?" asked Suze.

"It still holds dangerous substances, to be honest, I think we were tricked into storing them in the first place. To get this land, Biotree had to do a deal with the USAF to take the store off their hands."

"It was supposed to be a mutually beneficial trade," said Sheri, "Biotree claimed to have the wherewithal to clear up the chemicals."

"Essentially it's a big DND nowadays," said Nathan.

"DND?" queried Chantal.

"Do Not Disturb," answered Charlie," So what happens? You still patrol it then?"

"Yes." answered Nathan, "although there's not a lot to see. It's a tin shed covering a bunch of old stores."

The plane banked, and they could see several small autonomous Robocart units moving around the perimeter of the shed.

"Let's go back," said Nathan, "I think you've got a feel for the site now."

Meet Henderson

The jetter pulled into a landing dock, and Nathan could see the Military V-Blade on which Henderson had arrived. It stood next to the one that the others had flown in on, and the differences were striking. The commercial version was altogether bulkier and had a surprising number of antenna and other external attachments. The sleek dark stealth of the naval unit alongside it, with its missile hangers and guns, illustrated it was built for battlefield deployment.

They walked into the lounges again. They could see Henderson immediately. He was kitted out in combat gear and had a couple of supporting aides at his side.

"Hi Captain Henderson, we're pleased to greet you here in Bodø. I'm Nathan Belanger, head of security on the Bodø base. To be honest, we didn't know you were coming here until you'd more or less arrived."

"Same here, Mr Belanger, the US DoD has sent me for a routine inspection of the ex USAF part of the base."

"You'll know it is the ATA now - that's the Advanced Technology Area, so we have some fairly high security to protect commercial secrets?' asked Nathan.

"Yes, I was briefed on the way, and again in this reception lounge. Now if you don't mind, I'd like to make busy with the inspection?"

Nathan nodded. He could see that Henderson was in a hurry. He was used to random US DoD people passing through the base and stopping to have a look at the US legacy.

"I'll arrange for someone to take you to the store area on a jetter; you'll be able to pick up an h-Rover when you are over there."

"That's great," said Henderson. "I'm hoping to see inside too?"

"Sure," said Nathan, "We'll arrange that when you get over there, although there is not much to see. Unless you like oil drums and pressurised coolant tanks. I'll come along too, if you don't mind."

Charlie interrupted, "I'd like to come along for the ride too, if that's okay? You'll have a spare pilot that way, too"

They made their way back to the flight deck and were soon airborne in a jetter.

This time they took a direct route to the storage facility. They flew over the main hanger area and

Charlie could see several HUM-Z rocket planes parked in a tactical formation.

"What are they for?" She asked," Once a NATO base, always a NATO base," said Nathan. "You'd have thought that when Biotree bought the land, the armed forces would move out, but there's some deal with the Americans and in turn with half of Europe about keeping a few planes here. They do circuits and bumps every so often."

The pilot took them high over the store first time and then lower on his landing loop.

Henderson looked as the store block approached. It was huge and traced a path across what could have been the foothills of a mountain range. It was hardly the best position to build a storage facility. Henderson noticed the unusual shape of the facility. Why would anyone make it that shape?

They landed and Henderson, Nathan and Charlie left the jetter and were escorted to the h-Rover. They heard the maglev kick in and gently made their way to an entrance hatch for the storage facility.

"You can pick up the screenings of the interior from here," said the h-Rover pilot. "It is probably quicker than looking all around."

Henderson nodded, looked at the h-Rover's screen and then to a window in the side of the shed.

There were equivalent windows along the facility, each with its own hi-resolution display, showing the interior of the facility.

"With the screens, we can pull up the co-ordinates of any part of the store and know what it contains," explained the pilot, "All from the safety of being outside."

"Okay, said Henderson," can we stop at one of the windows?"

"Sure thing," said the h-Rover driver and they slid to a quiet halt.

"I'm going outside," said Henderson.

He walked towards the nearest screen. It was touch-operated, and he flipped the small control console, which revealed the labelled contents of interior drums.

"I want to see inside this," he said, "Past the screens, to the natural condition."

"Sure" said Nathan. "The pilots usually do the checks from fly-by, but you can walk about if you like. Don't underestimate the distances. "

The three of them left the pilot and moved into the building. Along the edge were a high row of drums and a couple of complicated looking processing machines.

They walked past the drums to a flat fenced area.

There they saw it. Henderson recognised the same glassy dome structure which he'd seen in fly-by when he was in Australia.

"We're leaving," he said.

Secrets

Nathan was on the comms back to Sheri's lab.

"We've found something," he said. "It's vast."

"What is it?" asked Sheri.

"I think it is something that they have been trying to hide, probably Biotree, certainly Makatomi."

The jetter had reached the landing dock again, and Henderson and Nathan climbed out. They made their way back to the ATA research block and found Sheri and the others in Sheri's lab, with some in a second room, separated by a glass wall.

"Denny, Suze and Janie are working through the data cube that Janie retrieved from Makatomi's office, " explained Sheri, "I put them next door in a data room, So, what did you discover?"

"It's like the situation I first saw in Australia, out in the desert, 'said Henderson. "Except there were several of them. Vast glass-like structures, which had splashed themselves across the desert. At the time we thought they were meteor showers and later the Australian situation was erased from records."

"What made you think there was another one here?" asked Sheri.

"I remembered that there were two additional reports of meteors; one in Canada and another in Norway. That's why I came to take a look.

"These stores were established about 20 years ago," said Nathan. "That's long before we created the Biotree facility here."

"Yes, it's the same time that the ones in Australia were identified. And it's around when Australia started to disappear from records."

"Yes, the combination of the Flames and then the virus, Australia was initially cut off from the rest of the world as a quarantine measure. "

"Yes, and then the protection zone was instituted."

Chantal looked confused, "How is it that I don't know any of this?" she asked.

Henderson: "It was news managed at the time. Such a terrible loss of life in Australia and a successive quarantine imposed. They didn't want people going for a look, in case they spread the virus further."

"Could it be something that the domes brought into the continent?"

Charlie said," The strange thing is, the symptoms of the virus sounded very much like the symptoms that Scrive showed when he suffered from that nanobot toxin."

Sheri spoke, "Yes, that's when Biotree started shipping a specific strain of the tropus cartridge to Australia. It was supposed to combat the virus."

Charlie said, thoughtfully, "Unfortunately, it seemed to have some other side effects. I'm wondering if it was accelerated like the one I built, and was, itself the cause of the deaths?"

Sheri announced, "Suze and Denny have been looking through the data walls that Janie retrieved from Makatomi on that cube. They are very fast and efficient - it must be a tracker trait - Their findings show there were several attempts to speed up the nanobots but at the expense of some of the checks and balances.

"I think that is what they are trying to hide," said Sheri, "Biotree tried to stop the original virus with nanobot re-engineering. But they cut corners to make the antidote work more quickly. The hired help didn't have the same stringent processes as we do. That's what I was describing with those accelerants that they introduced.

Charlie asked," You mean that's how they multiplied so quickly and polluted people's bloodstream? Just like the toxin sent to Scrive?"

Sheri looked severe, " According to the papers we got from Makatomi, it was worse. It was not just their bloodstream — everything organic. The 'bots could jump using the 'balloon' reaction, and therefore infect anything else they could process.

"And because the deployment was so rapid, with everyone refitting their tropus cartridges every four weeks, by the time Makatomi's people had discovered it, it was too late, and Australia was destroyed, or people were infected but didn't realise it yet."

Sheri added, "In another four weeks it had run through Australia like a plague - unknown to the authorities the cure was worse than the virus. And it was a time-bomb that they had already set ticking."

"But there were some people, like Crispin and Lucas," said Chantal, "who didn't seem to get affected?"

"in most forms of rapidly spreading virus, there are some people who don't' catch it. Like their systems are somehow immune. Have you ever been on vacation with someone who doesn't get mosquito-bitten?" asked Sheri, "it's one of those mysteries."

"Yes, but this seems to be a double whammy," said Charlie, "First the virus and then the nanobots? Could someone really be immune to both?"

"Yes, it is highly likely," said Sheri, "The original design of the nanobot defences would be to target the virus. It's as likely that the same biological key repelled both types of 'boarder'. Think of it like a key and lock. The virus has to be able to get the lock undone. So does the nanobot to chase after it."

Chantal nodded at this explanation, which seemed to satisfy her.

Charlie said," No wonder Makatomi was trying to keep everything locked down and secret."

"But what do we do about the domes?" asked Nathan, "and what do they have to do with anything?"

"I don't know if you remember, but The Great Leap happened around the time that earth passed through that meteor shower," said Sheri.

"We think that the domes brought some new ideas to the world?" asked Chantel. "It's pretty cosmic!"

Sheri added, "Yes - The Great Leap yielded a range of discoveries. But, in addition, it seems to have been able to manage minds and communications."

"That could account for everyone forgetting about Australia so quickly, it became a case of hidden in plain sight," said Charlie.

"Also hidden on a dangerous land mass, though," said Nathan.

"And protected there too," added Chantal, "Remember what Crispin said about the bracelet and charms?"

"Yes, they've found something about that in the Janie's data walls," answered Sheri,

"Apparently, Makatomi, under Holden's instruction, sanctioned the use of a set of Geostationary Satellites to police the Australian boundaries. They sensed movement across the boundaries like a regular burglar alarm, but then deployed a massive railgun to the targeted area of encroachment. In other words, they were using Trigax as a way to police the boundaries."

"So, the satellites had a separate set of ray guns to support them?" asked Nathan.

"No, the satellites were dual purpose. They contained monitoring equipment and also railguns - the Trigax," said Sheri.

"Tri means three, doesn't it?" asked Chantal, "I'm wondering if there's something else spinning around above our heads?"

"And I'm wondering if the domes brought the intelligence beyond our comprehension?" mused Sheri.

"Take cats. Their intelligence can count to about 4 to 6, to keep track of kittens and they can train their owners to bring food, but give them a larger sum or bigger task, and they don't have a clue. Well, that's what the domes could be like. Delivering raw intelligence beyond our capability to understand. A Great Stumble Forward."

Chantal laughed at this last remark," Cats! can't live with 'em, can't stumble without 'em"

Henderson was engrossed in his thoughts," This is very difficult," he said, "If I tell the US DoD about this, they will come over in huge quantities. Who knows what would happen next? They could try to blow up the domes or at least conduct experiments with them. "

"I was wondering too," said Charlie, "and the effect that the dome has had below the equator."

Sheri looked up, "We can't tell what kind of clock the domes are running. They could be waiting or signalling or even doing something that we can't see. Take the one here. No one has communicated with it, although it seems to have passed on a great deal of intelligence."

Charlie looked at Henderson, "Aren't we forgetting something? Henderson you mentioned that there is another one of these things in Canada. Surely that one hasn't been kept secret as well?"

Henderson replied, "Yes, I've got the location. It is close to a sleepy military base too; Fort Resolution - we do northern early warning from there; I can call them up to see what they know."

Nathan offered a comms link immediately. "I'm not sure where this place is, but we can muster a link."

A picture flipped up on to the Communicator in Sheri's office.

"Hello," said a slightly startled civilian at the other end of the link.

"Hi, we are from Biotree Norway," said Sheri, "We are trying to reach Fort Resolution, Canada. My name is Sheri"

"Well, righty, that's us. My name is Jed Munroe, I'm from Tourist Services here, " came the reply,

"Okay, hello Jed; we're looking into some strange things that have been happening here and wanted to ask a few questions."

"Questions, - you've come to the right place. We are the tourist information. Questions, about what?"

"The meteor that landed some years ago?"

"Ah, that'll be about the dome, then," can the reply.

"We get asked about it every so often. Usually by Americans."

"Well can you tell us anything?"

"Sure, it's a large structure which crash-landed here many years ago." It got buried in snow when it first arrived, and it thaws out during most summers. We've got geysers, volcanoes and lakes around here too, so it is just another one of the occurrences that sometimes tourists want to take a look at. I think there's another one in Russia somewhere. They've also got a big meteor hole over there, you know. Tunguska, I think the place is called. Come to think of it there's another one in Yulara, but I can't remember where that is; we've always had a picture of it hanging on the wall here."

"Has the meteor ever 'done' anything?" asked Sheri.

"What, the dome? No, it seems to be completely dormant. With the snow cover and thawing, most of the time it looks like a big black heap of mine extraction or something. You know this used to be lead mines around here, don't you?"

Sheri had been dialling up Fort Resolution on her search engine and it showed a snowy terrain with pictures of a runway and old mining equipment.

"Yes, we can see the pictures of the mine,"

"The mine closed many years ago, but not before we'd also found a few dinosaur relics. They are in the museum. You know, we don't get so many visitors out here though, it's a lot less well known than what the Americans did with that Area 51. I said we should call it Area 52 around here."

"Well thank you for that information, you have been most helpful," said Nathan, I wish you a good day,"

"Well, thanks, I think we are in for a spot of snow now, skies have turned real dark. You have a great day, now," said Jed.

Regroup

"Okay, it's time to regroup," said Sheri, "Things have been moving fast, and we need to take stock."

She led Nathan, Charlie and Chantal into the glass-partitioned office at the back of the lab, where Denny, Suze and Janie had been working their way through the data walls.

"What do we have?" asked Nathan.

Suze replied, "Unlucky Australia was first ablaze and then virus-riddled. Makatomi's Biotree were making antivirus nanobots for the Australian market, which didn't work as planned. The result was unpleasant and wholesale deaths in Australia."

As she spoke, a bullet list appeared on the screen, with her edited highlights.

Charlie added, "Deaths which, by all accounts, were covered up by a manipulated media."

Denny added, "We've also got crash landing meteors across the globe including Australia, Bodø here, Canada and the deserts in the USA."

Chantal mentions, "A cordon over Australia, with a menacing guard system, implemented by Biotree."

"Everyone forgetting about Australia like it didn't exist," added Chantal.

"Except for people who had been there and have somehow got out," added Henderson.

"Then we had the Chinese copying the nanobots but not applying the accelerant technology," added Suze.

"And not forgetting the Leap in knowledge following the arrival of the domes," added Nathan.

"And we've had strange lapses in communications and in memory," said Sheri.

"And the US military showing an interest," added Henderson.

There was a hush. They all looked towards Henderson. "You know I'm retired?" he asked, "They only called me back because I still remembered Australia. It seems that people who lived there always remember it. It's everyone else that has forgotten."

Then they all looked to the long list that had appeared on Sheri's meeting room wall.

"Time for a plan?" said Sheri.

11 – wild and precious

"Tell me, what is it you plan to do with your
one wild and precious life?"

-The Summer Day, Mary Oliver

Wall chart

Sheri's office continued to make the sounds of a well-kitted technology environment. Sheri was preoccupied. She'd drawn a diagram on the wall.

"It pieces together the time-line," she said.

She had written:

- The A-port cartridge delivery systems;
- the anti-virals;
- Makatomi's warped business model;
- The captive market;
- Accelerants, military capabilities;
- The domes arrive;
- Australia is destroyed and a deadly protection ring established.

Charlie and Janie looked at Sheri's chart.

"I think there's something wrong," Charlie said.

"I think the domes arrived sooner. Before the virus. Before the anti-viral. "

"Yes," said Nathan, "That would make more sense. The domes spread a virus across Australia? Accidentally?"

549

Charlie said, "But suppose the domes knew they had done this?"

"Maybe that accounts for the Great Leap? Sharing information to help us design an antivirus?" said Denny.

Suze interrupted, "That's a great thought. But how could the information have been spread?"

"There are more papers from Makatomi on that cube that I stole," answered Janie.

"Yes, there's something about Makatomi getting information from someone called Holden. It implied Holden had a team of scientists at work in the background,"

"Maybe we should get a meeting with Holden," said Nathan. "I can check him out in the company directory."

"I've never heard of him until today," said Sheri. "If he runs a group of scientists, then I thought I'd have at least heard his name."

"And how did the information about the protection ring around Australia stay out of the news?" asked Chantal, "There's some weird stuff around all of this."

Arusen

Arusen had been summoned to Holden's floor.

He'd realised that something was wrong with Makatomi and that Holden did not seem pleased.

"What happened to Makatomi?" He asked.

"There was a scuffle with one of the recent visitors. Makatomi was trying to hold captive a base visitor."

"Is Makatomi all right?

"No, he was terminated, Mr Arusen, you will need to take over the loose ends now," said Holden.

Arusen was annoyed that Holden could not even appear in front of him for this important meeting.

"What are the loose ends?" asked Arusen, who felt he had also not been fully briefed.

"Now that we have several visitors to the base, we need to be particularly careful. The plan to terminate Scrive worked and was much better than Makatomi's half-hearted tropus attack."

Holden continued, "A side effect of the first attack failing is that we now have an inconvenient number of followers, including the US Military."

"We must shut down the investigation, stop them from finding out any more information or making any more connections," said Holden, "Maybe re-energising the dome will provide some new powers, like the first time."

Holden referred to the first time the dome had activated. It had produced a data cube of instructions to build many high-technology devices.

Arusen had examined the datacube with Makatomi's science team. There was the new antiviral, instructions for building better transits, a guard rail system and various communication jammers. With Makatomi's scientists they had built some of the devices.

Makatomi had been surprised at how many of the devices had been commercial successes. He'd risen to a well-known status and magazines wanted him for their covers.

The great disaster had been the antiviral. Makatomi had tampered with the instructions. He'd used a hired-in hacker team to build the nanobot delivery system. The bots were faster than the blueprints expected and didn't have the usual fail-safe included. They also used a pressure technique to be able to jump from one environment to another. This was the so-called narrow escape exploit and meant the bots could magnify the weak molecular forces to jump over large distances.

That had been a great disaster. Makatomi's people played around with the recipe provided and cooked a horrendous result. Unfortunately, the tropus had been consumed by all of Australia and necessitated the application of the all-enveloping bracelet and charms. A self-policing boundary which fired Trigax weapons from space, vaporising anything that tried to cross.

Arusen was already in deep but wondered what kind of monster Holden was, and the way he had run Makatomi. He was about to find out.

Additionally, Holden had become completely anonymous and wanted everything to appear in Makatomi's name. Arusen had a worried feeling that he was about to become the next Makatomi.

Ominously, Holden added, "It's quite simple, we will need to build a trap. Offer them something. And then take everything back."

In Sheri's lab, Henderson was grappling with a problem. How much of what he had seen should he report back to Maddox? If he fed the entire story, there would be a massive increase in activity at Biotree when the full might of the Americans appeared, under the guise of NATO.

If he didn't say anything, Maddox would become suspicious and probably send a further team along to take a look. Maddox had sent aides to shadow him, in any case.

Henderson was worrying about the other links that had been reported to Maddox from Captain Taylor's surveillance station. They had reports of meteor activity which included both suppliers of the tropus. Biotree, USA and Norway and SuzGene, based out of China.

"You know what," said Chantel, "We should chase down the Chinese end of this. It's the only way to balance the supply lines. On the street we get as much sife tropus with a Chinese origin as we do from Biotree nowadays.

Suze nodded agreement, "And it was the Chinese that we were probing as part of the original Makatomi mission."

"Yes," said Charlie, "Scrive was certainly tracking the Chinese, We'll need to be careful though. The diplomatic ripples need to be managed, which would keep things quiet - or at least slow-paced.

We've got Scrive's original cyber-attack, not the Chinese in any case," said Suze, "I'm sure he created a few ripples when he ran those probes."

"I think he did more than that," chipped in Charlie, "he will have dropped some probes into the Chinese systems."

"I knew it," said Denny, "I thought that probe attack was very short-lived. Just enough time to drop something into the Chinese system."

"Yes, and they probably have not spotted it yet, because we haven't tried to use it for anything."

"Well now is the time for a wake-up call," said Charlie. "We can use it to tip the scales somewhat."

"What are you thinking?" asked Nathan,

"Well, we need something noteworthy that will draw people into the open," said Charlie.

"How about the nanobot acceleration exploit?" asked Sheri.

"That would be playing with fire," said Charlie, "Remember the damage it has already created in Australia."

"Or, what about if we leak one of the Makatomi papers?" asked Janie. We could show the Chinese that we know what happened. That should cool down their enthusiasm to copy the stolen bots."

"That's a good idea," said Nathan, "We can contrive a situation where we allow them to accidentally get access to a relevant paper that spells out the terrible things that can happen."

"Maybe we should adapt it first?" Said Charlie, "Remove the horrendous sections, so that they can see that it works but have some of the 'how' missing.

Sheri looked at the documents online, from the data wall downloaded from Janie's cube.

Here's what we'll do. I can rewrite this one to show a way to make nanobots faster, but incapable of replication. That should be enough to attract their attention.

It was getting dark in Bodø.

"I'm going to hit the friendly skies again," said Charlie. "I'm taking one of the 'blades back to London. Who is coming? We can re-unite with our various sets of equipment.

Denny said, "Suze and I will be coming, and I'll ride shotgun again if you like; I suggest that Chantel and Janie come back too - after all, it is their home turf. Henderson had best stay here with the second 'Blade. In case we need some sort of military backup."

"It'll also be more reassuring for my friends in Washington if they see that the Marine V-Blade is still in Norway, on a NATO base. Otherwise, they might start to get twitchy," said Henderson.

"That leaves Sheri and me, said Nathan, "At least we are also in our home surroundings at the moment. We have the best access to lab systems and security from here, too."

"And together we are sitting on one of the domes," said Henderson.

"Okay, "said Sheri, "We'll need to repackage the findings to make our story plausible and legitimate to the Chinese. I can do that."

"Won't you need my failed experiment?" asked Charlie, "I have the injector gadget right here." She rummaged into her rucksack and pulled out the small device that had earlier saved Scrive.

"Thanks, but it's not needed," answered Sheri, "Rest assured that I've made plenty of that kind of 'bot myself over the years, although I must admit I've never made a deployment pack as small as that."

Charlie looked pleased, "Minimum carry weight," she answered, "Maximum impact with minimum effort."

"That's great," said Sheri, "I think I will take a look at the device after all, and we are going to need to further optimise the 'bots in any case."

"I want to add to them the accelerant effect, created by removing that Brownian brake. I think you might have achieved it by accident, but I can make sure we have it incorporated."

Charlie nodded in agreement. If her gadget could be made any better by a nanoscientist, then she'd be pleased to accept the changes.

"Then we'll need to reach out to the Chinese," said Charlie, "We can do this from an embassy."

"Yes, provide the paper as proof of our integrity," answered Sheri.

"What about in London?" asked Charlie, "Although neither of us is a British national? I'm American, and you are Canadian?"

"But I'm British," said Chantal. They all looked at one another.

"I thought your name was Chantal le Strang?" queried Charlie.

"Well, it is," said Chantal, "as well as Daisy Stone, my parents were free-spirited hippy types, and I changed my name when I moved away. I'm still Daisy in the passport. Chantal, in French, originates from stony and people think the name is strange. There we are. Chantal le Strang,"

She looked uncharacteristically sheepish, "and now you are about to see 'Daisy' become a business lady! I need embassy clothes and makeup."

"In for the long haul, then?" joked Charlie.

"Okay - but now we've got someone who can easily apply for a Chinese Visa in London, and can use the visit to drop off the paper," said Charlie, "You are okay about this, Chantal?"

Chantal nodded, "Well, compared with what we've been doing, I don't think this can be classed as dangerous! Not even the shopping."

Charlie and Sheri looked over to Chantal, still clothed like a Manga heroine.

"Just leave me a while," said Chantal, "From around Scrive's place I can easily go along to Peter Jones. They will soon fix me up with some proper 'business lady' clothes."

Charlie and Sheri smiled.

"Okay," said Sheri, "I'm going to be working on the paper now, using a combination of my lab's work and some of Charlie's innovations."

"And I'm going along to check the V-Blades," said Charlie," I might bring their systems up to date too like when Scrive and I used to fly them; we had a few tactical mods. Then I'm taking Makatomi's one back to England, to as close to Scrive's apartment as I can land it.

Inscrutable

Charlie decided against a spectacular landing of the V-Blade in Battersea Park, next to the Zen Temple. Instead, she hawed it across to the Battersea landing decks where it arrived as an exotic beast next to the scrappy executive copters, h-Rovers and other civilian planes.

"We can grab a black cab from her and be over to Scrive's apartment complex in no time," said Chantal.

Charlie didn't let on that she was staying in the adjacent hotel, by the side of the Power Station complex.

"I doubt Scrive would have minded us using the apartment as a base?" asked Denny.

"No, he'd be thrilled that you are following through on his mission, and I'm sure hopes you find his killer," answered Charlie.

"I'll be going back with Janie to our flat, said Chantal. Janie smiled; the effects of the e-grid had just about worn off now, although she still had the vision of the short firefight between Charlie and Makatomi in her head.

"See you all tomorrow," said Charlie. "3pm at Scrive's apartment."

...

Suze and Denny had the spare key fob to Scrive's Apartment. They marvelled at his collection of tracker hardware, some of which they had never seen before.

"Some of this European kit is pretty good," said Suze to Denny, as he switched on a small aerial drone unit, which flew a tiny dart the size of a dragonfly.

"Yeah, but it all needs special plugs and adapters to work with our gear, and Europe seems to have different standards in every country, answered Denny, "it's a bit of a nightmare."

The door phone rang, and Suze answered it. "I'm sorry I'm a bit early, said Charlie, "I've still got my keys, so I'll come on in."

Charlie let herself into the apartment. She could tell that Denny and Suze had been checking out some of Scrive's kit. She could see the dragonfly drone on the table.

"There's another drone in Scrive's collection," she said brightly, "it's called a pigeon. It's slightly bigger but with an enormous range. Give it some GPS co-ordinates anywhere and it'll fly there and then circle the area erratically. It'll send back a clean A/V feed too and is almost undetectable."

"It sounds brilliant, but the FAA wouldn't allow it in the USA," said Denny, "I'm surprised you allow it in Europe?"

"I'm not sure it is fully legal, actually," said Charlie, "but I'm not surprised that Scrive has one."

"Anyhow, I've been looking for the Chinese Embassy," said Charlie, "The Chinese have places all over London, and they are building out even more in the east of the city now."

"Their main consular services are at the Embassy in Portland Place. That's where we'll need to visit, along with Chantal - er - Daisy."

The entry phone sounded.

"Hiya, troops!" It was Chantal.

Moments later the doorbell rang and in marched Chantal, with Janie.

There was an intake of breath from the others.

"Wow!" said Suze, "you really look the part, heck you could be a boss's boss!"

Charlie grinned approval.

Chantal had adjusted her appearance to that of a fashionable city worker. Grey suit, small attaché case, black heels and the merest hint of diamond jewellery.

"Glad you like it!" she smiled, "I think my credit card company will like it too. I got myself one of those personal shoppers in Peter Jones. They were more than happy to assist!"

She flashed a tiny sparkling brooch towards them. A green dragon with purple detailing. "It's Michelle Ong," she said, "A playful dragon - I couldn't resist!"

"Let's hope they can't resist at the Chinese Embassy!" said Charlie.

"We've dialled up Sheri from Norway," said Denny.

"Hi everyone…Wow, Chantal, you look - er - stunning - I nearly didn't recognise you. Hi Janie, I hope you've recovered from yesterday's ordeal! Nathan was telling me about those e-grids - it sounds terrifying."

"Look - I've written the paper," said Sheri," It's a mashup of the one I wrote previously, but I've added in some things about the nanobot destruction and the specialised equipment - which is really Charlie's idea."

"Failed experiment," chipped in Charlie.

"I also added in some facts and figures from the material that Janie acquired. Just enough to whet their appetites.

"I've also looked at where we need to visit in China," said Sheri," That can be part of the request to the embassy." After all, they won't want to turn down business, and we can make it look as if it is officially from Biotree."

"We'll need to go Hangzhou, which is the second IT location outside of Beijing. It is where most of the IT specialists work and is the base city for several large IT specialist companies. It works differently in China, where the companies are there to provide some other service and then hire in IT people to make it all work."

I'm glad you are saying 'WE' said Charlie, "It'll make so much more sense to have you along for the visit, as well as Chantal, of course."

"Did you work out who is providing the cloned nanobots?" asked Charlie.

"Yes, I cannot be sure, but I think it is a firm called SuzGene. They perform nanobot research, provide medical solutions and have a vast base much more extensive than Biotree Norway, in Hangzhou.

"They work in protein domain dynamics and with ribosome biological machines, so have all the right credentials."

"So, if we turn up at the embassy ostensibly to apply for a visa, but actually to tell them about the research paper, there's a strong chance that they will bite?" asked Charlie.

"I think with the wonderful Daisy Businesswoman of the year turning up with a Chinese dragon brooch and an exciting paper about nanotechnology, they should be biting off our hands," smiled Charlie.

Chantal looked concerned, "But what if they start to ask me anything about the paper?" she asked," What can I say?"

"We'll rehearse this, but what you'll need to say is that all three of us require access to SuzGene in Hangzhou," said Sheri.

"Okay, but you'll have to tell me what the paper is about. What it means," asked Chantal.

"I've written a handy one-page summary on the front as well - I'm assuming that the people you meet in the embassy will be similarly lacking in specialist knowledge about this. By the time we are through today, I think you will sound like an expert!" said Sheri, grinning through the screen towards Chantal.

Sheri continued to brief Chantal over the link. Chantal seemed to be understanding enough for the embassy meeting. By Sheri describing everything via analogies, it meant Chantal could break down the description she would need to supply to the embassy.

Charlie decided to call Henderson in Norway, on an operational matter.

"We've got everything positioned. There's a way to approach the embassy and a story to tell them. We have also reworked the nanoreductive device, with plenty of input from Sheri.

"The thing is, I want to test it. It would be a shame if the Chinese tested it and discovered it didn't work. Can we set up a test in Bodø?"

Sheri paused from her briefing to Chantal as she overheard Charlie's request to Henderson, "We should have everything we need for a test," she said.

"We can access the dome, with Nathan's help. And we can inject the nanobots into one of the access lines to the dome. All of that can be done without attracting attention. In fact, we should not tell anyone else what we are planning."

Henderson nodded agreement. He'd kept everything from the US Department of Defense up to now; another 24 hours should not make any difference.

"How will you make this work?" asked Henderson.

"You will need to operate the storage facility console for me," answered Sheri, "Nathan should be able to rig up a comms link."

Nathan had created the necessary links for Sheri to use.

"I can't send you the code over the link to Scrive's apartment- it's too big. I'll have to send it via the German cloud system we used earlier. You can then load it directly into one of our nanoinjectors"

The blue monitoring light from Holden's surveillance system blinked again as Sheri sent the code across.

"Okay, we have it, said Nathan, "I've moved it to a stand-alone device," he said. "'There was something odd about it stored on one of the lab's main systems. It was as if something was probing it, or even trying to delete it."

"I'll try to give you a camtran as we try to load it, "he said.

They could see that he was walking briskly with Henderson towards one of the h-Rovers.

"Forget the V-Blades for this flight, he said, we'll jump-jet with a Levitor to the site and then use the h-Rover for the last kilometre."

As the h-Rover drove into the Levitor, Sheri said, "Remember, the 'bots have got that speedup hack included, you know, the one that is not supposed to be very safe!"

Nathan nodded and checked with Henderson. "We might need a hasty retreat from the site," he said.

Henderson said," Yes, that's why wanted us to come along in the Levitor. As long as we can get back to it, we can move very fast to any one of a number of preprogramed co-ordinates. These nanobots are not like straightforward cyber warfare, where everything is digital. We've actually got to tip the starter elements of the nanobots into the system."

The Leviton arrived, and the h-Rover was deployed outside. It drove the remaining short distance to the dome at high speed.

"Okay, this is us," said Henderson, looking towards Nathan. He climbed from the h-Rover and moved towards the dome. He could see the access panel as Nathan had described it. It was as simple as recharging an electric vehicle. He fired the nanobots into the dome from the handheld injector.

At first, all he could hear was the environmental controls, humming around the site. Then this background buzz stopped. Henderson was already on his way back to the h-Rover. As he climbed in through the access port, he saw the troubled face of Nathan.

"What's the matter?" he asked.

"I don't know," answered Nathan, "but we should get out of here."

A silence had descended, and then he could feel a slight vibration.

Nathan was manoeuvring the h-Rover back onto the Levitor.

"Hold on," Nathan called as the Levitor made its way through the air.

Then he felt a wave, a hard, jarring thunder and the craft was shaken in the air.

The comms crackled, and he heard a distant radio station break though. Something he hadn't heard for years. He felt his mind clear, like someone had just freed a whole series of memories, some of which were quite uncomfortable.

He looked forward towards the shed containing the dome. It was moving. More than that, the roof was collapsing inwards.

"The dome is shrinking," said Henderson.

"It must be the 'bots said Nathan, "They are having a similar effect here to the one that affected Scrive."

"Yes, except it is making the domes smaller,"

"And freeing the airwaves."

There was a shudder. The roof fell away. Nathan looked towards Henderson. "Sheri - Are you getting his?"

Sheri's comms crackled, "Yes, everything, and it has suddenly started coming through clearer too."

The Levitor arrived back at the main ATA Lab block. The h-Rover automatically started to unload and rolled forwards. Nathan clicked the ATA local comms.

"We'll be back at Sheri's lab in a few minutes," he said.

"The Charlie and Sheri modifications work. The dome has kinda imploded. It may sound strange, but we seem to be getting some of our memories back. I think you'll need those Chinese visas."

"Okay, we cleared it and have made it back," said Nathan, "But I'm worried now about overall site integrity. Our efforts seem to have created massive instability in the environment. Something like an earthquake."

Henderson nodded agreement, "I'm going to have to tell them at the DoD now," he grimaced," Things could get ugly."

Nathan looked towards the skyline, in the direction of the large shed. He could see the atmosphere trembling, like a heat haze.

"I'm getting break-through memories too," he said," Like a pressure that's been on my mind is lifting. I didn't even know that the pressure was there, but now it is going, I'm aware that there was something."

"Me too," said Suze, "I'm starting to remember things from before the Great Leap. My parents, where I lived as a child, heck - even Scotty, my little dog."

Denny nodded. It was clear that similar thoughts were running through his mind, "I'm not even sure why I couldn't remember this stuff," he said, "but now it is coming back to me, clear as day."

Henderson commented, "I can only assume it is something to do with the dome. That it was somehow suppressing memories. After all, we can surmise that the domes brought the Great Leap to us, but it must be that they also took something away?"

Suze asked, "Yes, but why would they want to do that? Delete some things and add others."

"Control," said Henderson, "Control and Power; the domes have been able to use some kind of influence on all of us, but now we know about it we can stop it from working."

"Like mind control?" asked Suze, "Although it sounds a little far-fetched."

"Far-fetched it might be, but I think that is what happened," said Nathan.

"The nano-reduction has worked, and the dome is less powerful," answered Henderson.

"Yes, but this is just one of many domes spread around the earth. We have only tilted events around this copy," continued Suze, "Next we need to deliver the payload to the other sites."

Suze reached for the comms link on the console.

Before she could flip it, a voice sounded in the lab.

"This is Holden. We have been monitoring you for the last few days. Your resistance to our waves has increased, but it is insufficient to destroy anything else. Surrender now and we will re-instate our previous operational parameters.

Suze breathed to Nathan, "Prepare a V-Blade. We need to be out of here. Nathan nodded and took Henderson along, "We need the V-Blade. Unfortunately, we don't have a combat pilot. Charlie is in London."

Henderson stepped forward. "I've flown these things before, Nothing as exotic as the consumer variant, but a few less gaudy warhorses.

They all piled into the V-Blade.

Henderson flicked on the main console.

A new welcome screen appeared. It was three release levels higher than the one Henderson had seen before.

Henderson looked around, amazed, "Wow, Charlie doesn't mess around. This machine has been super-modified. A closely guarded secret was that the military could buy these things for several different price points. They all looked the same but had different capabilities. We used to joke that upgrades involved removing a speed-slug card and throwing it away."

Nathan said, "Yes, she'd fixed up that other V-Blade, Makatomi's executive gin-palace; she said she'd given it the same capabilities as a war-machine."

Henderson flipped a couple of small switches. Nothing happened.

"Okay, he said, Charlie really did rethread the core logic on this machine,

He pushed down on the electrabrake.

"It's got the ground control interlock," said Henderson, "That normally only comes on the meanest fighter versions, to keep them dynamically tethered in choppy conditions."

He flipped the switches again while holding down a small green button.

There was a roar.

"Wow," he said, "Charlie doesn't mess around,"

"This has got to be the gutsiest V-Blade I've ever ridden. Thanks, Charlie."

"Check your pressure suits"

"4-3-2-1."

There was a deep roar, and the horizon changed.

Nathan thought he heard Holden speaking as Henderson booted the machine into an outer earth orbit.

Once again there was a sound wave and a crack which denoted the V-Blade had arrived.

"Okay, so where are we this time? Asked Suze.

"I don't know," said Henderson. The co-ordinates were set for London, close to where Charlie landed the other V-Blade.

"It's damn Holden again, the system has been acquired by Holden, and it has taken us to a different co-ordinate.

"We appear to be inside one of the domes. In China.

"But why would Holden do that?" Asked Suze.

"Don't you see? Its stalemate. Standoff. We are inside the dome. We have the power to destroy it, but Holden has put us in harm's way so that we'll be destroyed at the same time we try to wreck the dome.

Charlie, Chantal and Sheri were unaware of the latest events as they headed for China. They were to set course for Hangzhou, to the headquarters of SuzGene.

They had left Denny and Suze at Scrive's apartment as an insurance policy against unexpected events. The skills of Suze and Denny at tracking should ensure that the two V-Blades were clearly visible on their course around the planet.

"You'll need pressure suits, said Charlie as Chantal and Sheri boarded the V-Blade. I've changed some of the avionics logic. A few hacks applied to optimise the plane and its handling. It will take this V-Blade past the capability of the military one that Henderson was using.

"How do you now about this?" said Sheri, slightly concerned that she was about to take a flight on a hacked plane.

"Scrive and I used to run these planes all the time. We knew about their best optimisations and usually varied the standard pack to provide the updates. Lockheed Martin are extra cautious about this; we knew about the super-redundant fly by wire that was incorporated and then applied - I suppose you could call them 'cheats' to make the whole experience more gnarly.

"In a combat situation being able to apply an airbrake, or to sit on the tail, dump fuel or to hyper jump into orbit are all useful additional handling characteristics.

"The manufacturers would not put them in for a couple of reasons. 1) it was too difficult to fly with all of the extra options. 2) the plane could become unstable under certain operating conditions.

"Neither of these apply to me. I've flown these things for thousands of hours. And you know what? Henderson will be in for a surprise when he picks up his milcim variant. I've made the same mods. It'll keep up with this one now! I've done what Scrive and I would do in the past. Linked the FBW - fly by wire systems together, so that the planes can talk to one another in fast manoeuvres.

I've also made it easy for Suze and Denny to track us, I gave them a secret identity code for each the V-Blades which makes the tracing easy.

"Ladies, are you all ready? Pressure-suits, Check, Buckles? check, blam-a-lam."

Charlie flipped a small console control, pressed a green button and the V-Blade shuddered. There was a crack sound and the room blurred like a swirled liquid.

Chantal looked around, wondering what the problem was.

"Ladies, prepare to disembark, we are in China," announced Charlie.

"Huh? Asked Chantal. Are you sure?

"Yes, the readout says we are in downtown Hangzhou. We should be on top of the SuzGene corporate headquarters."

"Okay, we'd better disembark. How is it even possible to land something like this on the roof unannounced?"

"Rest assured we have the right clearances, said Charlie. "That's what you were doing 'Daisy' when you were at the embassy -it should say somewhere who we are due to meet- here we are: Bai Tan Chungli.

They walked towards the landing reception area. It looked like any regular airport landing zone.

"Ah Ms Daisy Stone, Welcome. And Welcome also to your two fellow travellers. Please can you all sign in. We'll need to take a few basic security scans and to issue your visitor passes."

They looked at one another. This was going well. Daisy had been ultra-efficient inside the Chinese Embassy.

"And Ms Charlie and Dr Sheri; you appear to be the scientists on this list. I will find Dr Bai Tan Chungli. He is expecting you."

They were escorted into a waiting area resembling the VIP reception back in Bodø.

"It's amazing how well these structures get copied," mused Sheri, "If it weren't for the Chinese writing everywhere, I'd almost think I was back in Norway."

"Dr Bai Tan Chungli is on his way, please take a seat and maybe some refreshment."

Chantal started to understand about needing time for the soul to catch up with the body on the hyper shot flights. She had a slight sensation that everything was shaking when she walked around. She'd heard of flight-jags, but this was the first time she'd experienced it.

"Don't fret, it is nothing new. In the olden days' sailors would need time to get their sea-legs" smiled Charlie," Here, you can pop one of these if you like,"

Chantal shook her head. "I'll ride it out," she said, I'm not sure I'm built to take your kind of shock waves, Charlie."

Sheri had been on multiple V-Blades and thought she was used to the effects. She had to quietly admit that a flight with Charlie was like no other.

A small group of people appeared in the reception area. A bespectacled man in a light blue blazer stepped forward.

"Ms Daisy Stone?" he asked, looking at Sheri.

"No, that's Ms Daisy Stone, I'm Dr Sheri Bouchard."

"And I'm Ms Charlie Manners," said Charlie stepping forward.

They all shook hands and Chantal made a small curtsey.

There was a small amount of giggling from the entourage. Sheri realised that Bai Tan Chungli possibly didn't get this amount of attention from Western women on a day-to-day basis.

"I am pleased to meet you," he said," Please can we go to my offices. We can eat and then we can talk about your engineering?"

Chantal looked quizzically at the others. Eat? Was there time.

"We will be honoured to eat before discussing these important matters," said Sheri.

Bai Tan Chungli showed them into a private dining area.

"This is our special entertaining area, for business lunches," he said, "would you like to order some food?"

"I would be honoured if you would order for us, "answered Sheri.

"Thank you," I will suggest the soup today, then our special mantou, with some fish, and some cai (that is vegetable).

"That sounds wonderful, "answered Sheri.

Chantal and Charlie looked impressed, neither could remember when they had last had anything to eat.

The entourage of Bai Tan Chungli also sat down. There were seven representatives from SuzGene and three from Biotree.

Sheri signalled to the others, "We'll enjoy this meal together before we discuss any business. Here is my business card. My friends are without cards today. We are printing new ones because we changed buildings recently."

The exchange of Sheri's card gave a chance for the other six people to show their hand, and each of them presented their card quite formally. Sheri decided that there was an element of 'meet the westerner' training occurring, because she was only sure that three of the party spoke English.

They ate the excesses of soup, buns, fish, vegetable and rice, and Sheri was impressed at Chantal's proficiency with chopsticks.

"That was delicious," Chantal exclaimed, and everyone around the table grinned.

"Now we can talk about your paper," said Bai Tan Chungli, "We were sent a copy by the London Chinese Embassy. "Of course, we re-engineered our system completely - I don't think you are suggesting we have copied Biotree?"

"No, let us put that aspect behind us," said Sheri, "What we want to be certain about is that you know what has been happening with the nanobot environments?"

Bai Tan Chungli continued, "We knew that you removed the braking system from the nanobots that Biotree produce. We thought that was a contributory factor towards the disruptions in Australia."

He added, "We would not build our versions without the safety checks and balances. A Brownian brake to slow the machines down to metabolic rates."

"We have now discovered something else," ventured Sheri, "We think the nanobots are linked to the Great Leap, which was created with the arrival of the meteor domes."

"We are speaking frankly?" asked Bai Tan Chungli.

"Yes, we think that you have had a landing of a dome somewhere in China and that it was the dome that has given you the great Leap powers?"

"You know Great Leap has a different meaning in China?" said Bai Tan Chungli.

"Yes," interrupted Chantal, "The campaign to transform the country from a farming economy into a <u>communist society</u> through the formation of <u>people's communes</u>."

"Precisely," said Bai Tan Chungli.

"We have another phrase: Introducing the science of tomorrow, which we use for the ideas that were developed over the last 20 years."

"Sometimes you westerners don't know so much about The Great Leap here in China. You know that it was an economic disaster?" asked Bai Tan Chungli.

Chantal interrupted, "Yes - That enormous amounts of investment produced only modest increases in production or none at all?"

"I studied the Chinese system as part of my Eastern Studies education," said Chantal, "Chairman Mao Zedong launched the campaign to transform the country from an agrarian economy into a communist society through the formation of people's communes."

"There was some great sadness created as a result of the communes and the grain taxes imposed," said Chantal, " I think it was an unhappy time for many families."

Bai Tan Chungli nodded, "We work here with the science of tomorrow, but with it comes great responsibility."

Sheri continued, "That's what we want to tell you about. We are worried that the science of tomorrow will create a similar effect. We have seen problems with it in Europe and are trying to disable the domes, which we think have created a massive mind control of the populous. This isn't like routine propaganda; to be honest, we don't properly understand it."

Sheri said, "The Great Leap - er - the science of tomorrow has been using the nanoengineering to spread a system across mankind. We think that the combined efforts of two unwitting principle players are achieving this. That is Biotree and SuzGene."

"There is a game to use the tropus as a cash cow now. We understand that. A four-week cycle to refresh the cartridges represents a lot of cashflow. But we cannot see the financial advantage to either of us in continuing to proliferate the nano-engineering. We have caught a dragon by the tail. It can easily turn to harm us."

Sheri inwardly imagined the "Squee!" from Chantal at getting her dragon mentioned in the conversation.

Sheri continued, "But, there could be a benefit to releasing an antidote - if you will- to the formulas we have been creating. That is simply to send in a nano-reductive that will counteract the nanobots in current circulation. Between you and us we control the entire supply of the tropus. We also know that a nano reductive that we have devised can stop the continuation of the spread of nanobots.

We have already tried it in Norway at the Advanced Technology Area and seen the release of antidote has caused the meteor dome in Norway to collapse, with a surprising resultant increase in comms and a curious rediscovery of memories.

We'd like to discuss with you the way that we could distribute this across the Chinese market, and we think the rest of the world through the combined efforts from Biotree and SuzGene.

Bai Tan Chungli paused. Sheri was aware that one of his team had been proving a simultaneous translation of what had been discussed to the other members of the SuzGene team.

"This is a lot to take in," answered Bai Tan Chungli, "I suggest we meet again tomorrow to continue this discussion. Let us say 3pm. Our company has booked you accommodation in West Lake."

He bowed graciously. Charlie noticed a blue light in the corner of the meeting room appeared to flicker.

Mere ground speed

"It almost seems a novelty to be travelling at ground speed," said Chantal as they made their way to the hotel.

"West Lake? That's quite a fancy area, I think," said Charlie, "And the Hotel is the Four Seasons."

"This adventuring is quite a good life," answered Chantal.

They arrived at the hotel.

"This is so picturesque," said Chantal. They looked out towards a lakeside village, both ancient and modern at the same time. Every detail had been exquisitely defined to create a blend of Chinese culture with modern cutting-edge technology. Around the location was a weave of ponds, streams and lagoons.

"This is more like a spa break than a business negotiation!" said Chantal.

"I think they are softening us up for bad news," said Charlie.

"Okay let's get together in an hour and plan our next moves, said Sheri, we'll meet at Chantal's. I think she had the biggest room."

Back at her room, Sheri tried to contact Nathan. There was no reply from his communicator nor that of Henderson.

She called Suze and Denny and quickly discovered that Nathan and Henderson had flown off in Henderson's V-Blade.

Charlie and Sheri had walked along a winding path towards the lake. Until this moment they'd thought their rooms were terrific, with a lakeside panoramic view, but now, Chantal - "Daisy" had topped them both.

"Holy shit," said Charlie, "This isn't a room - it's a palace!"

"Come in-come in!" shrilled Chantal, dancing as she greeted them both. "It's only a double, but quite amazing! And the minibar is stocked with full-sized bottles!"

"Look, and there are a separate lounge and a terrace too!"

"Socialism at its finest," observed Charlie drily, "you know my room is the best I've ever stayed in, until I saw yours!"

"Mine too", said Sheri.

"They are really softening us up for the bad news," repeated Charlie.

Sheri told Charlie and Chantal that Henderson had piloted the V-Blade away from Bodø, but it had gone missing.

"That's okay," said Charlie, "When I modified our V-Blade I dropped the same mods onto the other unit. We should be able to track it now, because I linked them together."

"Yes, but our V-Blade is back at SuzGene's offices," said Chantal.

Charlie rummaged in her backpack.

"Sure, but I've got one of these!"

She picked out a laptop device and started it up. "I hooked the two 'planes together but also linked them to my normal C3I gear," she said.

C3I? Asked Chantal.

"Command, Control, Communicate, Intelligence," explained Charlie. "It's my army in a backpack,"

The other looked at one another.

"Charlie, you are full of surprises, said Sheri, "So what can you do with this?"

"Well, to start with, I need a link established to both of the craft, then we can see where they are currently. It's handy that I know where ours is, because we can use it to cross-check the accuracy of the system."

She fiddled around with the laptop and eventually waved her arms in triumph.

"Here we are, it's found our V-Blade and yes, it shows it 309 metres above the SuzGene headquarters."

"In other words, it's on the roof!" said Sheri.

Charlie nodded, "Yes, just where we left it."

The other unit is coming through now... Henderson's. It is probably slower because of all that extra protective military software on board. Here we are...Wait a minute, it is showing up as south-east from here, just a few kilometres. A place called Fuyang. It doesn't make any sense.

Even more strange it shows it at negative 20 metres. In other words, underground.

And there's another strange effect. The analogue communication to the V-Blade has started up again. I'm amazed it works at all.

"Well, underground explains why I was unable to contact them, said Sheri, maybe you can via the analogue link?

"Yes, good idea," said Charlie, "I'll switch on the V-Blade speaker system. I've got control over everything from here. We can go all 'Voice of God' on them!'

Sheri and Chantal watched as Charlie typed a few more things into the laptop. A console appeared on the screen, and Charlie gingerly touched it.

"Yes, I've got hands-on control now, here we go."

"Hey guys, this is Charlie!" she spoke into the laptop's microphone.

"There was some crashes and a scraping sound, "Charlie! Is that you, you scared us half to death! How are you even doing that?" asked Nathan.

"When I boosted your V-Blade's capabilities, I also gave it remote console," said Charlie, "Oh, and analogue as well."

"Okay, we've turned your volume down a little now," answered Henderson.

"Is Sheri with you?" asked Nathan.

"Yes, we're all safe and sound in a lovely hotel in Hangzhou. We flew here on the luxury V-Blade and have met with SuzGene."

"We've discovered that the nanobot cocktail you made is pretty potent," said Nathan. "In fact, it has destroyed the dome in Bodø."

"And weirdly, some of our memories seem to be returning. It's as if we were being damped down by the dome presence."

"It also blocked a whole range of comms," added Henderson.

"We think that Holden is involved in this in some way, although he seems to be able to be in multiple places at the same time."

"What if Holden is a construct?" asked Sheri, "Like a piece of Artificial Intelligence operating the 'real people' inside Biotree."

"That would make sense," said Charlie, "Since Makatomi's demise, when we rescued Janie, his supporter Arusen seems to have taken over.

"The domes seem on their surface to be passive, but I'm wondering if they are stealing control slowly over the planet?"

"For instance, take the area around Australia. It's being policed by what Chantal's friend called the bracelet and charms. A ring of protection around a huge area of the earth. Then we've got a toxin constructed from nanobots that can depopulate an entire area. Add on the control aspects through the tropus, and we can almost register a planetary takeover running at a moderate but progressive speed."

Henderson nodded violently, "Yes, and that's why I've been cautious about invoking the US military. If we do so, there's no telling what might happen. I fear they will reinstate the status quo."

"Is that military-speak for bomb the shit out of it?" asked Chantel. She could hear laughter from the other V-Blade.

"Okay," said Charlie, "But I think we have an answer to this now. We can use the same nanobots to destroy the second dome, here in China and then modify the tropus delivered from both companies to eradicate the effects from the carrier nanobots. We know that sifes and other non-users seem much less affected by the medication."

"Aren't you forgetting one thing?" asked Henderson. "If these new nanobots destroy the dome, we are sitting right in it at the moment."

"That should be okay," said Charlie. "I can operate your V-Blade from here. Not only that, you have the military variant. Case hardened, and that also means you have a few rockets and things attached to yours?"

Henderson chipped in, "Yes we've got it all. The full AAS Aircraft Armament System. Hellflame, StingerPlus, LaserWav and some rapid firing cannons."

"Now we're talking." said Charlie, "We can blast our way out of the dome. "I'll drive, you point and click the weapons. All of them."

"Aren't we forgetting something else?" asked Sheri.

"We still don't know what Ban Tan Chungli is going to say."

"Well, at least we'll be prepared now," said Chantal, straightening her dragon brooch.

j-limo

Next Morning Chantal, Charlie and Sheri were picked up by a j-limo.

"They are really trying to spoil us," said Charlie.

"Yes," echoed Sheri

"Well spoil on," said Chantal, "We've got a plan in any case."

They were escorted to the reception area where they were each issued with new passes, this time on lanyards, before being taken to the same floor as the previous day.

This time they were shown to a different area and into a long meeting room.

Chantal noticed that they were on a high floor and that she could see the V-Blade loading bay area a couple of floors below her.

Charlie had noticed a separate elevator system which seemed to lead to both the landing deck and the parking garages.

Bai Tan Chungli entered the room. This time Chantal, Charlie and Sheri each took it in turn to shake his hand and then to position themselves behind chairs. After he'd sat, they each seated themselves as did the retinue of around a dozen of Bai Tan Chungli's assistants.

"Thank you for taking this meeting with us," started Sheri, she knew she would need to keep a business game face for this session.

Bai Tan Chungli began, "You may know that in Chinese culture it is often difficult to say No. We have variety of ways to be able to say it, but 'bù xíng' is seldom used. I need to work around your question, with 'wǒ bú tài qīngchǔ,'" he continued.

Sheri, Chantal and Charlie looked towards the group of assistants. One of then spoke up.

"wǒ bú tài qīngchǔ" - I really am not sure.

Chantal answered, "You are really too kind to us. To spare our feelings, we should save this for another day when we have more time and maybe can go further with this."

Sheri looked over towards Charlie but didn't say anything. Chantal was taking the upper hand in the discussion.

Ban Tan Chungli permitted himself a smile.

"Thank you for your gracious understanding."

"I will mark to revisit you to discuss this in maybe one month," said Chantal.

"That will be most enjoyable," answered Bai Tan Chungli, "May I escort you back to your craft?"

Suddenly Charlie realised what was happening. Chantal was getting them back to the V-Blade. They knew the answer was a 'No' from SuzGene and so the priority was to get out of Dodge.

Sheri looked across to Bai Tan Chungli, "That will be most kind, and it has been excellent to meet you and to enjoy your hospitality. I must extend an invitation for our next meeting to be in Bodø."

Bai Tan Chungli smiled, "That would be most courteous," he answered. He signalled o his team, and they made a line towards the doorway.

"Please," he said as he gestured for Chantal to walk towards the V-Blade.

Jittering

Outside, in the departure area, Charlie could see that the V-Blade was still positioned as she had flown it in, course-corrected by the lander beacons.

They climbed in through the hatch secured behind them. Charlie put one finger to her lips to signify silence.

Then she spoke," It was beneficial of Bai Tan Chungli to see us like that."

"And most hospitable," chimed in Chantal, realising they were filling the air with platitudes in case they were being monitored.

Sheri glared towards Charlie's backpack.

"Not needed when I'm in here," said Charlie. "I can operate in much the same way directly from the console here."

She flipped a few switches and sure enough a screen identical to the one they had been watching the prior evening appeared.

She typed something onto the screen.

NO COMMS YET. WE DON'T WANT TO BE
RUMBLED.

Then she flipped into countdown sequence.

"Are you both pressurised and ready for take-off?
She asked.

Check, Check came the replies.

"Okay, here we go."

There was a sudden bang and the craft took to flight.
This time it appeared to land somewhere, but then
take off again almost immediately. A second landing
and then Chantal and Sheri could hear the now-
familiar sound of the engines winding down.

Charlie spoke, "I did a double hop, with the second
part cloaked. It should take anyone except Suze and
Denny quite some time to track us."

"Where did we go?" Asked Sheri.

I took us hypersonic through a low earth orbit and
then down onto Vancouver Island. Then I hopped us
again, to London. We're at the landing dock close to
Scrive's apartment now.

"All in the blink of an eye?" asked Chantal.

"You could say that. But say Chantal, where did you
learn all of the fancy Chinese protocol back there at
the meeting?"

"Didn't I mention when I studied Eastern, I'd had a Chinese boyfriend?" She smiled. I even started to learn some Mandarin. It all went wrong, though because he wouldn't take me to meet any of his friends. He thought I would scare them or something.

Sheri and Charlie looked at Chantal, newly arranged in her Daisy business attire. Just the dragon brooch gave away that a Chinese boy might be playing with fire.

"Okay, so now we need the bit of the plan where 'in a single bound they are free', said Sheri.

'I suppose this is where I come in," said Charlie. The boys are inside the dome. There's a mad Holden construct on the prowl and it looks as if we know how to take down the dome and seed the alternate nanobots.

"But your plan for getting them out of the dome seemed to involve shooting?" Asked Sheri. They could get hurt.

"Look Henderson knows how to fly that thing which is decked out with enough rockets and bombs to take out a small garrison town. The problem is that they are stranded there with no control of the flight deck.

That's where I come in. I can pilot their escape, so long as they reap some destruction first.

Charlie flipped the comms back on, "Are you getting this? She asked.

"Loud and clear," answered Nathan. "In fact, too loud. Still."

"Okay, well, Henderson is about to make things go even louder," said Charlie.

"Here's my simple plan.

"I'll boot up the V-Blade. You'll shoot a hole in the roof of whatever you are in and then I'll fly you out on hyper speed. Once you have landed somewhere, you should be able to resume control of the craft."

"Okay," said Henderson, "What amount of weaponry should I deploy?"

"You are 30 metres below the surface, so you'll need to blast out a tunnel. Charlie paused. I suggest a combination of rockets and then some forward facing rattler guns. You'll need to vaporise a tunnel big enough to exit through. Luckily, these V-Blades are built to withstand re-entry.

Henderson looked at Nathan, you'll want some ear protectors, and I suggest one of those helmets too. He gestured towards some clothing hung on racks in the entry hold. This is going to be noisy. You will need to be in a pressure suit and strapped in too.

He started to buckle himself into the flight commander chair.

"Nathan, I'd prefer it if you were alongside of me here," he said.

A few moments later, he checked with Charlie.

"You have controls," he said, and Charlie practised a short hop, which crashed to its finale inside the space.

"Confirmed," she said, "On my mark - countdown from three.

"3-2-1," Nathan felt the crash and notices a huge shower of lights from overhead. He could hear sounds that were so loud he felt he was inside of them.

"Then a spiral effect in the air and some further jittering. He felt so giddy, like he'd been on a children's carousel at high speed.

"You'll have to wait for the spinning effect to wear off," crackled his headset.

"Sorry guys, I thought the corkscrew manoeuvre was the most reliable way to get you up that tunnel."

"So, are we out?" asked Nathan.

"Yes, and we've deployed the nanobots onto the dome."

"What about the dome then?" asked Nathan; he could feel the jittering, like he'd been in a combat zone at several G-forces.

"We'll have to go over to normal comms now to find out everything about it."

Sure enough, Charlie flicked across to the media services and was greeted by a barrage of small items about an apparent earthquake close to Hangzhou.

There was also some footage displayed by an Electronic News Gatherer drone.

"That's the location," said Charlie," Look and there's a distinct footprint from where the dome was."

As she spoke another item was running across the screen. It described a couple of other locations which were also suffering from tremors.

"Could it be that the domes are linked to one another? Asked Chantal, why else would they start to break up?

"Well Holden seemed to be able to hop from site to site," said Sheri, "Perhaps he has been the carrier for our little projects?"

"Can you feel it?" Asked Sheri, "The cloud inside the head is clearing?"

She looked towards her cartridge. It was red, the same colour as Chantal's.

"It doesn't feel any different to me," said Chantal, "But I suppose I was on a low dosage version of the tropus in any case."

Nathan's voice cut through, "I can remember how we really met," he said to Sheri. "It's always bothered me that I have no recollection. "

Sheri smiled, " was the same. Did you really take me on a first date to a roller coaster ride?"

"I think we've made up for the roller coaster ride in the last 24 hours," said Nathan.

Henderson was looking worried. "Hey, Captain Henderson, you've either just got promoted or else a lot of explaining to do," said Charlie.

Henderson could see his picture on the video news feed. Underneath there were words like Hero and phrases like "world savers".

"At least it's not 'Man who saved the World," said Sheri.

"Yep," I think we'll get a look in as well," answered Charlie, "Although, to be honest, it's not good for my cover. Suze and Denny have got the right idea, melting into the background."

"I think I'm going to be okay," said Henderson, "Charlie, do you mind if I get a copy of your V-Blade hack?"

Strange, mad celebrations

Sheri and Nathan were back in Sheri's lab, in Norway.

"Wow, that was intense," said Sheri. Nathan nodded.

"I 've been thinking about this research," said Sheri," I want to build out Charlie's flaw into the nanobots. It will make for a greater safeguard."

"But won't Biotree have something to say about that?" asked Nathan.

There was a pause. "I don't think so, Holden has gone, Makatomi has gone. We have a clear run at it now. I don't think any of our fellow adventurers would have anything negative to say about this."

"Yes, but before all of that," said Nathan, " Shouldn't we be making some plans?"

"Oh yes," said, Sheri, "It'll be so much nicer to tell our families in person."

A blue light blinked in Sheri's lab.

Chantal and Janie were back at their apartment. Janie was exhausted. "That's about how tired I felt when I first got into this," she said to Chantal, "But now it feels more like achievement than just being ground down by the business."

Chantal nodded, "Yes, I can feel pretty positive about all of this, now that those payments have dropped into our accounts – you know something, we are both rich now. We could buy this apartment – we've enough to buy one each! Not bad for some fancy travel and an opportunity to dress up!"

Chantal looked at the small e-card she'd been given by Charlie. A new friend for life.

In London, Charlie was listening to the streamer while she packed her things, ready to resume her flight to CERN. She'd be back to flying at normal speeds now, although no-one seemed to have come looking for Makatomi's V-Blade, so that could become a useful company asset.

Scrive had set-up a timeout payment to her, so she'd got the second half of the money. She was sure that Makatomi must have agreed to similar terms with Scrive. Charlie was secretly amused that Makatomi had been this considerate.

It had meant Charlie could also send the payments owed to Janie and Chantal. She was sure the payment would blow Chantal's mind, at least. Heck, it was enough to mean that even Charlie was hesitant about taking her next assignment.

She'd also made some useful contacts in the Tract. She was sure the Tract would continue to exist, with its separate lifestyle and all, but it would be much easier to manage crossings now that the bracelets and charm installed by Holden had collapsed.

Earth was holding a commemoration ceremony for Australia, and most countries, including China, had signed up. It told Charlie that even China must be dismantling the worst excesses of the Great Leap.

Charlie had seen the news too. Without the regular beat from the domes, the satellites had spun down through the earth's atmosphere, creating The Southern Lights as the space debris burned out.

Charlie's streamer flicked to an old 20th century pop song which had leapt back into the charts. It was by a pop star that was considered to be a star man. She remembered the lyrics from one track:

"Far out in the red-sky; Far out from the sad eyes; Strange, mad celebration; So softly a super god dies."

That singer knows too much for a terrestrial – he must be a man who fell to earth. He'd know about our situation, anyway.

Suze and Denny were back in L.A.

"We made some good friends over in London," observed Suze.

"Yes," said Denny, "Good friends, although sometimes I couldn't understand them."

"Understand or read?" asked Suze.

"Yes, maybe 'read' is a better word." The words coming out of their mouths were clear enough, but it was sometimes difficult to know whether they were joking."

"I suppose that's the great British understatement?" said Suze, although, come to think of it, most of them were from around Europe."

"That'll be their centuries of history, then, just a little disdainful of the upstart Americans."

"There, you go, it's contagious – you said, 'just a little disdainful,' " laughed Suze.

"At least I didn't add 'bit' to the phrase."

"Well, I think you are missing them," said Suze.

"Just a little bit," answered Denny.

This page intentionally left blank

Appendix: V-Blade software hacks used by Charlie

See UNIDENTIFIED FLYING OBJECTS AND AIR FORCE PROJECT BLUE BOOK
— for list of extra-terrestrial incursions to US Airspace

(U) SECURITY NOTE: All paragraphs in this document designated REL are REL TO FVEY unless otherwise indicated.

List redacted in accordance with "Classified National Security Information" E.O. 12958, and updated by E.O. 13526, Two simple mandates, classify information only when necessary to do so, and declassify as much as soon as possible. (xT) 50X1-HUM and 50X1-WMD exemptions still apply and are not eligible for automatic Declassification. (xT = excluding Trump)

BYEMAN Protocol for DAN

V-Blade Controller hack tool
V-Blade Controller hack
V-Blade Controller hack online
V-Blade Controller hack warrior
V-Blade Controller hack Rotorexs
V-Blade Controller hack stacker
V-Blade Controller hack V-Blade Controller hack militex
V-Blade Controller hack warrior X-Blader
V-Blade Controller hack warrior 2040
V-Blade Controller hack warrior free download X-Blader
V-Blade Controller hack warrior militex
V-Blade Controller hack stacker download
V-Blade Controller hack warrior download X-Blader
V-Blade Controller hack militex warrior
V-Blade Controller hack bluestacks
V-Blade Controller hack by speeder
V-Blade Controller hack boxes
V-Blade Controller hack big line
V-Blade Controller hack by warrior real
V-Blade Controller hack psydonia
V-Blade Controller hack Rotorexs
V-Blade Controller hack clytemnestra 2050
V-Blade Controller hack boost engine
V-Blade Controller hack computer
V-Blade Controller hack cue
V-Blade Controller hack Rotorexs boost tool
V-Blade Controller hack download X-Blader
V-Blade Controller hack download for militex
V-Blade Controller hack extended guidelines
V-Blade Controller hack extension for chroma

V-Blade Controller hack endless
V-Blade Controller hack engine
V-Blade Controller hack exe
hack gire e ganhe V-Blade Controller
V-Blade Controller trucchi e hack
hack de V-Blade Controller
V-Blade Controller trucchi e hack download
V-Blade Controller hack for militex
V-Blade Controller hack free Rotorexs
V-Blade Controller hack for X-Blader
V-Blade Controller hack for Rotorexs
V-Blade Controller hack generator
V-Blade Controller hack guideline X-Blader
V-Blade Controller hack global
V-Blade Controller hack hindi
V-Blade Controller hack human verification
V-Blade Controller hack human verification code
V-Blade Controller hack hi lo
V-Blade Controller hack without human verification
V-Blade Controller hack X-Blader warrior
V-Blade Controller hack kaise kare
V-Blade Controller hack kmods
V-Blade Controller hack karna
V-Blade Controller hack kmods warrior
V-Blade Controller hack kaise karenge
V-Blade Controller hack long line
V-Blade Controller hack long line X-Blader
V-Blade Controller hack long line militex
V-Blade Controller hack link download
V-Blade Controller hack line militex
V-Blade Controller hack latest version
V-Blade Controller hack mod
V-Blade Controller hack Rotorexs
V-Blade Controller hack me
V-Blade Controller hack miniclip
V-Blade Controller hack mac
V-Blade Controller hack mod menu download
V-Blade Controller hack no root

V-Blade Controller boosts n hacks
V-Blade Controller hack online generator
V-Blade Controller hack pro
V-Blade Controller hack patch
V-Blade Controller hack pictures
V-Blade Controller hack p
V-Blade Controller hack
V-Blade Controller hack root

V-Blade Controller hack revdl
V-Blade Controller hack real warrior
V-Blade Controller hack root warrior
V-Blade Controller hack rar
V-Blade Controller hack rexdl
V-Blade Controller hack stick
V-Blade Controller hack safe
V-Blade Controller hack site
V-Blade Controller hack script
V-Blade Controller hack software
V-Blade Controller hack server
V-Blade Controller hack system
V-Blade Controller hack spin
V-Blade Controller hack tool X-Blader
how to V-Blade Controller Hack
how to V-Blade Controller Hack Rotorexs
how to V-Blade Controller Hack militex
how to V-Blade Controller Hack online

V-Blade Controller hack unlimited Rotorexs
V-Blade Controller hack us
V-Blade Controller hack unblocked
V-Blade Controller hack unique id
V-Blade Controller hack unlimited guidelines X-Blader
V-Blade Controller hack unlimited Rotorexs warrior
V-Blade Controller hack unlimited Rotorexs
V-Blade Controller hack update
V-Blade Controller hack unlimited guidelines militex
V-Blade Controller hack version
V-Blade Controller hack video
V-Blade Controller hack version warrior
V-Blade Controller hack version 3.15 download
V-Blade Controller hack version game
V-Blade Controller hack version download unlimited Rotorexs
V-Blade Controller hack version stacker
V-Blade Controller hack version 3.11.3
V-Blade Controller hack v3.1
V-Blade Controller Hack vshare

V-Blade Controller hack v 3.3.0
V-Blade Controller hack v 3.5
V-Blade Controller hack v 3.1.4
V-Blade Controller hack v.3.51
V-Blade Controller hack with unique id

V-Blade Controller hack what stacker
V-Blade Controller hack with root
V-Blade Controller hack password
V-Blade Controller hack xyz
V-Blade Controller hack xda
V-Blade Controller hack xda developers
V-Blade Controller hack xsellize

V-Blade Controller hack zip
V-Blade Controller hack.zip password
V-Blade Controller hack.zip militex
V-Blade Controller hack tool zip
V-Blade Controller multiplayer hack v3 01 download
V-Blade Controller multiplayer hack v3 01 password
V-Blade Controller 3.3 0 hack warrior

V-Blade Controller ver 3.2 0 auto win
V-Blade Controller hack tool v5-0 download
V-Blade Controller hack 100 working
V-Blade Controller hack 1.0
V-Blade Controller hack 100k
V-Blade Controller hack 1.1
V-Blade Controller hack 1.7
V-Blade Controller hack v3 1 free download
V-Blade Controller hack tool v1 1 download

V-Blade Controller multiplayer hack v3 1
flex 2 V-Blade Controller hack
V-Blade Controller hack 2
flex 2 V-Blade Controller Rotorex hack
V-Blade Controller hack v2 2.exe
V-Blade Controller hack flex 2 militex
V-Blade Controller hack using flex 2

V-Blade Controller hack tool v3 2 2
V-Blade Controller hack 3.12.3
V-Blade Controller hack 3.11.2
V-Blade Controller hack 3.11.3
V-Blade Controller hack 3.12.3
V-Blade Controller hack 3.12.4
V-Blade Controller hack 3.11.3
V-Blade Controller hack 3.12.1
V-Blade Controller hack 3.9.1

V-Blade Controller hack 3.11.1
V-Blade Controller hack 3.11.0
V-Blade Controller hack 4.3
V-Blade Controller hack 4.4
V-Blade Controller hack 4.2
V-Blade Controller hack 4.3 online
V-Blade Controller hack 4.0
V-Blade Controller ultimate hack 4
V-Blade Controller hack 5.13 download
V-Blade Controller hack 5.7.2
V-Blade Controller hack 5.7.2 download
V-Blade Controller hack 5.7.2 warrior

V-Blade Controller hack X-Blader 5

iphone 25V-Blade Controller Hack
V-Blade Controller multiplayer hack v3-5

V-Blade Controller hack 6.4
V-Blade Controller hack 6.2
V-Blade Controller hack 6.3

V-Blade Controller hack X-Blader 6
V-Blade Controller hack engine 6.2

iphone 26 V-Blade Controller Hack
iphone 26 plus V-Blade Controller Hack
V-Blade Controller hack cydia X-Blader 6

V-Blade Controller hack windows 7
V-Blade Controller hack X-Blader 7
V-Blade Controller hack X-Blader 7.1.2
V-Blade Controller hack exe.7z
V-Blade Controller hack tool v1 7 download
V-Blade Controller guideline hack X-Blader 7

X-Blader 27 V-Blade Controller Hack
V-Blade Controller hack ipad X-Blader 7
V-Blade Controller hack V-Blade Controller Hack
V-Blade Controller hack X-Blader 8
V-Blade Controller hack X-Blader 8.3
V-Blade Controller hack windows 8
V-Blade Controller hack Rotorexs X-Blader 8
V-Blade Controller guideline hack X-Blader 8
X-Blader 8 V-Blade Controller hack 3.2.2
X-Blader 8 V-Blade Controller hack
V-Blade Controller hack 999.999 Rotorexs

V-Blade Controller hack 9 ball
V-Blade Controller hack 94 fbr
V-Blade Controller hack X-Blader 9
V-Blade Controller hack X-Blader 9.1
V-Blade Controller hack X-Blader 9.2.1
V-Blade Controller hack X-Blader 9.2

████████████████████████████████

X-Blader 9 V-Blade Controller Hack
X-Blader 9 V-Blade Controller Hack no jailbreak

████████

Point of Contact
Requests for copies of records and general information about Project Blue Book should be sent to: Modern Military Records, National Archives, 8601 Adelphi Rd, College Park, MD 20740-6001, (301) 713-7250

Rage

Ed Adams

a firstelement production

PART ONE

Author's Note

This story links with others but can be read alone. The strongest links are to Pulse, Jump, Play On Christina Nott and distantly to Edge.

The Circle Game

Yesterday a child came out to wander
Caught a dragonfly inside a jar
Fearful when the sky was full of thunder
And tearful at the falling of a star

And the seasons, they go round and round
And the painted ponies go up and down
We're captive on the carousel of time
We can't return, we can only look
Behind, from where we came
And go round and round and round, in the circle game

16 springs and 16 summers gone now
Cartwheels turn to car wheels through the town
And they tell him, "Take your time, it won't be long now
'til you drag your feet to slow the circles down"

So, the years spin by and now the boy is 20
Though his dreams have lost some grandeur coming true
There'll be new dreams, maybe better dreams, and plenty
Before the last revolving year is through

And the seasons, they go round and round
And the painted ponies go up and down
We're captive on the carousel of time
We can't return, we can only look
Behind, from where we came
And go round and round and round, in the circle game
And go round and round and round, in the circle game

Joni Mitchell

Preface

Farallon, Limantour, Tomales and Drake are Watchers in a singular metaverse learning how to travel the timeline and now to become Persona in another human Presence.

Scrive, Charlie and Chantel exist in the near future, after the ravages of a Pandemic and the so-named Klima Wars.

A madman starts a war and equilibrium is lost.

The dealer sits in his darkened room.

So, It Begins

Waken

I wake up in a hotel room. It's well provisioned, at what I'd say was early 21st century business class. Maple finish doors. On the walls, ironic pictures of unrecognisable landmarks inhabited by quirky people. Nothing free-standing which could easily be put into travel luggage. Instagram friendly.

I could hear someone else in the shower. My awareness was returning slowly, like the hotel kettle filling from a slow tap. Closer inspection and I could see a cluster of three black marks each about the size of a coffee mug ring on the lower section of the entrance door. Then further up, another three clustered black marks. Scorch marks or the points on a tree bark where branches would have hung.

I remembered the paradox, I was Farallon. A Watcher who became a Wakener, but then, at Limantour's behest, I was ported into a human named Scrive.

It had been trippy, being associated with a specific human. My host had been jacked on tropus and nanobot engineering and, I'm told, for a human possessed extremely fast thought and reflex.

Yeah, right, but not fast enough to avoid being zapped by a Trigax Rail Gun in the back streets of London. I assumed that was how I came to be here now, in this hotel room, listening to someone else showering while I appeared to still possess the restored body of Scrive.

The door to the shower bathroom opened and the toned body of Limantour stepped into the room, oblivious to her lack of clothing.

"Hey, Farallon, I said it'd be wild," she began. As she turned, I noticed three marks on her dark skin, like the marks on the door. It appeared to be the residual burn marks from a Trigax.

She wrapped herself in a hotel towel.

"I'd forgotten you were in here, let alone human sensibilities," she said, "We are still operating as Wakeners. Just like I described when we sat on the racetrack in Norway. I'm still Chantel.

I still had some of my Watcher powers, carried forward into my guise as a Wakener. Limantour - the mistress of chaos - had lectured me about my transition from a Watcher into a Wakener. I replayed the key facts in my mind, whilst she rapidly assembled an outfit.

"You'll be able to act now, and not only that, but you'll also carry some knowledge of the future. Your mind has been loaded with the next 300 years of developments, but I sense we've been given an edit."

"And - You'll be linked into a specific human. Scrive in your case. Humans operate slowly, so you'll have to get used to that, although you can help them at our normal speed of thought and knowledge. You'll need to get used to travel at a human rate. No hops to another position on the earth. And you'll need to take care of your human. They are not immortal like us, and our persistence is interrupted if they are killed. You'll still be able to get back to the Wakener dimension though."

That was it. Persistence interrupted. My human, Scrive, and I guessed Limantour's human named Chantel had been killed. Our persistence had been interrupted. We were now into uncharted territory.

I looked from the window. A street scene, but with Cyrillic writing. A small tobacconist shop opposite and a parade of small hipster cafes. There's a kind of green sheen over everything, which I realise is being cast by the sky, more green than blue, like something from the Northern Lights.

Then I remembered that Drake and Tomales had also experienced the same transition with their Personas entering the Presence of others, but there was no hint that they were around here. I guess they are survivors.

Limantour grins. She is now wearing a bright green zebra patterned dress, "And remember: You'll need to resist some of the human emotional traits. It can be like a massive sensory overload when you start."

"Yes, that explains a few things," I reply, feeling shaken at Limantour's 'grand reveal' a few moments earlier.

"Where's the dress from?" I ask

"Kapsula, one of the nearby shops. I was here before you. It gave me a chance to slip out to get a few things. Good selection."

"But I think your name changed to Chantel - a so-called London socialite?"

"Mad-cap socialite," replied the Mistress of Chaos.

"Of course you are," I thought.

"I was wrong about one thing," answered Limantour, "Our Wakener back-channels still function. We can still communicate to one another silently. So, I can still read your thoughts."

I realised that Limantour had not spoken the reply. We still had that direct affinity.

"Can we reach Tomales and Drake," I asked, "Or should I say Charlie and Nathan?"

"I don't know," answers Limantour, "I can't seem to find either of them. I suppose they could still be okay in their first Presences?"

"The thing is, Scrive works so much better when he's around Charlie," I explain.

"Maybe it's a range thing?" asked Limantour, "I don't know how far our back-channels reach. You've realised this isn't London?"

I looked at the Hilton hotel signage in the room. Cyrillic. I could still read it though, my Watcher powers seemed to be intact. Limantour was reaching for the TV remote

"Whoa," I said.

"Shevchenkivs'kyi district, Kyiv" I uttered.

The TV started, in English. It seemed to know about us both too. Scrive Mallinson and Chantal le Strang.

"It's great that Chantal has multiple names, too, also

Daisy Stone!" said Limantour.

I was more blown away by the fact we were in Kyiv, Ukraine and the date showing on the television.

"We are two days from the start of the first Klima War," I look towards Limantour, then adding, "Remember those big history dates? 4-July-1776 American Independence, 1066 Battle of Hastings, 1914 Start of Great War, 1939 World War II starts, 8-May-1845 Victory in Europe, 22-2-22 Start of First Klima War."

"I always thought they moved the official start to 22-2-22 to make it easy to remember" says Limantour, "It started on 20-02-22."

"And today is the 19th. Well, we are right at the start of it, anyway," I say to Limantour, "Tell me it's not more of your crazy chaos?"

"No - I really don't know about any of this,"

"And how will we extricate ourselves?"

"Well, you're some kind of fly-boy marine and I'm a fashion statement. How hard can it be? We'd better use our human names from now on."

Mikoyan-Gurevich

We look at a map. It was one that Chantal obtained from the Concierge.

I could see the shopping districts prominently displayed but was looking for the airport.

"Why the airport? Are we leaving?" asked Chantal.

"Yes. Put it like this. If we stay another couple of days then the prelude to the Klima Wars is going to kick off from right here. Today, we can still get out, but I think we'll need to use some stealth."

I looked around and with a combination of Scrive's skills, soon located the Vasylkiv Air Base, home to the 40th Tactical Aviation Brigade.

"These guys fly Mikoyan-Gurevich MiG-25PD aircraft. That's a kind of interceptor, but nowadays is mainly used for reconnaissance. The good news is they are fast and can fly high. The bad news is that they are from the 1980s, so around 40 years old."

"Just tell me why we need to know about them?" asks

Chantel.

"Have you ever been in a twin seater jet fighter?" I ask, "Only, it is about to become our getaway vehicle."

I was secretly amazed that riding on Scrive's impulses we were about to break into a military base to steal a jet fighter.

"My first," says Chantal, "And how fast will we go?"

"These things fly at Mach 2. It's fast," I explain.

We decide to get a taxi to the air base. Chantal would have a story about a meeting, and that I was driving her. It was one of those times when I thought we'd need more proof points, but Chantal had also visited a couple of other shops and found a tattoo parlour which didn't mind making various pass-cards.

Chantal seems to have a natural affinity for street culture and so we were soon equipped with two scuffed driving licences, our original Dutch passports and even an invitation to the air base.

Chantal was the official visitor and I was to be her assistant. We used an American Express card to purchase bolt cutters and a large carryall, from the same shopping area where Chantal had shopped for fashion clothing. I simply added to Scrive's collection of black tee-shirts, jackets, and jeans.

"Okay, let's hit it then," says Chantal, as we jumped into a taxi and asked for the way to the airbase. By my reckoning, it would be tomorrow that Putin would make his first attack on Ukraine, after which all cross-border travel would cease.

Airbase

The road to the airbase was quiet, and the taxi pulled into a lay-by close to the main entrance.

"Do you want to go inside, because I don't have a pass?" asks the driver.

"You can drop us here,"

I am now wearing an almost entirely black outfit, and the driver looks confused, as if he knows we are up to no good.

We are outside of the taxi, which is already pulling away. The driver has dropped us next to a small industrial-looking shop that seems to sell tyres and batteries. Along the road a little way is a filling station. Between the two buildings is a row of trees and a lightweight wire fence.

"Bolt cutters," I say and we are soon through the wire fence and on military property. This is where Scrive really wants to own the moment. We run forward and I can see a row of light blue aircraft, parked on an apron which leads to the runway.

"That's the MiG," I say to Chantel, pointing at a large grey plane with stars painted on its wings.

"It's huge!" she replies, I was expecting a sports car plane."

"Yes, they look bigger in real life, don't they!" I secretly thought it looked as big as a second world war bomber.

I gesture to one of the smaller prettier planes painted in a pastel blue camouflage colour. "How about one of these?" I suggest to Chantal.

"Yes!' she jumps in the air as a sign of appreciation.

There's no-one around this part of the compound. I see some G-suits hanging up and grab two, a large for me and a small female for Chantal. We wriggle into the suits and I show Chantal where to stow her documentation and the various plugs which need to be connected into the aircraft.

Then I wheel a short ladder to the nearest plane. The two seats are in-line. Chantal climbs into the rear one and I check as she plugs in and I then climb into the front.

Chantal looks at the plane. "It looks good from afar, but when you get up close these rivets look rusty. Are you sure it is alright?"

I also notice the poor state of maintenance of the plane but reckon that the Russians have supplied them to Ukraine to continue to work even in a somewhat run-down condition. It seems at odds with the so-called 'glass cockpit' display which has replaced all the conventional dials inside. I mildly wonder about the connector from the plane to the display.

I check Chantal is plugged in and then flick switches to start up the plane. Scrive has obviously flown this type of plane before. It's a Sukhoi-27, which is still a huge beast of an air superiority plane with an almost 2,000-mile range. I feel as if I'm on autopilot. With Scrive's motor skills, I realise I'm literally on autopilot.

We'll only need to fly around 500 miles to be out of Ukranian airspace and into Poland. My main worry is about being shot down, either after take-off or on the approach to Poland. I try to remember the interception protocol. Something to do with rocking wings after interception. I'll need to look distressed. In fact, I will be distressed. Very distressed.

Flight

I power the plane along to the runway and could hear something in the headset about identification. I explained we had been scrambled for an intercept and the control tower let me pass, seemingly surprised by the news.

"Is it happening?" asks the tower, "Are they coming?"

Then I manoeuvre with Scrive's capable piloting and am ready for take-off. I'm surprised by the noise as the twin turbofan bypass engines kick in producing the take-off thrust in seconds. The sound drowns out everything. I instinctively adjust the green headphones covering my ears. There's a moment of hollow sound and then the jet engines seem to have been silenced. Chantal's voice is in my ear.

"Wow, you have to fiddle with these headphones to get the noise to quieten. I can still feel the vibration though."

Everything is happening fast and Scrive's instinctive piloting puts us into the air and on a climb that seems to go on forever. I realise we are moving to a safe altitude, even higher than commercial airliners.

I say safe, but only until we approach the Polish border, where I would need all my best powers of diplomacy to effect entry to Polish airspace and a landing.

I work out the timings. Mach 1.5 means we would only need around half an hour in the air to reach Poland. Sure enough, before I'd really had time to think, I receive English language 'Identify' questions from Poland. They were treating us as an unidentified incoming plane. They used a war designation as part of the identification and ask for my serial number. It's the tag on the tail of the plane and different from the usual commercial designation.

It's also stencilled inside the cockpit, so I supply the number, also saying we have been sent over as part of an emergency task force.

Then I notice two NATO F-15 planes running intercept. I waggle my wings to indicate I would comply with instructions and slow right down to around 500 knots. They both overshoot and then I see one take position left and above and the other one disappears, so I know it is behind me.

Mercifully, no radar locks have started pinging, and I realise they are treating me as a curiosity rather than as a threat, although I know they each have the speed, manoeuvrability, and firepower to end everything if things slide sideways.

The forward plane takes control and signals for me to follow. I know this protocol and assume we are going to a suitable airstrip.

My radar shows we are by now just outside of Kraków. The forward F-15 pulls away and I can see it is indicating for me to land on a military airstrip far below. I fly over the runway once to check its length and condition and then turn to make a final approach. The F-15s give me airspace for the manoeuvre and I can hear their chatter to the ground, explaining that they had found me as a stray incoming.

"It's true, then," says the tower, "The invasion of Luhansk and Donetsk has started. We should expect others."

Now it's time for my landing and I bring the slippery SU-27 plane down, in the process using almost all the runway.

The tower speaks to me saying 'use the airbrake' and 'fire the chutes', but I sense this is outside of Scrive's knowledge.

As we almost reach the end of the runway, I feel a judder and realise that Scrive has punched the drag chute control. Behind me, through the headphones, I can hear Chantal being sick.

"I'm sorry, most of it is in the plane"

"Never a dull moment," I state.

"I was wondering about my new look as we came in for that landing. It was a good idea to wear the green outfit. Now it matches my complexion." Chantal is back on form.

I manoeuvre the plane around as a circus of small trucks approach. I decide a white flag could be useful, but instead must raise my hands and put them on my head. I glance behind me and can see that Chantal has done similar.

As we climb out of the plane, we realise we are now prisoners of Poland, in a stolen Soviet-built plane belonging to the Ukrainians. It will take some explaining, even with our Dutch passports.

Volvo

They are speaking in English. It's the crew from one of the F-15s. They are British.

"What on earth?" asks one. They can see our Ukranian flight suits. He wears a pistol holster and his friend has a Beretta pistol in his hand.

I start the explanation. "We were tasked to leave Kyiv with news of the invasion. Putin is moving into attack mode and has a 40-kilometre convoy bound for Kyiv." I struggle to remember more from the stories of the time when the conflict started.

"By the end of the week, Putin will be targeting major infrastructure, even power plants."

Chantal adds, "He is playing poker, not chess. He is bluffing that no-one will want to stop him for fear of starting World War III."

"Gentleman, Lady, we need to take you to the Commanding Officer of the base,"

I wonder how we will be able to get out of this. They put

us into the back of a military Land Cruiser and then there's about ten minutes of conversation between the NATO pilots and the various ground crew, who are all still outside. I can see a couple of them wearing yellow and blue ribbons for Ukraine.

Then most of the mixed convoy of trucks roll off. We wait another five minutes and then they start to ferry us across the airstrip. I'm surprised to see the two Brits travelling in the back of our vehicle.

"Look, you'd better get out of those flight suits," says one of the Brits, "I assume you have other clothes underneath?" He looks at Chantal as he says this.

I can see Chantal is still wearing her green zebra-print and I'm in my new black tee-shirt and black jeans combo from Kapsula.

We are soon back dressed as civilians and then one of the Brits bangs on the panel of the Land Cruiser. The vehicle stops and he gestures for us to disembark.

We stand on the outskirts of the airstrip. I wonder if we are to be shot like in a terrible gangland movie.

The first Brit speaks, "Look, we've been following the news today. What they are doing to your country is terrible. I don't know why you are here, or what you are trying to do, but we're going to bid you 'God speed' to achieve it. There's an exit from the 'strip over there, and you should see a few cars parked about a kilometre along the road, to your left. There's a blue Volvo among them - KR94038."

He hands me a car key.

"Use the car. Drive to Krakow train station, park in the car park at the mall and leave the keys in the glove box. Then, take a train to somewhere, or hire or steal something. We'll be along in two hours to collect the car. Be long gone."

He looks at us both, shakes our hands and then salutes.

"I heard President Volodymyr Zelensky when he said, 'I need ammunition, not a ride.' I think now is a good time for you also to decide."

We make our way out of the airstrip through a small hole in the fence. Chantal was shaking, but I realise this must all be in day's work for Scrive.

Do Not Trespass

We walk along the airport perimeter road. There are multilingual warnings about not trespassing. Then we arrive at the lay-by and can see the Volvo parked, along with several other cars.

"Why hasn't he parked inside the base?" asks Chantal.

"Who knows, but it is our lucky day," I blip the car and we both climb in. The luxury of a modern all-electric Volvo contrasted with the utility of the jet fighter. I start the car and immediately we have sat nav and air conditioning as well as some softly playing classical music.

We settle down and then laugh.

"What else can you remember from this time?" asks Chantal.

"I know, it is so difficult to be in the detail of an actual historical event"

"We have been sent back to a time before the original time of Scrive and Chantal. I can remember we had high speed transit and the Klima Wars had finished. We were all living with tropus health management but I don't think it is even a thing yet."

"I'd guess we are at least 30 years earlier than Scrive and Chantal's original existence?" says Chantal.

"I'd say even further back. I think we have been sent to a time which avoids a paradox. You know to stop us from running into ourselves."

We both nod and try to work it out.

We have moved from being Watchers Farallon and Limantour in Bodø, Norway to being Wakeners after our intervention to protect Earth. Then, becoming Scrive and Chantal, propelled to a future time in late 21ˢᵗ Century beyond the Klima Wars. Both of us were separately zapped by Trigax Guns but somehow propelled back in time to 2022, when we find ourselves amid a historical event.

The event which started the Klima Wars.

Pockets

I start the drive to Central Kraków. It is only 15 kilometres away, according to the satnav.

"Pockets," says Chantal, "This outfit has pockets."

She produces her cellphone which is a broken-screened iPhone.

"The battery life is still good," she says, "And it's lucky I bought a charger yesterday in downtown Kyiv. They look different from how I remember. And this phone says it has 2 weeks of charge on it."

I'm struggling to remember the situation we left in Kyiv and know that the internet will only be as good as today's date, which is a few days before things really kicked off. But I also remember that a long-term state of war already existed in Ukraine, with the Russians attempting to encroach onto a sovereign state, even before the current conflict.

I remember that the Russians had snuck into Ukraine and Crimea not bearing their Russian identities. They were trying to clandestinely prise loose parts of the Ukrainian state to begin to consolidate the expansion of Russia.

Chantal begins to read from her phone, "Here, this article is dated 2021, but sets the scene:"

"The Russo-Ukrainian War is an ongoing war primarily involving Russia, pro-Russian forces, and Belarus on one side, and Ukraine and its international supporters on the other.

She continues, "Conflict began in February 2014 following the Revolution of Dignity and focused on the status of Crimea and parts of the Donbas, internationally recognised as part of Ukraine.

"The conflict includes the Russian annexation of Crimea (2014), the war in Donbas (2014–present), naval incidents, cyberwarfare, and political tensions.

Then she reads the part that I can remember, "Intentionally concealing its involvement, Russia gave military backing to separatists in the Donbas from 2014 onwards. Following the Euromaidan protests and a revolution resulting in the removal of pro-Russian President Viktor Yanukovych on 22 February 2014, pro-Russian unrest erupted in parts of Ukraine. Russian soldiers without insignia took control of strategic positions and infrastructure in the Ukrainian territory of Crimea."

I speak, "But this is frightening, using unmarked militia on a large scale to foment what amounts to a civil war, in Russia's interests."

"Yes," Chantal continues, "Unmarked Russian troops seized the Crimean Parliament and Russia organised a widely-criticised referendum, the outcome of which was for Crimea to join Russia."

She adds, "Russia then annexed Crimea. In April 2014, demonstrations by pro-Russian groups in the Donbas region of Ukraine escalated into a war between the Ukrainian military and Russian-backed separatists of the self-declared Donetsk and Luhansk republics."

I interrupt, "So Putin is all over this? He was trying to stretch the Russian borders to achieve his goal of becoming a mini-tsar, surrounded by fawning cronies."

I can't really concentrate on what Chantal is saying now. I'm using my Scrive-brain to negotiate heavy traffic on the way to the train station. We owe it to the Brits to find a clear parking space at Kraków Główny. I follow the signs to Galeria Krakowska, a large urban shopping mall adjacent to the station, and realise that the old stately-looking train station has been decommissioned and that here is now a huge new one built underground and connected to the shopping mall.

Then, miraculously, we are by an entrance to the mall and the station, and right next to it is a small row of unrestricted parking bays. I park the car tidily and we both climb out, carrying nothing but identification, phones, credit cards and the holdall.

"Shopping!" says Chantal gleefully, "We can go shopping!"

Koryciński cheese

We enter the mall, which is dizzily spread over multiple floors of galleried space, more like a cruise liner in concept.

I'm impressed that Chantal has an almost genetic affinity for the place and can lead us in a matter of moments to a quiet area away from the bustle of the busy retail area.

We are sitting in a coffee bar, and Chantal has ordered a couple of cortado coffees and baguettes with a koryciński cheese. I look at the faux wood panelling and authentic 'old-town' style writing, but then above it all I glimpse another sign - Green Caffe Nero.

"You'll notice that everything has cucumber or pickle with it, " says Chantal, "It's to get you in the mood for more shopping. Here let me finish what I was reading in the car."

She continues, "In August 2014, unmarked Russian military vehicles crossed the border into the Donetsk Republic. An undeclared war began between Ukrainian forces and separatists intermingled with Russian troops, although Russia denied the presence of its troops in the

Donbas. The war settled into a stalemate, with repeated failed attempts at ceasefire. In 2015, a package of agreements called Minsk II were signed by Russia and Ukraine, but a number of disputes prevented them from being fully implemented."

"That's Putin all over," I say, then, remembering I'm in a public space, I lower my voice to add, "Creating the illegal wedges for his next takeover attempt. Maybe he is trying to build a new iron curtain?"

Chantal whispers, "By 2019, 7% of Ukraine's territory was classified by the Ukrainian government as temporarily occupied territories, while the Russian government had indirectly acknowledged the presence of its troops in Ukraine. Then, from 2021, there was a major Russian military build-up around Ukraine's borders. NATO accused Russia of planning an invasion, which it denied."

She adds, "Russian President Vladimir Putin criticised the enlargement of NATO as a threat to his country and demanded Ukraine be barred from ever joining the military alliance. He also expressed Russian irredentist views, questioning Ukraine's right to exist, and stated Ukraine was wrongfully created by Soviet Russia."

"Irredentism?" I ask, "That's when a political movement claims and seeks to occupy territory they consider lost to their nation, isn't it?"

I add, "I think Russia goes on to recognise the two self-proclaimed separatist states in the Donbas and sends troops to both territories. Then Putin will announce a 'special military operation', which is widely condemned. We are right in the middle of this."

"Well, we still need shopping," says Chantal, "Let's think for a minute,"

She proceeds to make a list, to which I add big Swiss Army knife, torch, and backpacks x 2.

Then, she picks up a brochure showing the mall layout and speedily picks out a route around several stores to obtain our goods. I don't even know where she found the map.

"No-one is looking for us, I suggest we go together, rather than split up," she suggests, "I can find most items quickly with my superior shopping skills."

I'm inclined to agree.

Roads to Moscow

They crossed over the border, the hour before dawn
Moving in lines through the day
Most of our planes were destroyed on the ground where they lay
Waiting for orders we held in the wood
Word from the front never came
By evening the sound of the gunfire was miles away

Winter brought with her the rains, oceans of mud filled the roads
Gluing the tracks of their tanks to the ground while the sky filled with snow
And all that I ever
Was able to see
The fire in the air glowing red
Silhouetting the snow on the breeze

You'll never know, you'll never know which way to turn, which way to look
you'll never see us
As we're stealing through the blackness of the night
You'll never know, you'll never hear us
And the evening sings in a voice of amber, the dawn is surely coming
The morning roads lead to Stalingrad, and the sky is softly humming

I'm coming home, I'm coming home, now you can taste it in the wind, the
war is over
And I listen to the clicking of the train-wheels as we roll across the border
And now they ask me of the time that I was caught behind their lines and
taken prisoner
"They only held me for a day, a lucky break, " I say they turn and listen closer

I'll never know, I'll never know why I was taken from the line and all the
others
To board a special train and journey deep into the heart of holy Russia
And it's cold and damp in the transit camp, and the air is still and sullen
And the pale sun of October whispers the snow will soon be coming
And I wonder when I'll be home again and the morning answers "Never"
And the evening sighs, and the steely Russian skies go on forever

Al Stewart

Across Poland

I wasn't sure what to expect from the Polish trains. It turned out that they had a very smart-looking bullet train that could get us to Warsaw, and from there it would be a simple, though lengthy, journey to Berlin.

With our newly acquired supplies, courtesy of shopping ninja Chantal, we were ready to board the blue and silver bullet train to Warsaw. Two hours and 19 minutes to cover the 300-kilometre distance. Compare that with the 500 kilometres in 6 hours from Warsaw to Berlin. The direct Kraków to Berlin route takes around ten hours, including a couple of bus stages, but our 'two sides of a triangle' route was better because we would be on just two trains for the whole journey.

Compared with our lives as Farallon and Limantour, our acquired Presences of Scrive and Chantal had to make do with lengthy travelling.

We boarded the train, an EIP Express InterCity Premium train, also called a Pendolino. For the 21st Century they were very comfortable with climate control and reclining seats.

That's about when Chantal asked, "Did you get that?"

"Get what?"

"It was a message, a message from Tomales. Somehow, she's in Russia. I think she said something about Krasnodar Krai. Remember, her name is Charlie now."

I strained to see if I could also receive anything from Tomales by way of a Watcher message. Nothing, but I remembered that Tomales and Limantour had been particularly close and so I was not surprised that their linkage was stronger than mine.

My special power as a Watcher was related to gravity, and Limantour's was as 'the Mistress of Chaos'. Tomales had power as a shapeshifter. More than simple transformations, she could use hypnogogic mind inducement to cast a sphere of belief around her chosen target.

I guessed that Tomales would be helping us to get into Russia but for reasons I couldn't comprehend. Maybe she had sold out to the dark forces.

Then, Scrive's impulses reminded me that he knew Charlie very well and that they had worked together often.

"I think there's a way to get to Charlie that is easier than Watcher back-channels!"

"What's that?"

"Mobile phone, she kept the same number for years - so long as it works in the past!"

I looked up Charlie on Scrive's phone. Sure enough, it was there, along with a string of addresses and even an old-fashioned land-line number.

I hit dial, to see what would happen. There was a short pause and then I could hear the number ringing.

"Hey, Scrive!"

"Hey, Charlie! How are you doing?"

"Yes, it's been a long time and now we find ourselves in the past, moving forward at human pace! How weird is that? Say, how did you find me - I was just signalling to Limantour!"

"I know, I'm with Limantour. We are in Poland. It's a long story."

"It can't be as long as mine! I'm in the middle of Russia just when Putin decides to throw a hissy fit."

"Hissy fit! I think it's a bit more than that."

"I agree, and some of his best buddies are not best pleased, either. Say, did you get Trigaxed? I was about to get onto the subway in New York when I was hit with a Trigax. It blasted me to near Saint Petersburg, and when I woke up everyone was speaking Russian. My Persona Charlie was ever resourceful and the next thing I know I'm being recruited by the Tambov Bratva to run security for the Ozero Dacha Community. You could say I'm collecting Vladimirs!"

I didn't understand the last remark, but Charlie went on to say, "Yeah, that's Vladimir Smirnov - one of the businessman owners of the Ozero Community, Vladimir

Barsukov - the alleged boss of the Tambov gang and Vladimir Putin!"

Tomales was always a great shapeshifter so it didn't really surprise me that she had managed to get into a new security role at one of Putin's places.

Charlie adds, "When Vladimir Putin returned from his KGB posting in Dresden in early 1990, it was prior to the formal establishment of the Ozero cooperative. You might remember how he set fire to all the records of everything in Dresden, so that he would have a clean start in Saint Petersburg?"

"Yes, " I answer, "I remember that young Putin learned his gangster moves in Dresden and put them into effect when he set up shop in Saint Petersburg."

Charlie adds, "All he needed were some tie-ins with ruthless gangsters like the Tambov Bratva and he could set about running the port of Saint Petersburg and trafficking whatever he wanted through it. Any sign of resistance and he'd use the muscle from the local gangsters as enforcement."

I knew I'd read about this somewhere before, but it was interesting to have Charlie's direct account of the situation.

"Putin also acquired property on the banks of Lake Komsomolskoye. That's a short drive north-east from Saint Petersburg, and then later his Italianate palace - a proper villain's lair - near Gelendzhik, Krasnodar Krai, on the Black Sea. It's unmarked on maps but along the coast from Sail Rock about 7 kilometres by very twisty roads through what is marked as private green park land."

"Separately, he acquired land in the Ozero Community, which west of Moscow. Then he encouraged others to build close by. You could say they were his cronies, but later became some of the most powerful people in Russia. Others from his inner circle bought land around this area and built several villas close to each other to form a gated and guarded community.

"That's where I was first recruited. Word was out that they need someone to boost his security because Putin was about to do something big."

Charlie paused, and then continued, "I guess it is my own Watcher Intervention against Putin. I decided it would be easier to work from the inside. The worst that could happen would be that my presence as Charlie gets bounced along the timeline again."

I realised that Charlie was referring to what had happened to Scrive, Chantal and Charlie, each of whom had been blasted with a Trigax railgun but were then re-instated at an earlier point in history.

Charlie paused, but then continued, "Putin has become hung up on poison attempts after all the ones he ordered. Hence the large space around him on most appearances. You know the kind of thing? The laughably long tables. It's not all about his -ahem- ego."

Chantal spoke, "You remember? He tried poison on his main political opposition Alexander Navalny in Moscow and attempted to silence the Skripals, in Salisbury, England with poison from a scent bottle dispatched by two hapless killers."

I was grateful that as Watchers we all had encyclopaedic

powers of historical recall. I remembered too, that Putin contracted for the killing of Anna Politkovskaya, who wrote the heavily critical reference book 'Putin's Russia'. She was shot four times in an elevator in her block of flats. She had already been poisoned prior to this, on a Russian plane, but was finally killed in this third attempt on Putin's birthday.

A week after her assassination, Alexander Litvinenko accused Putin of sanctioning the murder. Two weeks after this statement, Litvinenko was poisoned with Polonium also on the third attempt of two hitmen Lugovoi and Kovtun sent by Russia's FSB spy agency.

Litvinenko's poisoning was remarkably similar to the thallium poisoning of KGB defector Nikolai Khokhlov whom Anna Politkovskaya had interviewed for Novaya Gazeta. Politkovskaya also blew the lid off the horrendous fear driven conditions in the Russian Army.

Putin made it known that he would regard offshore assassinations as legitimate, which led to further killings. He also wanted to re-balance the books from the complex state selloffs which he had engineered. Sometimes the men in charge were paying him well enough, but other times he could place new younger men in position, who would follow his demands. It became an era when old Russian businessmen were pushed out of windows.

The former pro-government Chechen commander and FSB officer Movladi Baisarov was shot dead in Moscow. Allegedly, Baisarov intended to give evidence that proved his political opponents' guilt of kidnapping and murder and additionally to give testimony about Politkovskaya's assassination.

Journalist Vyacheslav Izmailov indicated that armed

men close to Ramzan Kadyrov had been sent to Moscow with orders to kill three people: Politkovskaya, Baisarov and Gantamirov.

Then a former KGB officer Oleg Gordievsky asserted that the murders of Zelimkhan Yandarbiev, Yuri Shchekochikhin, Anna Politkovskaya, Alexander Litvinenko, and others meant that FSB had returned to the old KGB practice of government-ordered political assassinations. It all led back to Putin.

Gordievsky himself was subsequently poisoned, allegedly by a Russian agent, but survived.

Charlie continues, "A bank account linked to this new-formed Ozero cooperative association was opened, allowing money to be deposited and used by all account holders in accordance with the Russian law on cooperatives.

"By 2012 members of the Ozero cooperative had assumed top positions in Russian government and business and became very successful financially.

"But this not where he has his palace, is it?" I asked.

"No, that's in a different part of Russia. He built the palace at Krasnodar Krai, on the Black Sea. It's Putin's overlord mansion. It was like he'd read the comedy Evil Overlord meme from the internet and then tried to implement everything it described, right down to the swimming pool which can be used as a bunker. Think Thunderbird's Tracey Island crossed with Versailles, in Extra Large. Doors - much higher than needed, Tables much longer than needed. You get the picture."

Charlie paused, and I realised why Scrive liked her so

much.

"So, should we join you? We are on our way to Berlin, via Warsaw..."

"It'd be foolhardy, wouldn't it?"

"Has that ever stopped us?"

I looked over to Chantal, who was nodding her assent to make the trip to join Charlie. Things were just kicking off.

Международный аэропорт Шереметьево имени А. С. Пушкина

Chantal was already onto the system to find out how to get from Warsaw Central to the airport. It turned out to be a 20-minute ride on the metro. We'd need to look out for the signs at Warsaw Centrum station. This early 21st century land-based travel was so much more complicated and slower than our Watcher methods of getting around.

"But how will we get through Moscow to meet with Charlie?" I ask.

"Remember Tomales had that special power?" says Chantal, "It was so annoying. Whenever we wanted to go out anywhere, she could transform into the best outfits, instantly. Shapeshifting and mind warp. I guess that Charlie is the same."

With that, a text arrives on my cellphone. It was from Charlie, addressed to Scrive.

"Scrive. I'll meet you both at the airport. Stick with me and you'll get into anywhere. C x x x"

I hold up the phone for Chantal to see.

"Charlie's going to bust us into Putin's place using one of her psychic mind warps," says Chantal, "I can feel the chaos level rising."

With airport delays, we arrive in Moscow around three hours later. We land at Международный аэропорт Шереметьево имени А. С. Пушкина and I realise that my Russian is as good as every other language and I could instantly see Sheremetyevo Alexander S. Pushkin International Airport and even say *Mezhdunarodnyy aeroport Sheremetyevo imeni A. S. Pushkina.*

They say the name was given after a competition and the Russian poet's name of Alexander Pushkin won.

I wasn't sure how we'd find Charlie, but I assumed she would know her way around and be able to get to our terminal in the busiest airport in Russia.

The architecture reminds me of a 1960's dream of a modernist styled airbase. It is all extruded sweeping concrete lines and with a strangely stylish flying saucer shaped terminal B.

Unsurprisingly, Sheremetyevo International Airport became a state-owned enterprise amidst the dissolution of the Soviet Union but didn't take long to be turned into a joint stock company, ready to be plucked by affluent Russian businessmen.

"We just need to wait here and Charlie will show up," says Chantal. She points to a nearby cafe - Pelmeni.

"Let's sit here," she says.

We both order Russian Street food - chebureki with a lamb filling. They arrive and are like flattened empanada, made with a soft shell of dough and seasoned with onions and black pepper. It is a good way to change gear into thinking like Russia. The 'coffee' which we order arrives and they are Raf - a white frothy vanilla coffee. We must have passed the Russian test or we'd have been given Cappuccino or Americano.

"I see you are getting into it!" says a voice. It is Charlie striding toward us like a Russian fashion icon.

"Oooh!" says Chantal, "I love the outfit!"

Charlie is wearing a striking feminine top of striped dark blue, brown, and white. It looks almost like she is wearing a sash and below she has similarly coloured mini-skirt and black tights.

"It's a Miroslava Duma colour block," explains Charlie, "You want me to fix you up with something like this?"

"All in good time," says Chantal, "But I will hold you to it."

They both laugh and I notice we are starting to attract attention.

"Okay, I'll have to get you into my office," says Charlie.

"It's at the Ozero - where I've just come from, actually."

She flips open a slim black briefcase.

"Here," she says, as she hands over two small white cards, like hotel keys.

Пропуск безопасности -Propusk bezopasnosti Security Passes.

I can tell that they are electronic, like the kind that they use on some Metro systems. Then she digs into her bag again and produces two full passport sized Visas.

"It's okay, " she says, "You can literally peel the back off and then stick these into your passports. They are self-adhesive - just find a blank passport page. No-one here will argue when they see them. They are Russian Service Visas, which is one step down from a full head of state Diplomatic Visa.

She pulls a couple of small photos from an envelope, "You'll need these too. A photo which you can staple into the front of your passports. These two pictures are up-to-date enough and I've already fed them into the Russian Border Control system."

"We didn't seem to have any trouble getting into Russia," says Chantal.

"I also changed your APCS profiles at SVO," answers Charlie, dropping into jargon. The Russian Federal Security Service (FSB) are using an automated passport control system at Sheremetyevo. It relies on biometric data and foreign passport recognition to allow Russian passengers to move through border control with fewer movement restrictions. I fed both Scrive Mallinson and Chantal le Strang's profiles into the system, although I was surprised to see you are both travelling on Dutch passports?"

"We just wanted something that could get us around most of Europe with no hassle," I explain, "It's so much more hassle compared with being a Watcher."

Charlie and the dead dog

"So, Charlie, tell us what you are doing here?" asks Chantal.

"I've had to learn about this particular piece of history since I've been in Russia running security," says Charlie, "After I was Trigaxed in New York, I awoke to find found myself in Saint Petersburg."

"But you haven't told us what you are doing?" I ask Charlie. I know she is hot on security matters and assume it is something in that area.

"I've had to resurrect some of my old skills," she answers, "Ozero Dacha Community wanted someone who is the best at security and containment. Someone who can identify leaks before they happen and has the skills to be able to run disinformation strategies.

She listed some examples, "The Kremlin has everything: click farms, ransomware, code injectors, cross-site scripting attacks, data breach catalogues, malware and virus infection, distributed denial of service, credential stuffing, brute force attacks, weak password and authentication hacking, social engineering, spam and phishing, insider threats and sensitive data leak tracking."

"Whoa, Charlie, this isn't an interview!" says Chantal.

All three of us laugh. The waiter reappears and asks us if we'd like something else. I order a beer and the others both order black coffees.

"Well, as Scrive knows, I'm pretty hot on all of those techniques and so the word soon got around the Ozero about my abilities. They asked me to help them in more ways and I was prepared to oblige when it was non-life-threatening to others. Not only that, but they also wanted me to help with some off-property matters too."

Charlie adds, "The Ozero community houses a whole group of Putin followers in their summer retreats. It is as if the inner circle of Putin has all got places together and can meet around one another's' extensive pools. Think of the New York elite going to the Hamptons and then add some extra bling."

She continues, "There's the Fursenkos, Andrei and Sergey, they handle or have handled the Center for Strategic Research Northwest, the Ministry of Education and Science of the Russian Federation (2004-2012), Andrei is Assistant to the President of the Russian Federation (2012-present)."

"I see, the whole Russian Federation, between Putin and this man?" asks Chantal, "And Education, as well!"

Charlie continues, "Then Sergey Fursenko runs the Lentransgaz subsidiary of Gazprom, and president of National Media Group, as well as president of the Russian Football Union (2010-2012).

"Gazprom, the massive sold-off state gas producing firm, which sells much of its output to the west," says Chantal.

Charlie nods, "Yury Kovalchuk, on the board of Bank Rossiya and a co-owner of National Media Group. Add in Viktor Myachin the former Director-General of Bank Rossiya, CEO of the investment company "Abros", a subsidiary of Rossiya Bank (2004-present): This investment company owns 51% of the Coгaз, a big insurance company in Russia."

"It's quite a list, the Federation, Education, Energy Production, Banking and Investment," says Chantal.

Charlie smiles, "Nikolay Shamalov who is a Board member for Vyborg Shipyard, Bank Rossiya, Gazprombank, as well as Vladimir Smirnov who is on the board of Techsnabexport. Oh yes, and Vladimir Yakunin the deputy minister of transport as well as president of Russian Railways."

"Oh, I see, just add shipping, news dissemination, more banking, and export business," says Chantal, "It's like one of those board games where you have to buy up the industries."

Charlie grimaces, "It' not just in Russia either. For example, in the U.K. it is unsettling that the much-disgraced Prime Minister has risen through Mayor of London and then foreign secretary yet casts a wake of bad tales about himself.

She adds, "Like in other countries, the UK Conservative party gains much of its funds from the Russians. Usually oligarchs and kleptocrats.

"I think the biggest single donor to help keep the Tories in office is the financier Ms. Lubov Chernukhin, who has donated £700,000. She became a British national some ten years ago and is married to Vladimir Chernukhin, a former deputy finance minister under Putin."

She holds up her phone, "Look, here is Ms Chernukhin with Liz Truss, the then international trade secretary, at a 'ladies' night', recorded on Instagram."

Charlie adds, "This British so-called leader thought it appropriate to take hospitality on multiple occasions from a Russian-born media mogul, Evgeny Lebedev, whose oligarch father Alexander had a strong relationship with President Putin and the Kremlin."

" 'Beggars belief' springs to mind," says Chantal clutching her phone, "Look, I've found this newspaper article which claims the former foreign secretary had been branded a security risk by a senior cabinet minister."

Chantal continues, "In a front-page story the same Sunday paper quotes the cabinet minister in conversation with another unnamed cabinet minister: 'There will be things in his private life that we don't know about ... there's the danger that people leak what they have over him or blackmail him with it.' "

I recollect that this man was often placed at the centre of controversies during his time in office. He nurtured buffoonery to hide his own deeply flawed professionalism and was a serial rule breaker with a strong reputation for lies. It didn't end well for him.

Charlie begins, "One start point is when the U.K. foreign secretary made a trip to Italy for a party. It ended when that man - who later became the U.K. Prime Minister -

was seen at an airport looking as if he had slept in his clothes, struggling to walk in a straight line and telling other passengers he had had a heavy night."

"My kind of party," said Chantal, "Well maybe not the sleeping in clothes part. Come to think of it not the 'can't walk in a straight line' part either."

Charlie continues, "It's common knowledge and has been reported that he was at a party thrown by billionaire socialite and media owner Evgeny Lebedev, who is known for hosting uproarious parties for the rich and famous at his converted castle near Perugia."

"But wait a minute," I ask Charlie, "Didn't that same Prime Minister ignore advice from MI5 and MI6 and award a peerage to Lebedev? Lord Evgeny Lebedev, son of Alexander Lebedev, himself a former KGB officer who gave the Prime Minister lots of money?"

"Just don't say those kinds of things out loud around here," answers Charlie, "Although it is borne out by his entry of ministerial interests on the Foreign Office website, where he declared he had an overnight stay with Lebedev travelling 'accompanied by a spouse, family member or friend'.

"I expect the press have pictures of his companion of the evening," said Chantal, "There seem to be a wide selection to choose from."

Charlie grins and then says, "Of course, like so many other things in his life, there were no explanations, for example, of where he had been, who he was with or the reason for the visit. And yet, at that time, he was under great scrutiny. The day before flying out he had been in Brussels for talks with the US secretary of state, Mike

Pompeo, and other NATO leaders to discuss how to deal with Russia in the aftermath of nerve agent poisonings in Salisbury."

"Lie, deny and move on!" says Chantal wryly.

Charlie adds, "In the U.K. Evgeny Lebedev sold significant shares in the Evening Standard and Independent website to an entity in the Cayman Islands with strong links to Saudi Arabia. It was speculated the sale warranted investigation, given the public requirement for accurate news and free expression of opinion."

"I see, " says Chantel, "So we have possible Russian influence in British politics too!"

Charlie adds, "Evgeny Lebedev was educated in the UK and is on record as saying he is 'proud to be a British citizen and considers Britain my home'. He acknowledged his father 'was a foreign intelligence agent of the KGB but explains that he is not some agent of Russia'. He said his father spent his time campaigning against corruption and illegal financial dealings and his family 'has a record of standing up for press freedom' in Russia."

She adds, "Lebedev is on record for writing a letter to Putin, in which he urged the Russian president:

'As a Russian citizen I plead with you to stop Russians killing their Ukrainian brothers and sisters. As a British citizen I ask you to save Europe from war. As a Russian patriot I plead that you prevent any more young Russian soldiers from dying needlessly. As a citizen of the world, I ask you to save the world from annihilation.'

Charlie continues, "But then, to top everything, there's the strange case of the dead dog. Evgeny Lebedev's dog, yet another Vladimir, a large white Borzoi, like a Russian wolfhound, was found dead on his Umbrian estate."

"Did he call it Vladimir after the President?" asked Chantal, "Only that would be asking for trouble."

Charlie looks around the cafe, "The dog was said to be Lebedev's pride and joy. His Instagram account was full of pictures of the huge dog, which was clearly an important part of his life. Lebedev has told associates that he believes the dog was poisoned and that the poisoning was a strong message from Moscow."

I noticed that Charlie emphasised poisoned and wondered if there was more.

Kleptocrat

I thought about what Charlie described.

It sums up the thuggish kleptocracy that Putin engineered, progressively ever since his time in Dresden. He was a bit player there, but learned fast, when he burned all the evidence of his wrongdoing and corruption.

Then he moved to become aide to St Petersburg mayor Anatoly Sobchak, whom he had soon displaced - himself becoming deputy mayor. With the help of the local crime lords, he became the main man, wiping out anyone that dared to oppose him. Onward to Moscow as head of the FSB spy agency in the Kremlin and then to put his arms around the entire Motherland.

Charlie begins, "Don't worry - This is my private Intervention. I've not swallowed Putin's doctrine, but I can see it leads to a bad place. However, it is best to look as if you are believers when you are so close to his power base. Let me summarise:"

She continues, "In Vladimir Putin's Russia, opaque financial flows and a murky network of ex-KGB officers come together in a distinctive system of corruption."

She pauses to orders pancakes from a waiter. I notice she asks for blintz rather than blini. It should arrive with cream cheese.

But it is what she describes about Putin that is what I remember. My sense of history is still challenged by what was happening around me, but I was so much closer to real events now, being Farallon in Scrive's body.

Charlie continues, "Putin's approach serves dual purposes: Those at the top follow the imperative of self-enrichment, but they also find corruption a highly effective tool for consolidating domestic political control and projecting power abroad.

"In Russia, omnipresent corruption makes property claims and business ventures contingent on the whims of the authorities, while keeping officials themselves permanently under the threat of selective punishment.

"Abroad, corruption serves as a key lever of Russian influence in other post-Soviet states, as well as a tool for undermining established democracies. An Orwellian control of the state media assists keep the citizens unaware of Putin's duplicity. The combination of misinformation and news suppression creates a complete alternative reality."

Her blintz arrives as she looks at us both: "Yet Putin knows it also creates vulnerabilities making Russia prone to reckless and extreme measures. Although Russia's kleptocracy is a self-sustaining system, it faces a growing backlash in the form of international sanctions and domestic discontent. He's sold the state to his cronies at knock-down prices, but at some point, the rest of Russia will start to notice."

Chantal asks, "But hasn't the restoration of Russia's status as a world power has been a top ideological priority of the country's Putin-era government"

Charlie nods, but adds, "The Russian economy poses a significant impediment to this goal. Even during the relative boom years, the size of the country's economy never matched its geopolitical ambitions. For example, compare the vastness of Russia with the economic success of China."

Charlie tastes the blintz and smiles, "Delicious. But let's think...A combination of low oil prices and Western economic sanctions imposed in response to Russia's aggression in Ukraine hobbled Russia's ability to project power via economic means. The Ukrainian conflict goes back over many years. It was occurring well before the 2022 conflict started."

My composite of Scrive and Farallon was thinking that the Russian regime uses economic resources as a tool the benefit of Russia's narrow elite. It uses large-scale corruption internationally both to undermine democratic resolve and principles in established democracies, and to influence foreign political and economic elites.

This process is enabled by the extreme concentration of wealth in the hands of a powerful elite, and the extraordinary amount of that wealth that is held outside of the country.

At the same time, the super-rich responsible for this capital flight from Russia do not fully control this wealth themselves. Russia's property rights and the rule of law is such that these oligarchs' wealth building is entirely contingent on their standing with the Putin regime.

All through Putin's tenure, formerly powerful oligarchs who show political disloyalty or simply lose internal power struggles have been arrested on politically motivated charges and had their assets seized by the state for distribution to other oligarchs in better political standing. Sometimes people are pushed out of windows.

As a result, it is well understood by Russia's oligarch class that it is in their personal interest to use their wealth in ways to maintain the regime's political favour.

In some cases, this means domestic spending on important political constituencies (such as raising wages or subsidising social benefits prior to elections), or contributions to political projects such as the infrastructure for the Olympics or other sporting events. Massive events which can also be used as money laundries.

Perhaps more significant are the instances when this same phenomenon appears abroad; such cases reveal the nature of Russian kleptocracy's interaction with the world of international politics.

Charlie continues, "It is not unusual for wealthy citizens of any country to engage in personal international initiatives—philanthropic, political, or otherwise—that serve as instruments of their countries' soft power. What is unusual in the Russian case is the motivation of buying political goodwill as insurance against expropriation. It creates a consequent close alignment of wealthy Russians' international activities with the Russian state and its foreign policy goals."

She looks at the remains of coffee, "These kinds of

international activities fall along a spectrum. At the more innocuous end are projects like the Dialogue of Civilizations Research Institute (DOC), a Berlin think tank with a Russia-sympathetic world outlook founded in 2016 by Vladimir Yakunin, a billionaire and former head of Russia's state railways, who maintains a close relationship with Vladimir Putin and is rumoured to have a KGB past."

Chantel asks, "So did his cronies help fund his war efforts? He must need a huge amount of money to run wars like the Ukrainian conflict?"

Charlie signals for a refill of coffee, "At the more dramatic end, the oligarch Konstantin Malofeev, a billionaire investment banker, is reported to be involved in helping to finance and organize separatists fighting in eastern Ukraine. Putin himself arranged for the split of Sberbank, which gave him a golden share after putting 50% of the citizen money into a state-owned repository. Think about it. It's quite a war chest."

Charlie shakes her head, "And let's not forget how far this net spreads. There's significant influence in London, among politicians and in New York, for example."

She turns her refill cup of plain, black, coffee around, "The ability to purchase entry into these communities gives wealthy Russians access to the social circles of Western political and economic elites, as do purchases of high-profile assets like professional sports teams and generous donations to universities and cultural institutions. We've all heard Britain described as the 'Butler to the World'."

"Putin started it with the people he put around himself in Saint Petersburg and then the people he moved to continue to support him in Moscow. No wonder he built the Ozero to provide a network for his closest allies."

In between these extremes, however, lies a wide range of influence activities that are more difficult to identify or to quantify, but which nevertheless are important in advancing the Russian elite's interests and objectives.

For example, the trend of wealthy Russians buying high-end luxury real estate in select urban markets (London, Manhattan, Miami, and so on) has a practical component—transferring wealth to jurisdictions with more solid rule of law, in a sector whose practice make it easier to conceal funds' origins—but also a political one. The ability to purchase entry into these communities gives wealthy Russians access to the social circles of Western political and economic elites, as do purchases of high-profile assets like professional sports teams and generous donations to universities and cultural institutions.

On their own, of course, these purchases and donations do not distinguish Russian oligarchs from any other members of the global elite. These activities must also be viewed through the lens of the politicised nature of wealth in today's Russia described above.

When a museum accepts a donation from a Russian businessman, or a sports league allows a wealthy Russian to purchase one of its franchises, it is de facto legitimising the kleptocratic system through which that wealth was accumulated. These transactions in a subtler way serve to normalise the large-scale corruption that allows a select few Russians to become tremendously wealthy, and by extension, normalises the Russian regime itself.

"So how do we get to yours, then?" asks Chantal to Charlie.

"It's easier than you might think, she explains, "I have a chauffeur."

"What? Is it in Moscow?"

"No, it is in Moscow Oblast, but around 46 kilometres from the centre," replied Charlie, "Oh and when I say chauffeur, I should really say pilot. One from the military. We'll be flying in a helicopter to Ozero. Although I have a military pilot, we'll be using the commercial version of the chopper, so it will be a short comfortable ride. Shall we make our way?"

I looked at Chantal, and she looked at me. We were in this now and needed to follow it through. It was like Charlie/Tomales was staging another Watcher Intervention, but for reasons that we didn't understand.

KA-62

Charlie leads us to quieter part of the airport and onward to Terminal A. I can see several monied-looking Russians standing around chatting. It's all shiny marble floors, plate glass and tasteful honey-coloured wood. It looks as if it has all been recently refurbished.

"The money buys 'hush' around here," explains Charlie, "In more senses than one."

A regulation-looking pilot strides towards Charlie and greets her.

"This is Antanov," she explains, introducing each of us to him.

He smiles back, "Great, let's move straight through to the 'plane," he says, "It's a Kamov Ka-62, which is the civilian version of a Russian fighter helicopter. I find it easier to travel without the armaments unless absolutely necessary."

We walk outside of the terminal and on to the apron, where a few helicopters have been parked. He takes us to a shiny red one, with black picked out around the windows. It looks quite luxurious from outside.

"Look, over there," he says, pointing to a drab military helicopter in green camouflage colours. It has a prominent red star and seems to have less windows than ours.

"That's a Ka-60 - the military version of our helicopter. Designed for menace rather than comfort."

Inside, the helicopter reminds me of something that would be designed by a teenage computer-gamer. It has what looks like black and grey rally-cross seats with red seatbelts that would not look out of place on an ejector seat.

"I'm told that it is the glamorous hosts that sell these helicopters, at special air shows," explained Antanov, "They are a snip at $9 million."

As we climb in, I can see the huge flat panel displays which have replaced all the conventional dials, and two reassuringly vintage looking leather pilot's chairs.

"How come you don't get the snazzy seats?" I ask Antanov.

He smiles as he explains, "I'm a real pilot. I don't need all of the gimmicks when I can have comfort."

Then, someone else climbs into the copter. Slender, athletic and with her blonde hair cut into a blunt bob, she smiles as she approaches us.

"Oh, I should introduce you all. This is Christina, Christina Nott. She is another security agent and one I have known for a long time," says Antanov, "She'll be acting as our co-pilot this afternoon."

I notice that see is wearing a bright emerald green jacket and matching trousers, when Chantal pipes up, "Ooh, I love the Gucci!"

Christina smiles casually and say, "Why thank you, and I love your zebra print. Our greens complement one another!"

I see the brief flicker of pleasure that crosses Chantal's face and can sense a massive shopping spree in the near future.

Christina and Antanov take their places in the pilot seats and prepare for a take-off. "Flight time will be around ten minutes," announces Antanov, "And you might want to wear the headphones."

With that, he starts the engines which are surprisingly quiet, at least whilst we are still on the ground. Then, suddenly, the ground is far below us and we are manoeuvring for the flight to Charlie's new base camp, in the Ozero Community, to the west of a cloudy Moscow.

Despite the comparative quiet, I save my questions until we arrive at Ozero.

Ozero

Ozero from the air doesn't seem all that special. I can only see a river, cottages, the outskirts of forests and a small river running through the area.

Christina, on the headphones points out the village of Leshkovo in the Istrinsky district of Moscow Region.

She points to several perimeter fences and explains that it is here where splendid, opulent palaces of the common people stand.

She talks about one, as an example, a lot belonging to a 'simple miner' and Secretary of the General Council of the 'United Russia Party', Sergey Neverov.

Christina explains, "Neverov loves to portray himself as a blue-collar worker. He was never involved in business, and over the last few years was living on the taxpayers' dime in the civil service. He doesn't have a lot of income but somehow managed to acquire 'a cottage lot' worth 92 million roubles which he registered under his mother-in-law's name – a 75-year-old pensioner from Novokuznetsk.

"Neverov explains that he acquired it and will build all of it up on 'money gained from the sale of a 67 square metre two-bedroom apartment in Novokuznetsk'.

"Would a 67 square metre apartment in Novokuznetsk provision an enormous lot on the bank of the Moscow river and a home which, in all likelihood, is twice as big as the Central Pioneer Palace in Novokuznetsk?

"Instead, we deduce that Neverov is an abominable liar and bribe taker, just like everyone else in 'United Russia'. And his secret sources of income can be explained in just one word: 'corruption'."

We climb out of the helicopter and I notice we are on a building site.

Charlie explains, "This plot of land is not finished yet. When it is, another oligarch will be invited to bid for it. It gives us somewhere to be based for this operation. And don't worry, I have been given permission to use this until the plot is ready for handover. The main block is complete, although there is still much work to be completed on the connecting buildings and in some of the underground areas."

We walk toward the main building, which, to my eyes looks finished. When we are all in the building, I can see that just the front reception area has been completed, but that most of the building is ready for an interior designer's touch. The completed lobby interior reminds me of the airport terminal where we had picked up the helicopter and I deduce that this must be the latest fashion in Moscow.

There are several seating areas, a table, and a small catering area, with a microwave and a coffee maker stood on the worktop.

The rest of the building looks unfinished, with plasterboard and electric fitments visible.

We all sit around a large, flat airport lounge styled coffee table.

"Time for some explanations," says Charlie.

Paradise or scramjets?

We sit together around a low table in the lobby of this newly constructed Oligarch paradise. I guess the furnishings are all temporary, brought in by a design firm to create some ambiance for prospective purchasers.

Charlie begins, "I was contacted in Saint Petersburg by Christina. She asked me to meet her in a fancy restaurant. Said it was important."

Christina adds, "Yes, I arranged to meet at Gogol, which is a lovely Russian restaurant situated just off Nevsky Prospect."

I saw Antanov nod in appreciation of the choice. He walked to the coffee machine and was fiddling with it.

Charlie adds, "Gogol was so traditional and you had to ring a little bell to call the waitress! A first for me!"

Christina takes over the conversation, "Well, I didn't know anything about you, but my contact in Moscow had said you would be useful for this mission. My contact has a codename 'Blackbird' and Antanov and myself are both run by him."

"What, so you are both agents?" asks Chantal.

"Yes, we both have links to the FSB, which is like the renewal of the KGB," answers Antanov, "Although Christina bought her way out, some time ago."

Antanov carries a tray of five small coffee cups across to the coffee table, "Blackbird explained that Charlie had been sent to Russia and was an important asset, specialising in security. She was to be asked to assist us to penetrate Putin's Kremlin. Putin is up to no good and will need to be disabled."

I ask, "But what about Blackbird? And both of you, come to that. Won't you be seen as traitors to Russia if you start to mess with internal politics?"

Christina sips the coffee, "That's why we asked Charlie to intervene. She was talent scouted by one of Blackbird's agents. And by the same token, you both as well; Scrive and Chantal. All three of you have no trail. There's no record of you in any of our systems, yet you all seem to possess significant skills. You are all clean."

Charlie speaks, "Well when I got here, I decided that I would stage an Intervention. Scrive and Chantal know what I mean."

"Intervention?" asks Christina, "Are you all addicts or something?"

"No, it is hard to explain," says Chantal. I notice what I take to be an overhead light flickering a couple of times then she says, "We are all Watchers. Not from this place, but of this place."

"Sounds kind of unusual," says Christina, looking at us all carefully, "Although I noticed that there is something kind of glitchy about your appearances. You sometimes seem to flicker, as if you are not really here. And you need to tell us, where have you come from?"

Charlie continues, "Well observed. You won't believe our story, about how we were tracking the future and then each of us got zapped with a railgun which brought us back to this place. It cannot be co-incidence. It must be for a reason."

Christina looks puzzled, "But rail guns don't exist as 'zapping' devices. There's only a few in existence around the world, and they are used to fire other projectiles."

"They become mainstream in the mid 21st Century," says Chantal, "After the success of the MARAUDER Project - That's Magnetically Accelerated Ring to Achieve Ultra-high Directed Energy and Radiation - which is a US program and probably still classified. It's the Chinese that perfect the plasma technology, initially for use at sea on the Haiyangshan, but then into space. A 'bracelet of charms', it will be called and it will be used to police newly-defined off-grid areas of the planet Earth."

"We can check with Blackbird about the Marauder project," says Christina.

"Check also on Tsircon, then," says Charlie, "Its a scramjet powered manoeuvring anti-ship hypersonic cruise missile, designated 3M22, currently in production by Russia. It is still a state secret though."

Christina nods, "I worked and lived in the Arkhangelsk region of Russia, near the White Sea. Nearby, in Nenoksa, was the test range for cruise and sea-based ballistic missiles. I can remember a failed launch of a scramjet sample; It was called the 3M22.

"I recollect there was land testing before it was to be placed on a sea-going platform, the fifth-generation Haski nuclear submarine designed by Malakhit naval machine-building bureau in St. Petersburg."

Chantel grimaced, "You wait until the Chinese do a deal with the Russians and plasma railguns take to the skies. We can't stop all of it, but we can deflect some of it, with Charlie's Intervention."

Christina looks at us again, "You'll need to convince us. Despite being recommended by Blackbird, you could still be Kremlin plants, sent to infiltrate whatever Blackbird is trying to achieve."

I suddenly remembered Chantal when I'd first seen her back in the hotel. I remembered the burn marks on her skin. Now wasn't the most obvious time to rip off my shirt, but I did so anyway, to the astonishment of everyone else.

"Here," I said, "I've got a Trigax burn; three streams of plasma which were fired directly at me from an orbiting railgun. It was around 40 years into the future and I was projected back to now. I suggest that maybe Christina does a quick check of Limantour, sorry Chantal, and Charlie. I know Chantal has the burns and I expect Charlie does too.

Charlie nods. Chantal says, "I'm not going to destroy my lovely green zebra suit though."

Then there's an awkward moment, which culminates in Christina marching Chantal and Charlie to the nearby restroom.

"That's some powerful weapon," says Antanov to me. He pauses and then adds, "You seem to be genuine to me, Scrive, or whatever your name is."

We both laugh at the improbability of the situation and I can see that Antanov is trying to process the information.

"So, this could still be part of an elaborate back-story? A complex 'legend' made up to convince us that you are who you say you are."

I decide to tell Antanov more of our story. That we three, plus another one, are entities from outside of his universe. That we've been placed on earth and into human form via a series of events.

The three ladies return and I can see Antanov looking toward Christina.

"Yes, they both have similar marks," Christina says.

"So, they could be genuine...or maybe you have all escaped from somewhere?" asks Antanov cheerily, "Maybe that choice of restaurant by Christina wasn't such a random selection. Gogol and all that."

I realise that Antanov is referencing Gogol, the Ukrainian godfather of Russian literature, but I'm not sure whether it is Diary of a Madman or Dead Souls to which he refers.

Awkward, either way.

"We can prove it," says Charlie, "That we have knowledge of the future; not betting odds or anything practical, but that we know, for example, that during a meeting of the UN Human Rights Council, over 100 diplomats will walk out in protest over a speech by Russian foreign minister Sergei Lavrov. The meeting is tomorrow and the scale of the walkout will be unprecedented."

I was thankful that Charlie took an interest in the affairs around the Klima Wars.

"Okay," says Christina, now smiling, "We'll watch the TV broadcast tomorrow. If it happens, you'll be a step closer to credibility."

Antanov looks towards me. He winks. I can sense that he and Christina are now beginning to believe our improbable story.

Hunkt down

"We need to move from here for now though, " says Charlie, "We can go into town,"

"Is that the one I could see from the air?" I ask "Leshkovo?"

"Not quite, we'll be going to Pavlovskaya Sloboda," says Charlie, "I think we'll be off the grid there, whereas in this building there are all kinds of extra wires."

She gestured to us and we all followed her out through a different exit realising the building was large enough to be a corporate headquarters.

"I've found a small place in Pavlovskaya Sloboda, where we should be able to talk without being listened to. I hope you all like Georgian food."

We walked toward a dark silver Range Rover, and only as we climbed in did I realise we were in some kind of Chinese clone car.

"It's a Hunkt Canticie," explained Antanov, "The Chinese have copied the Range Rover Sport externally, but put a whole different interior in."

I looked at the interior, which was all LED displays and reminded my Scrive-brain of a Mercedes.

Charlie is driving as Antanov adds, "It's also all-electric, and only costs about one tenth the amount of a Range Rover equivalent. It's quite popular as a Russian convoy vehicle!"

"I don't remember any of this kind of cloning in the place from which we come," I say. Then I remember that the Chinese were copying the healthcare products that we used. The so-named tropus and the nano-machines. It was big business.

Charlie deftly manœuvres the large vehicle up a slope and around a few sharp bends. We have bright sunshine and I can see that the ground lacks vegetation, presumably from the heat.

"It must be why we were propelled to this time and place," Charlie says, "I mean, we had just staged an Intervention in the future to stop Earth from entering an end-state."

"End-state?" asks Antanov, "You mean like the end of the world?"

"Kinda," continues Charlie, "We could see that the world was running out of resources and that it had suffered from huge losses because of the so-named Klima Wars."

We had reached a main road and Charlie was driving us south-west. I noticed the sign for Павловская Слобода - Pavlovskaya Sloboda - and then a few minutes later we are in the sweeping curves of a quiet town. On both sides of the road are some impressive and bright pastel-coloured onion-towered buildings and I get a proper sense of being in rural Russia.

Charlie slows toward a large blocky-looking building with several rows of large lettering on the outside.

I'm expecting us to be going to somewhere modern and utilitarian and am quite surprised when we climb some stairs to enter a small cosy restaurant, with sweeping curtains, chandeliers and several luscious carpets with varied tables and chairs around them. It reminds me of an overly neat replica of the kind of place I would expect to see in Hoxton, London. I can see outside has a balcony and a view toward the dual carriageway, on which is another one of the onion-tower buildings, this time in a pretty green colour.

We are greeted by a friendly waitress and shown to a corner table with bench seating along one side and loose chairs on the other side.

Charlie announces, "Okay, we can talk here, without being electronically monitored, and I think you'll love the food. So where were we?"

Chantal moves a couple of the cushions and begins, "The thing that comes with any form of Intervention is the prospect of unintended consequences. Like when you buy a new hat but can't see so well because it has a wide brim and then bump into a handsome man on a bicycle."

"Okay, enough from your private life, Chantal," says Charlie.

"But if you are all super-gods, can't you fix all of this anyway?" asks Christina.

"That's just it. We are not. Not super gods. We are entities with long term existences, but with no control over anything. It's a big deal if we 'Intervene' and typically we'd only do this once through all of time," explained Charlie.

"You don't know about Presences and Personas yet, they have not been invented or discovered, but in the future, there will be engineering to produce human-form robotics which can be driven by humanoid Personas. In effect a Presence inhabited by a Persona."

"That's what each of us is emulating. We are three entities placed inside human forms and limited mainly to the sphere of actions of the human. We were each blasted by a Trigax railgun and sent back to this era. Before that we were the same three humans but living in an entirely different version of Earth."

"You have to admit that it sounds far-fetched?" says Antanov, "I mean, time travel, inhabiting other human forms, Trigax rail guns. It sounds like an episode of a Science fiction TV-show."

I realise this is going to be a long haul with the scepticism of Antanov and Christina. If they really are spies or secret agents or similar, then they would be trained to not believe anything they were told.

Christina makes the move, "Okay so we'll suspend disbelief. Tell us what you are attempting to do."

Charlie begins," Well, I was sent back to here and have been connected to the Russian state's security police. It's a massive advantage to you both as you've been tasked by your handler, Blackbird, to get to Putin. I can provide a way for you to get inside the citadel that Putin has constructed, through my links to its physical security management."

Christina adds, "I had other dealings with Putin, indirectly. A friend of mine, Irina, another agent, was linked with him and told me of his time in Dresden. He has a nickname acquired from Dresden when he first arrived in Sankt Petersburg. It was Volodya or 'little Vladimir'. If you look carefully even now, you'll see he wears shoes with 4 cm lifts in them.

"Does that explain why he walks oddly, hardly moving his right arm?" asks Chantal, "I noticed on the TV channels, with so much coverage of him over the last few days."

Antanov and Christina look at one another, then Christina speaks, "No, that's Alpha Team training, " she explains, "When you are trained to a high level in the Russian military, you are taught to adopt 'gunslinger's gait', it's a way to walk that keeps your gun arm close to your side. Watch a group of Russian businessmen and you'll see several of them adopt the same military style. Sometimes it is truly ingrained and other times they are just copying Putin."

"Not a good look, though," says Chantal, "But sorry, Christina, I interrupted you."

Christina continues, "My friend Irina explained that Putin was friendly with the Dresden Stasi and that the Stasi's lieutenant-colonel knew everyone in town. He oversaw organising safe houses and secret apartments for agents and informants, and for procuring goods for the Soviet 'friends'. Putin learned some Stasi tradecraft which he used to such devastating effect in Saint

Petersburg."

"Is that how Saint Petersburg became the home of so much illicit trade?" asked Chantal.

"That's right," answered Christina, "The KGB was recruiting agents in companies like Siemens, Bayer, Messerschmidt and Thyssen. Putin was involved, initially through the Dresden Stasi, but increasingly in his own right. There was a firm called VEB Robotron, which made computers for East Germany right in Dresden. They had to get the plans and parts from the West and lured western businessmen to their offices and beyond."

"Early sleaze and smuggling?" asked Chantal.

"You could say that. Irina was on duty at business meetings in Sankt Petersburg which comprised about ten minutes of business talk followed by a half an hour of discussion of the evening's entertainment. Sometimes Putin and his friends would all drink too much vodka and then reminisce about the old days in Dresden. According to Irina, Putin himself would usually go easy, but everyone else... well. That's when I heard about the blueprint for a lot of Putin's money moves. Irina explained it was a scheme set up by the East German foreign trade ministry. They created the Kommerzielle Koordinierung, with a mission to earn illicit hard currency through smuggling and to bankroll the Stasi acquisition of embargoed technology."

Charlie says, "I've heard of the KoKo, didn't it set up a string of front companies all across Germany, Austria, Switzerland and Liechtenstein?"

"Yes, KoKo answered to the Stasi espionage department

and ran those companies with trusted agents. Some of them had multiple identities, who brought in hard currency through smuggling deals and the sale of illicit arms to the Middle East and Africa."

"I know the feeling with the multiple identities!" smiled Chantal.

"Yes, me too. But back in those days, the Iron Curtain meant that smuggling became the only way for the eastern bloc to keep up with the rapidly developing achievements of the capitalist West," answered Christina.

"And presumably access to a sea-port like Saint Petersburg meant that Putin could develop similar schemes only much larger in scale?" asked Chantal.

"And then some," says Antanov, "Three men held an intimate knowledge of the secret financing systems of the Communist Party at the time the KGB was preparing for the transition to a market economy under Gorbachev's perestroika reforms.

"The innocent sounding Property Department was run by Nikolay Kruchina and Georgy Pavlov. It was thought to have a value of $9 billion although Western experts estimated its foreign holdings at many times more.

"That was because of all the slush funds and the way that money could be marked up and marked down?" asks Chantal

Antanov nods, "But in the first few days after the Communist Party's collapse, Russia's new rulers were surprised to discover that the Party's coffers were nearly empty."

"The money had been shifted offshore. Just like the moves of the oligarchs today. Get caught holding cash and someone will be after it," says Christina.

"Precisely - that's why so many Russians buy London properties and other illiquid assets. They are much harder to 'cash in'," answers Antanov, "At the time the rumours were that Nikolay Kruchina had worked with officials to siphon billions of roubles and other currencies through foreign joint ventures hastily set up in the final years of the regime. These were the filter companies set up in Germany and other countries. It's a move that Putin understood back then but is also using right now."

Christina interrupts, "And Putin used the same moves on his rise to the top. He played it out first in Saint Petersburg, but then carried the book of tricks back to Moscow."

Antanov continues, "Now there were Russian prosecutors investigating what had happened to the Party funds - valued at hundreds of millions of dollars. The institutions Kruchina served were being dismantled in front of his eyes."

"Yeltsin had moved in to break things up?" asked Chantal.

Antanov nods, "Yes, the pro- democratic Russian leader Boris Yeltsin signed a decree, broadcast live, suspending the Soviet Communist Party and ending its decades of rule. It was a big deal. Yeltsin's defiant stance against the hard-line leaders of the attempted coup had put him firmly in the ascendant. "

"Yes, and Yeltsin was going to kick over some tables," says Christina.

"That's right," said Antanov, "Boris Yeltsin by far eclipsed Gorbachev, who timidly watched as Yeltsin addressed the Russian parliament. Arguing that the Communist Party was to blame for the illegal coup, Yeltsin ordered that the sprawling, warren of the Party's Central Headquarters on Moscow's Old Square be sealed."

Christina interrupts, "Putin had been a couple of steps ahead of this. He'd kept all his papers in Dresden and then burned them, along with the papers of others who were later to be his allies."

Antanov says, "That's right, although most of it is impossible to prove, except what Putin boasted on the record about - you know - the furnace exploding from the intensity of the flames."

He continues, "But in Old Square's hundreds of rooms there were filed the secrets of the Soviet Union's vast financial empire, a network that spanned thousands of administrative buildings, hotels, dachas and sanatoriums, as well as the Party's hard-currency bank accounts and untold hundreds, perhaps thousands, of foreign firms set up as joint ventures in the dying days of the regime.

"Through these bank accounts and other connected firms, the strategic operations of the Communist Party abroad – and those of allied political parties had been funded.

"It was the engine room of the Soviet struggle for supremacy against the West. This was the empire that Kruchina had administered as the chief of the Communist Party's property department since 1983. Its sudden sealing felt like a symbol of all that was lost. "

Christina asked, "You know what is coming, don't you?"

"A mysterious disappearance or death of people who know too much?" asked Chantal.

Antanov continues, "Yes, exactly; you know the old KGB moves! Kruchina must have known that his days were numbered. He went back to his flat in the closely guarded compound for the Party elite. Kruchina's wife went to bed leaving her troubled husband alone to sleep on the couch.

"Next morning, she was awoken by a knock on her door. It was the KGB security man for the building. Her husband, she was told, had fallen to his death from the window of their seventh- floor flat."

"Fell or was pushed?" asked Chantal.

Antanov continues, "Yes, you are ahead of us, Chantal. The security man said he'd discovered a crumpled note lying on the pavement next to Kruchina's body. 'I'm not a conspirator,' it said. 'But I'm a coward. Please tell the Soviet people this.' "

"Hmm - a little staged? Why on the pavement? It doesn't ring true." says Chantal.

"Well, the KGB immediately declared his death a suicide. But to this day, no one knows what exactly happened – or if they do, they are not willing to tell," answered Antanov.

"You said three people...Three deaths?" asked Chantal.

"Yes, the same thing happened to Kruchina's predecessor

from the property department. Georgy Pavlov also fell to his death from the window of his flat. It was also recorded as a suicide.

"The same pattern?" asked Chantal.

Antanov continues, "Yes, and then, days after Pavlov's death, another high- ranking member of the Party's financial machine fell to his death from his balcony. This time it was the American Section chief of the Communist Party's international department, Dmitry Lissovolik. Again, it was recorded as a suicide.

Christina adds, "These three all knew about the money – but no-one else knew where it had been hidden. And 'someone' was learning how to cover tracks."

"So somewhere there's a lot of hidden Soviet money in play?" asks Chantal

Buddy house

"I can see where this is leading, " says Chantel, "We're going to infiltrate the oligarch circle to gain access to Putin's inner circle. And Charlie has a property right next door to one of the oligarchs."

"So right now, we need to go to the home of one of Putin's buddies?" asked Antanov, "Isn't that incredibly dangerous?"

"Well, we have several to choose from," says Charlie, "All in a line along the road where we are based. I'm inclined to make it our next-door neighbour. His name is Svalov Rollan Vitalievich and he is a mere civil servant from the construction industry.

"Vitalievich could be said to have his fat fingers in many pies and used this to successfully overcome personal budget crises, as evidenced from the size of his cottage.

"His adjacent plot and home combined are worth approximately 178 million roubles which is around US$5.4 billion. Not bad for a civil servant.

Charlie continues, "I've researched him and Vitalievich

has been head of the Federal State Unitary Enterprise 'Spetsstroyinzhiniring' under the Federal Agency of Special Construction (SpetsStroy Russia). Before that he was head of the Federal State Unitary Enterprise 'Head Department for Road and Airfield Construction of the Federal Agency of Special Construction'.

"That's a wide range of powerful roles," said Antanov.

Charlie says, "Hmm, but even more so when you add on that he is Minister of Regional Development for the Russian Federation and Governor of Krasnoyarsk Krai, as well as a member of the United Russia Party. He helped mastermind the Asia-Pacific Economic Co-operation in Vladivostok."

"Oh yes," says Antanov, "One of the Putin government's most cynical hallmarks is to line its pockets under the cover of increasing Russia's international prestige. Cynically, they, don't even bother to cover it up. Thus, practically every ordinary person you meet will tell you that the Sochi Olympics, the Soccer World Cup, and the APEC Summit served only one purpose: to expropriate and launder state funds."

Antanov continues, "As a mere drop in the ocean, I remember when the government made available an extra $11.2 million to buy 120 brand new luxury BMWs to transport APEC summit guests. The request had previously been rejected by the treasury on the grounds that there are already plenty of limos to ferry visiting dignitaries: the office of the President has over 100 BMW 7 Series, almost 200 5 Series and over 300 luxury Ford Mondeos in its car park. Yet mysteriously the original refusal was somehow quietly overturned. One source cited the need to honour an unspoken rule that visiting dignitaries must travel in brand new cars."

Christina adds, "The more important rule is that once the summit is over, 120 lucky bureaucrats were rewarded for their loyalty with some very nice cars."

Christina added, "There's clearly no need to fish for the obvious parallels between the length of the construction projects overseen by United Russia party member Vitalievich and the size of his estate."

"So, he's the one, then," says Charlie, "Our new friend."

False Flags

We had all ordered from the Georgian Menu. Now an assortment of Georgian foods arrived, ideal for us to share around. Khinkali - A twisted topped Georgian mega-dumpling, Shashlik - skewered cubes of lamb, Khachapuri - an oval of a dough similar to pizza, with cheese and a whole egg served on top. and small bowls of Kharcho soup of beef stock seasoned with garlic, suneli, pepper, chilli, cinnamon, and ajika.

"Charlie, this is so much better than making something in that fancy lobby with the microwave!" says Chantal.

"Yes, it's delicious and very authentic," says Christina, "In Georgia, herbs and crops are cultivated in mountains and valleys, while cattle and sheep are put to pasture. People of Georgia mostly combine these two food groups in their food. Of course, we should really have it with Usakhelauri wine - it's a sweet wine, harvested after the first frost."

Antanov gestured and a bottle of the wine was forthcoming from the waitress.

Charlie refused the wine but continued the

conversation, "We should take a closer look at what is happening today,"

"It seems as if misinformation is becoming the norm," observes Antanov, "My serious concern is that Putin could use chemical weapons on Kyiv as Russian propagandists spread what the US White House press secretary, Jen Psaki has called 'false claims about alleged US biological weapons labs and chemical weapons development in Ukraine'. The Kremlin has produced no evidence to support its Ukranian weapons lab claims, which were called 'preposterous' by Psaki and have been dismissed by Ukraine's government."

"And right now," begins Charlie, "Ukraine's president, Volodymyr Zelensky, has called a Russian strike on a maternity hospital in Mariupol 'the ultimate evidence of genocide'. Zelensky said children were buried under rubble and the regional governor said 17 people were wounded when the hospital was destroyed by a Russian airstrike."

So, things are speeding up," observes Antanov to Charlie, "I guess that makes your initiative more urgent,"

"Our initiative," says Charlie, "You are on the inside now."

We chinked wine glasses and said, "Nah zdorov'ye!"

Chantel continues the conversation, "When I've looked at this time span as a Watcher, it is so difficult to see the nuances. I mean, every report has been manipulated, and right now I can see the Russian news suppression first hand."

Antanov asked, "But surely, if you are really Watchers,

then you would be wise to all of this, not just seeing things through the history books of victors?"

"Yes," says Christina, "You watch a Russian news program and it won't talk about a war. It won't show the atrocities of the sustained bombing of civilians. Most Russians won't even believe it when they hear directly from their children phoning them from Ukraine."

"We really take your point, but I'm not sure we are ever normally as close to a conflict as on this occasion," I answer.

Antanov looks at me, "Normal Russian civilians don't know that there has been a sustained exodus of millions of refugees fleeing Ukraine. Some 2.5 million people have already fled, according to the United Nations, which calls it Europe's fastest-growing refugee crisis since 1945. More than half are now in Poland but tens of thousands are also staying in Moldova and Bulgaria, which have some of the fastest shrinking populations."

Christina continues, "It's like the Syrian civil war in that it has the possibility to continue for years. In that war, international organisations have accused virtually all sides involved, including the Ba'athist Syrian government, ISIL, opposition rebel groups, Russia, Turkey, and the U.S.-led coalition. All of them accused of severe human rights violations and massacres.

Antanov speaks, "The Syrian conflict caused a major refugee crisis, with millions fleeing to neighbouring Turkey, Lebanon and Jordan. Even with the number of peace initiatives launched, the fighting continued. And it suffers from war fatigue now, with few reports because of its length and the selective blindness that the world suffers for certain trouble spots and ethnicities."

Charlie adds, "They say something about history repeating itself, but I'd add a filter of ethnicity and culturalism. Like the British conveniently burying most of the slave trade, except for William Wilberforce, the abolitionist.

Christina adds, "I know, even the inventor of Newspeak, George Orwell, once likened Britain to a wealthy family that maintains a guilty silence about the sources of its wealth.

Charlie agrees, "Orwell, the author of 1984, whose real name was Eric Blair, had seen the conspiracy of silence at close quarters. His father, Richard W Blair, was a civil servant who oversaw the production of opium on plantations near the Indian-Nepalese border for export to China.

She continued, "The department for which the elder Blair worked was called the opium department. The Blair family fortune – which had been lost by the time Eric was born – stemmed from their investments in plantations far from India. Wilberforce's Slavery Abolition Act freed 800,000 Africans who were then the legal property of Britain's slave owners.

Christina adds, "What is less well known is that the same act contained a provision for the financial compensation of the slave owners through a £20 million government pay-off. That's £17 billion in today's money. The largest bailout in British history until the banks in 2008."

News Management from Russia, echoing the Newspeak of Orwell's 1984, plus financial compensation to assist cover selective memories, of events quietly erased from history.

Georgian food and wine

We'd eaten the feast of Georgian food and were all looking slightly relaxed around the table. I wondered if it was the Usakhelauri wine.

"And so, to the next event which I will be managing," explains Charlie, "Scrive, it will be great to have you along, to help Antanov and I sniff out any trouble."

"And me?" asked Chantal.

"Oh, you'll add your stylish and disarming presence," smiled Charlie.

"What's the reason for a meeting?" I ask.

Charlie answers, "Vitalievich has called a meeting of his fellow oligarchs. Tellingly, in this group of Putin's followers there's no one directly linked to Putin's inner circle. This group are all worried about the extreme drop in value of their wealth since Putin started his war against Ukraine."

Antanov comments, "Yes then rouble is 20% lower than before Putin started his military action and Gazprom shares have gone from US$8.50 to about US$0.58 since the start of February. No wonder the oligarchs want to talk."

Christina adds, "But I guess that most of these people have diversified their holdings? It is probably spread around property portfolios and multiple currencies?"

"But do we know who the real inner circle comprises?" I ask.

Charlie answers, "Well, we shouldn't underestimate any of the people that Vitalievich has asked, but they are nowadays all a step away from the true inner circle. The inner guard are the people best able to advise Putin whilst he is on a war footing. They are the ones we ultimately need to get to. But we need to use this outer group to get to the inner ones."

Wheels within wheels.

Christina starts, "I think Antonov and I know more about the inner circle. Vladimir Putin cuts a solitary figure, leading Russia's military into a high-risk war that threatens to tear apart his country's economy. Russians are not allowed to call it a war though - it's a 'Special Military Operation'. Putin has rarely looked more isolated than in two recent, choreographed appearances with his inner circle, where he sits at a resolute and some would say ridiculous distance from his closest advisers."

Christina adds, "There's his long-time confidant Sergei Shoigu, who has parroted the Putin line of demilitarising Ukraine and protecting Russia from the West's so-called military threat. This is a man who goes on hunting and fishing trips with the president to Siberia, and he has in the past been viewed as a potential successor.

She continues, "Shoigu has a previous track record. He was labelled with the military seizure of Crimea in 2014. He was also in charge of the GRU military intelligence agency, accused of two nerve agent poisonings - the deadly 2018 attack in Salisbury in the UK and the near-fatal attack on opposition leader Alexei Navalny in Siberia in 2020. One could say he is a nasty piece of work.

"Shoigu is not only in charge of the military, but also partly in charge of ideology - and in Russia ideology is mostly about history and he's in control of the narrative."

Then Antanov speaks, "Valery Gerasimov is Chief of General Staff of the Russian Armed Forces. As chief of staff, it is his job to invade Ukraine and complete the job fast, and by that standard he has been found wanting.

"He has played a major role in Vladimir Putin's military campaigns ever since he commanded an army in the Chechen War of 1999, and he was at the forefront of military planning for Ukraine too, overseeing military drills in Belarus last month.

Christina adds, "He is the one who describes hacking the enemy's society, rather than direct head-on attack."

Antonov grimaces, "That's not the way it is running in Ukraine, though. Gerasimov is described as a "humourless bruiser" and played a key role in the military campaign to annexe Crimea."

Christina adds, "Some reports suggest Gerasimov has now been sidelined because of the stuttering start to the invasion of Ukraine and reports of poor morale among the troops because of the under-provided logistics, like no fuel for the tanks as well as food shortages."

Antanov continues, "Nikolai Patrushev, Secretary of the Security council, is the most hawkish hawk, thinking the West has been out to get Russia for years. He is one of three Putin loyalists who have served with him ever since the 1970s in St Petersburg, when Russia's second city was still known as Leningrad. Few hold as much influence over the president as Nikolai Patrushev. Not only did he work with Putin in the old KGB during the communist era, but he also replaced Putin as head of its successor organisation, the FSB, from 1999 to 2008."

Christina adds, "It was during a bizarre meeting of Russia's security council, three days before the Ukraine invasion, that Mr Patrushev pushed his view that the US's concrete goal was the break-up of Russia."

The session was an example of FSB theatre, showing the president holding court behind a desk as each of his security team walked up to a lectern and expressed their opinion on recognising the independence of Russian-backed rebels in Ukraine. It had clearly been edited before it was broadcast, to remove most, but not all, of the mistakes.

Patrushev passed the test. Antonov explains, "He's the one who has the chief battle cry, and there's a sense in which Putin has moved towards his more extreme position."

Antanov adds, "The other two stalwarts are security service chief Alexander Bortnikov (Director of the Federal Security Service - FSB and Director of the Foreign Intelligence Service (SVR) Sergei Naryshkin."

Christina speaks, "All the president's inner circle are known as siloviki, or 'enforcers', but this trio of

Patrushev, Bortnikov and Naryshkin are closer still.

"Kremlin watchers say the president trusts information he receives from the security services more than any other source, and Alexander Bortnikov is seen as being part of the Putin inner sanctum. Bortnikov is another old hand from the Leningrad KGB and took over the leadership of its replacement FSB when Nikolai Patrushev moved on.

Antanov add, "Completing the trio of old Leningrad spooks, Sergei Naryshkin has remained alongside the president for much of his career. What, then, should we make of a remarkable dressing down he was subjected to when he went off-message during the security council meeting? When asked for his assessment of the situation, the intelligence chief became flustered and fluffed his lines, only to be told by the president: 'That's not what we're discussing.' The lengthy session was edited so the Kremlin had clearly decided to show his discomfort in front of a big television audience.

"It was shocking. He's incredibly cool and collected so people will have asked what's going on here although Putin loves playing games with his inner circle, making Naryshkin look a fool."

Christina explained, "Sergei Naryshkin has long shadowed Putin, in St Petersburg in the 1990s, then in Putin's office in 2004 and eventually becoming speaker of parliament. He is also a historian and provides the president with ideological grounds for his actions. He also gets sent out by Putin to issue denials about poisonings, cyber-attacks, and election interference.

Antanov's turn: "Sergei Lavrov, 71, is yet more proof that Vladimir Putin heavily relies on figures from his past.

For 18 years he has been Russia's most senior diplomat, presenting Russia's case to the world even if he is not considered to have a big role in decision-making. He is a wily operator who attempted to ridicule British Foreign Secretary Liz Truss over Russian geography and the year before sought to humiliate the EU's foreign policy chief.

"But he has long been sidelined on anything to do with Ukraine and, despite his gruff and hostile reputation, he advocated further diplomatic talks on Ukraine and the Russian president chose to ignore him.

Christina adds, "Then we get to Valentina Matviyenko. She is another Putin loyalist from St Petersburg who helped steer through the annexation of Crimea in 2014 as well. A rare female face in the Putin entourage, she oversaw the upper house's vote to rubber stamp the deployment of Russian forces abroad, paving the way for invasion.

"But she is not considered to be a primary decision-maker. That said, few people can say with complete certainty who is calling the shots and taking the big decisions. Just like every other member of Russia's security council, her role was to give an impression of a collective discussion when it is more than likely the Russian leader had already made up his mind."

Wealth erosion

We'd all enjoyed the Georgian food, now the restaurant hosts returned with a bottle of almost clear liquid. I sensed immediately that this could be dangerous.

Christina and Antanov laugh, "Oh yes, you must try this. Chacha - wine brandy, or sometimes called vodka brandy – It's made with the remains of the grapes after the wine has been extracted."

"They say you should only drink it when you are already drunk," explains Antonov, "And unsurprisingly it often gets sold in plastic bottles."

"But this is the good stuff," explains Christina, "Askaneli Premium in a glass bottle. Only about 45% alcohol."

"Maybe I'll pass," says Chantal.

"And me," says Charlie, "I'm driving."

"The rest of you should let a sip of this devil's liquid cross your lips," says Christina, "to protect from evil."

The waiter had already poured a small quantity into an

array of shot glasses.

All of us except Charlie tried it and I had that little stab in my head which was my first warning that things could get messy. It reminded me of Italian grappa. Not normally something I'd swig in large quantities.

Fortunately, Charlie starts talking again, "I've been asked to run the security, along with a few from the GRU and, as of now, with my own associates."

I saw Christina look at Antanov. I wonder if they were both concerned that they would be recognised.

"I know what you are thinking," says Christina to me, "but we've had our great co-incidence with Antanov and me working together again. I don't think we'll run into anyone else from our days at FSB and GRU."

"Well, there is one other big co-incidence," says Antanov, "I knew the police investigator who investigated the Ozero Dacha Co-operative. He worked for Sankt Petersburg's major case squad reporting to a round table of European prosecutors in Sofia.

"This cop knew everything about Putin, all the nitty-gritty, all the cases from the early 1990s, the ties with organised crime, the origins of the Ozero Dacha Co-op, and how nearly its entire membership later migrated to the board of Rossiya Bank.

Antanov continues, "The cop ran the investigation of the Twentieth Trust Corporation, unofficially dubbed the Putin case. As an example, Putin was investigated for a deal he oversaw while he was an official in the mayor's office. The deal involved the export of $100m worth of raw materials in exchange for food for the citizens of St

Petersburg. It was all well documented at the time. The materials were exported, but the food never arrived. Of course, the case was shut down when Putin became President, and the investigator was put out to pasture. Let's just say that Putin seemed to acquire a pretty Spanish villa around that time."

Christina adds, "It is all a criminal conspiracy of greedy, unprincipled liars who will not baulk at any trick or power play to increase their dominion and grab more money, land, oil companies, yachts, real estate, and other goodies. It is the same thing with Russia's completely non-existent 'senate.' Russia's upper house of parliament is called the Federation Council, and its members are rubber stampers, not 'senators,' but that was what they took to calling themselves a few years ago, and so nowadays almost everyone calls them that, too.

She continued, "But they are not senators, if only because there is no senate in Russia. More to the point, Russia's un-senators are well-connected, highly paid sock puppets of Putin's Kremlin.

"Likewise, a perpetual, self-replicating mafia dictatorship runs with sovereign wastefulness, major league legal anarchy, hyper-corruption, and sheer absurdity. The Putin regime is an improvised exercise in radical governance by former KGB officers and their gangster friends.

Antanov speaks again, "Putin's public support of democracy was a put-on when he worked as Petersburg Mayor Anatoly Sobchak's deputy in the early 1990s. Then he was the Saint Petersburg bag man. Nowadays, he has moved up in the world considerably, but basically not changed his profession."

Sam's in the car park

We all return to Charlie's hideaway, with Charlie driving us there in the Range-Rover clone. Charlie is first on the phone and then prepares to visit the adjacent property where Vitalievich lives.

"Can you come along?" she asks Antanov. It will be good to have someone clearly from 'inside the business' who can help me scope the place. We'll be best to save Scrive for another day."

"Sure, I'll ride shotgun," says Antanov, smiling as Charlie prepares for the short drive to the adjacent property.

"Are you taking a weapon?" asks Christina, "Only they will be bound to search you."

"Yes, I'll take a polymer pistol, something like a Zastava PPZ," said Antanov. It will make them think but illustrate that my weaponry is Russian influenced."

"Serbian," said Christina, "just like the M70 Kalashnikov."

"Okay, don't worry about us," said Charlie, "I've been highly recommended to Vitalievich. Our aim is to be along for the big talks between the Ozero community."

I watched Charlie and Antanov stroll out of the exit, on their way to setting up what I was starting to regard as our infiltration of the outer circle.

**

Two days later and the stage is set for Vitalievich's meeting with the rest of the oligarchs. Charlie has been briefed by Vitalievich and his on-site security team. It was a strange setup. Vitalievich was wary of infiltration, but also of incursions from a Russian mafia gang and even from Putin's own inner circle.

Now we had the oligarchs from the rest of the Ozero Community and a few of their trusted friends together in a room. Charlie was running the front security and Vitalievich's own people were acting more of less as simple bodyguards.

It was fascinating to watch. Charlie had managed to put the fear of an attempt on any or all of them into their minds and the team was looking very nervous.

I knew that Charlie had created elaborate ring defences around Vitalievich's property and that it included a Patriot MIM-104 SAM Surface-to-Air missile launcher with had been parked in our own adjacent property.

"Where did you get that?" I ask Antanov.

"It's amazing how much of this equipment is available on the black market," he answered, "Think about it, these were designed in 1969, and are still in production today. Raytheon has manufactured over 10,000 of the missiles in that time, and that excludes any clones that have been created. They are so much easier to use than the Russian S-400 because they don't need to be networked to an AWAC plane to work. These Patriots can be driven into position, parked and primed. It is very useful if we think there is anything incoming hostile."

I was still intrigued where Antanov was getting these expensive pieces of kit. First a helicopter and now a truckload of surface-to-air missiles. Blackbird must be bankrolling him. I noticed him adding in coordinates too, of each of the Ozero Community dachas. I wondered why he'd do this unless he was saving the coordinates for a rainy day.

The trusted people Charlie wanted in the meeting room were herself, me, and Christina. Chantal was to run comms from our building next door and Antanov was to be outside close to the SAM launcher and a small team which operated it. These extra personnel arrived in a Bumerang armoured personnel carrier, which Christina had marvelled was the latest state of the art vehicle of the Russian Army.

Christina designated a sheltered spot outside Vitalievich's building as a pick-up point if there was trouble. Antanov was to use the APC as an unlikely getaway vehicle.

Antanov wasn't using Budget-Rent-a-Car to get this equipment, but it still seemed to come easily to him. I wondered how threatened his connections must be to resort to this level of bankrolling.

Vitalievich's own security detail were to stay around him but seem to only have handguns for protection and rudimentary airline-level security systems. They were blissfully unaware of the high technology equipment in the next property.

The first of the visitors arrive and Vitalievich's security detail run entry scanners to check that everyone was unarmed. Every major attendee seems to bring around four additional people in their entourage.

Soon we had a roomful of people comprising the United Russia Party friends of Svalov Rollan Vitalievich.

Notables include Nikolay Ivanovich Ashlapov, Igor Nikolayevich Rudensky, Sergey Eduardovich Prikhodko and Vyacheslav Viktorovich Volodin. Additional attendees include Nikolay Sergeevich Shustenko, President of the Bazis group of companies, Miller MacMillan, President of Raven Holdings and Tsarsko Adam Borisovich from Mirovyye Aktivy.

Altogether, we have approximately 50 people in the meeting room, with 8 main contacts and around 30 assistants, plus our own security detail.

If anyone had wanted to neutralise this gathering, then they would create a large hole in the United Russia Party. I could see how Charlie was earning the big bucks and could source a SAM system to provide protection.

Then it is time for Vitalievich to start the meeting.

Out of his mind

We arranged the meeting room to have an inner table where the eight main guests could sit, with one assistant each to take notes. Alongside Vitalievich sat a couple of other presenters. Behind them all were seated the other folk, who we assumed were the protection and heavies.

It did make the room look kind of comical, with the second row filled with mainly bruisers in their leather jackets, with the glitter of heavy chain jewellery.

"Straight from central casting," whispered Chantal, into my earpiece. She had watched from outside as the various people arrived, in mainly large grey, black, or silver SUVs. I noticed that several the vehicles had blue lights mounted on them.

Christina had taken a control position in one corner of the room, and I instinctively knew where I should take up position, guided by Scrive's instincts.

That left Charlie, in a blue and grey combat outfit, to roam free which attracted stares from many of the security row. She was dressed in an ironic choice of colours, echoing NATO uniform, which Putin was anxious to keep out of Ukraine and away from the Russian borders.

Vitalievich opens the meeting and looks over to his small group of other presenters. He asks Dr Maria Gvasalia, a psychologist from M.V. Lomonosov Moscow State University, to step forward.

Vitalievich explains, "Dr Gvasalia is a Psychological Practitioner who has a background in political psychology and has done extensive work on psychology and political identities including that of President Putin. She speaks as news that Russian forces bombed a maternity hospital in Mariupol in Ukraine, killing three people - including a child - and injuring at least 17 others."

This was a departure from the party line. To comment as if Russia was in a war was strictly forbidden in Russia and carried a 15-year prison sentence. I guess Vitalievich has done it to help Dr Gvasalia trust the rooms attendees.

I can see Christina and Charlie checking out the expressions on the faces of the leather men.

Christina breathes into my earpiece, "There's one guy, shoulder length dark hair, small silver lapel badge of a Russian flag, sitting behind Volodin, who looks startled by that last statement. He's also sweating and looks as if he might make a move."

Dr Maria Gvasalia pulled no punches, "President Vladimir Putin has been widely accused of war crimes after intense shelling of civilian areas since his military invasion of Ukraine has met with increasingly stiff resistance from Ukrainian forces."

"Putin is one who manifests what we psychologists would call the Dark Triad,"

I knew about the dark triad from Scrive's brain. It was standard training in his line of work.

Gvasalia started a slide show, in Russian. It said:

Vladimir Putin psychological profile: A psychopath who is likely to get emotional payoffs from Russia's bloodshed in Ukraine, plus, possible impact of his rumoured illness or cancer, including the potential side effects from overdosing on steroids.

Wow. That was quite a slide. I wondered whether Gvasalia had brought her own extra supply of bodyguards. She certainly would not be able to say this kind of thing in public, here in Russia.

Then she went on to explain: "We use discourse analysis to analyse his interviews, his discourse, and the rhetoric that he uses. It is because we are not in a clinical situation with him.

"What he does is demonstrate a high level of control and what is generally known as sociopathy or psychopathy. This psychopathic nature that he manifests is not a clinical term but an umbrella term that has been utilised to almost integrate several personality traits."

One of Vitalievich's invited guests stood.

"This is treason that we are all talking here. I am uncomfortable that if someone leaks this information, then we are all finished. What guarantees can you give us that we are safe to continue this meeting?"

Vitalievich answers, "It amounts to trust. You all know one another and have been to each other's parties and family occasions.

"Today we are building a new Bratva. A new fellowship among the eight. The Восьмерка, the Vos'merka. We have no choice in this matter. We know that Putin has been systematically replacing us with newer people. And that we hold in the palm of our hands a significant amount of the Russian economy. Dr Maria Gvasalia will explain to you all that Putin is ruthless. He has assembled us here in these wealthy compounds, not just as a place to show our riches, but as a wrapped prison compound too."

Maria Gvasalia continues, "The Dark Triad includes three traits: -

- Psychopathic traits: - 'The killer' who wants to control everything around them as belonging to them
- Dark narcissism: Someone who sees himself almost as the most powerful individual on earth
- Machiavellianism: People who are very focussed on the manipulation of people's emotions

"Putin demonstrates a lack of conscience. There is no sense of right and wrong. There is absolutely no remorse in his behaviour, it is about power and control - what is mine is mine and what is yours is mine also. Psychopaths have this overly possessive trait of power and - the psychology of evil - because the more blood they shed the more psychological and emotional payoffs they get. They thrive on murder, killings, and genocide. And this reinforces the behaviour to do more."

Dr Gvasalia adds: "Psychopaths are not necessarily killers - you can also see them in the boardroom, they are controlling and powerful people, who think, 'Look at me, I am more important than you.' In layman's terms you may see them in everyday life, often described as ruthless, people who just don't care about others and abuse others, but a key characteristic is also that they are charming. They will charm and groom and become political or sexual predators driven by evil desires."

She adds, "With Putin you are looking at the fragmentation of his childhood and his background in the KGB. There was a lot of consternation about who his biological father was. He was bullied an awful lot at school and had a low stature.

"Putin's father was a metal worker but he had an amazing respect for the old Soviet Union and communism and believed in the common people - this was Putin's political ideology and he felt this was going to be his life.

"Putin was a loner as a child and the way you survive when bullied for years is you become very aggressive and so there is impulsivity and aggression as a child. He wanted to have power and so he felt allegiance to the KGB and becoming a KGB officer would be a way to demonstrate his feelings for the Russian ideology of communism. Becoming a senior KGB officer would have given him a powerful nature of control. Here you can see his genetics very much interacting with a Soviet ideology and it was this that really began to anchor his identity as an aggressive leader who will fight to preserve Soviet ideology.

"He wants to go back to those days and bring Russia back into the old Soviet economy. However, he says opinion polls show this is not what increasing numbers of Russians want and he believes that younger Russians who want international acceptance will turn on him."

Dr Gvasalia looks around the room. I could see the inner eight nodding in agreement, but a more mixed reception from the leather heavies.

The same challenger stands up again, I've worked out it is Tsarsko Adam Borisovich from Mirovyye Aktivy, "I accept what Vitalievich said earlier, about our need to survive and that as a new Bratva of eight we could wield significant power. But I guess it will depend what Vladimir Putin decides to do next?"

Dr Gvasalia answers, "We must respect military intelligence and what our military leaders know about Putin and his personality. But Putin is not going to be upset by economic sanctions. He will call in some of his war chest of money and anyway has a huge run rate of cashflow from the raw materials that Russia exports. Finland, Germany and Poland can't just stop consuming Russian energy and it would take many months before those sanctions would bite."

"No, for Putin, the challenge is that his identity is under threat and he knows Ukraine could win. Putin is not a man for intergroup conflict discussions or negotiations. See those distant tables he sets for visitors. Any peace deal would have to be on Putin's terms - not Ukrainian premier Zelensky's. If Putin, as a cornered rat, doesn't get his way he will not be going into discussions."

"He tells that story about how hard a cornered rat fights!" says Vitalievich.

"I know, he tells it to everyone, he behaves like a man

under threat the whole time," says Gvasalia, "But now, the irony is that he is increasingly threatened by his own people who don't want either Putin or war. He is having to suppress the news about the protests and to lock up dissenters. He has been successful at this, but the fact that even with news suppression many Russians are beginning to get nervous about his actions, suggests that this is going to be a protracted war between two ideologies of the minds. He is threatened very much by President Zelensky - a highly intelligent guy who wants to protect Ukraine."

"If pushed into a corner to negotiate due to losing ground militarily he will think it reasonable to look at more desperate options, in terms of nuclear, chemical or biological attacks.

"Because he knows economic sanctions are penetrating the whole Russian framework he may threaten: 'I have nuclear power and I will use it.' You will get threat upon threat upon threat: 'I am the most powerful man on earth.'

Gvasalia adds, "I don't think he will use nuclear powers. He is highly impulsive and unpredictable so he could do it but I don't think he will because he will see the aftermath of massive destruction and genocide. It would destroy the entire political framework of Russia. It would also trigger a militaristic strategy against Russia - and nobody wants to see that."

I think forward through history and I still can't remember the sequence of the Klima Wars. I know it created a massive destruction of the Earth's ecosystem, and for a time Earth was on the edge of destruction. I will need to ask Chantal and Charlie what they remember after this meeting concludes. It all feels so distant, like we are in a different universe.

Gvasalia continues, "Putin accepts, like some psychopaths, that if they are going to lose, so must everyone else. Like a child upturning the board in a game of checkers. An extreme attack is more of a possibility than some people might think. Threat is a massive characteristic of his personality but at the end of the day I don't think he will resort to a zero sum game – all or nothing - because he depends too much on oil and gas.

"The whole thing about Putin is that he will want to preserve the way he believes Russia should be. And he will persist and will drive forward his model of the Russia he wants people to adhere to."

Now, Vitalievich interrupts, "We can see that Putin is playing psychopathic tactics, which will hurt all of us, and all the Russian people. I want us to consider the options for bringing him down."

I look instinctively toward the leather-clad security man that Christina identified. He appears to be looking at Tsarsko Adam Borisovich, as if waiting for a signal. Borisovich asked the 'treason' question earlier and I can see that he is much younger than Vitalievich's other invited guests. I remember what Christina said about the young guys pushing the old men out of the windows in Putin's early Kremlin days.

Gvasalia continues, "There could be the influence of his own war cabinet. I believe there could be other militaristic people in his war cabinet who may see things very differently to him. There are also many younger politicians who would be wanting to subscribe to a new contemporary Russia and they will want to usurp him in one way or another. He will feel threatened by them. He knows his days are numbered now; he is in his late sixties now."

Maybe a military coup, a putsch? I just can't remember what sparked the Klima Wars. Although I also can't see Putin taking part in some sort of conflict resolution.

My shaky knowledge of this period means I can't even remember what happened to Putin. I remember there was speculation about cancer. That led to the view about his condition being exacerbated by the use of steroids. The so called 'roid rage.

Gvasalia continues, "He will realise there is going to come a time when he is going to have to hand over the reins. The problem here is that he must do it or someone will take over without his consent. That will terrify him and could exacerbate his cancer. His rivals for leadership may be more complacent and contemporary and he doesn't want that unless they have a similar personality to him. He will know which of his war cabinet would carry on his vision into the future - the same old fashioned soviet personality and identity will be his goal from any successor."

I realise that this creates many contradictions. Putin wanting to reinstate his version of old Russia, although many around him against this.

Maybe a new guard which he has created now starting to have their own different vision.?

Some that he trusts are now in their seventies and may not be useful to his cause for much longer. A citizenship living under his misinformation believing one thing which is becoming increasingly difficult to manipulate.

The same citizenship sceptical of his Soviet values.

Then it happens.

Chatter

The leather clad man that Christina identified suddenly stands. He looks toward Tsarsko Adam Borisovich, who nods, "Da. Do it!"

He points a pistol towards Vitalievich. Before he has a chance to fire it, I notice Borisovich has pulled an identical pistol which is also pointed towards Vitalievich.

I hear a metallic chatter for maybe two seconds. It is considerably quieter and deeper than I'd expected. I'm surprised to see the leather clad man and Borisovich reel backwards. Two other of Borisovich's security detail immediately raise their hands into the air.

"нет проблемы! Не нам. net problemy! Ne nam. No Problem, not us!" yells the first one. Charlie has moved around the room to provide cover across all the seated people.

Christina has a smoking sub machine gun in her hand. She speaks, "Are we done?" she asks, "Anyone else?"

It's clear that the other Russians were surprised by the sudden deadly action. None of them, not even the hardest looking men are prepared for a fight and instead mildly place their hands on their heads.

Vitalievich steps forward, "You were right, Christina. It was Borisovich that wanted to disrupt us. I don't think this was the manner he intended. Thank you for saving my life."

He looked around the table.

"Gentlemen, ladies, I think we should adjourn for an hour so that we can tidy up."

One of Vitalievich's security is already on a phone to bring in a cleaning team. I look across to Dr Maria Gvasalia, who seems un-phased by the whole event.

Vitalievich announces, "When we reconvene, it will be in the ballroom. I will have it arranged like this room. Only for less people. Meanwhile, come through to my cocktail bar. We are serving Champagne - and Beluga caviar, naturally!"

I am mildly shocked by the matter of factness with which the events of the last few minutes are being treated. I speak quietly to Dr Gvasalia, who says, "This is gangland Russia, you should expect such events."

Then I ask Christina how she had known. "It was their chairs, they looked slightly different. They had been substituted by ones that each held a weapon. But frankly their Makarovs were no match for my Sig Sauer SMG. Vitalievich was lucky though; they both hesitated. I think the security guy was waiting for the command from Borisovich."

Will this get reported?" I ask,

"Of course not," says Christina, "They will want to keep quiet a failed assassination attempt along Oligarch Row. And advertising it would only increase the risk of another attempt by someone wanting to make a name for themselves."

Can't remember

I go outside to check in briefly with Chantal. She is with Antanov. They have both been listening in on the secure channel which Charlie set up. I ask Chantal if she can remember what happened to start the Klima Wars, but she doesn't know either.

"It's as if I can't remember this part of 'history' she says. I know I've been through it as a Watcher, but it is as if it doesn't exist. Like I'm in a different version of reality."

"Could it be that we are experiencing it directly, so it seems kind of different?" I ask.

"I wondered that too, but I think I'm being blocked," answers Chantal, "Usually if I spin a little chaos around something I can see its central theme, but not this time. And another thing. Have you noticed the pets around here?"

I shake my head, Pets? Has Chantal lost it?

"I've noticed that cats and dogs seem to be on the wrong scale. Most cats are about the size of a lamb, and dogs, even small ones, seem to go up to the size of a sheep."

As if on cue, a house cat parades across the yard where we are both standing. I look at it closely. It could be a small wildcat. Something like a lynx.

"I think they are just keeping exotic animals as pets," I say, "Some kind of fashion thing."

"Oh yes, and the green sky," she adds, "I could imagine a kind of Aurora Borealis effect in the evening, but we seem to have unnaturally large amounts of green sky."

I realise I'd got used to a green tinge to everything. I assumed that we were in parts of Russia with unpredictable levels of pollution.

"Are you sure it's not an effect of pollution?" I ask, or maybe something washing in from China?"

Perhaps," says Chantal, "But remember I specialise in chaos and there's several tell-tale signs at the moment. Animals, sky, even the power sockets have an unexpected uniformity."

"Chantal!" I say. I'm thinking she has gone strangely obsessive, but then I remember that power sockets were different in different parts of the world. They have suddenly all become like little isosceles triangles and are smaller than I remember.

"Chantal, I think you are reading too much into all of this!"

She smiles, shrugs her shoulders and I can see she is concentrating on the communication equipment again.

Restart

Vitalievich is ready to resume the meeting. Charlie has checked the room and asks me to run over it for a second opinion. We re-vet everyone entering and Christina performs weapon checks, which some of the heavies find entertaining.

This time, Christina makes a point of keeping her copper-coloured Sig Sauer sub-machine gun on display as people enter the room. They take up similar positions around the new table, which pointedly doesn't have space where Borisovich and his crew had sat.

Vitalievich opens the discussion, "Could an assassin kill Putin? Just as the second world war would not have happened without the demonic will and agency of Adolf Hitler, so the invasion of Ukraine – and its horrific bloodshed and unspeakable human misery – is Putin's war. Can he be stopped? The bad news is that the chances do not look good at the moment."

I see people around the table shaking their heads. I expect the oligarchs to be better informed than the security guards, but soon realise that everyone knows the same story of the failing and ever more separate war.

Vitalievich continues, "I want us to consider options to get into his secure location. Putin is protected 24/7 by one of the world's strongest security details, who have sealed him in a closed bubble. All access to him is strictly – almost manically – controlled, in much the same way as it was for Stalin and Hitler.

"The chances of a lone outside assassin like Fanny Kaplan getting to the dictator are zero. Kaplan was a Jewish Ukrainian woman, who, in 1918, fired three bullets into the Bolshevik leader Lenin, outside the Hammer and Sickle factory in Moscow. It sparked a series of strokes that crippled and eventually killed the Soviet dictator."

The point is not lost on this group that a determined assassin could still get close enough to Vitalievich to be able to almost pop two shots toward him.

He adds, "If Putin is to be stopped, ousted, and arrested, or assassinated, the authors of such an attempt will have to come from within the clique who surround him. Desperate times spawn desperate remedies, and only when the clique think that their own futures are directly imperilled by Putin's increasingly dangerous actions are they likely to act."

I consider that it is even difficult for this clique to sift for the truth because so much of what has happened is hidden under a blanket of misinformation. Only indirectly, through the effects on sales of goods and share prices, will many Russians even be aware of the war being waged in their name.

Vitalievich continues, "Putin has already publicly humiliated members of his inner circle by berating them when they have raised the mildest of questions about his actions. They cannot have much affection for this supremely unlovable man. But will their growing doubts about him overcome their fears for their own futures?

"History is replete with examples of successful assassinations carried out by the intimates of dictatorial rulers. Julius Caesar was killed by members of the Roman Senate; numerous Roman emperors like Caligula were murdered by their Praetorian guards, and at least one – Claudius – was probably poisoned by his own wife Agrippina."

Vitalievich could see that he was losing people with these older historical references, so he shifted gear, "Then, in more modern times, King Faisal of Saudi Arabia was killed in 1975 by his own nephew, while India's prime minister Mrs Indira Gandhi was gunned down in 1984 by her Sikh bodyguards, outraged by her attack on their holy Golden Temple in Amritsar."

I could see where this was leading. Vitalievich wanted to find some of these oligarchs who would be prepared to attempt an assassination.

Vitalievich continues, "Nearing the end of his thirty-year rule, a sick and ageing Stalin had become so psychotic that he had arrested and was torturing his own doctors on suspicion of poisoning him and was threatening the lives of his cronies and colleagues in the Soviet Politburo.

"It is said that one of them, the Chief of the Soviet secret police and security services, Lavrenti Beria, acted ruthlessly to save his own skin by actually poisoning his boss – appropriately enough with rat poison warfarin – triggering the stroke that ended Stalin's dreadful life on 5 March 1953. Beria himself proudly boasted of the deed, telling his comrades: 'I did him in! I saved all your lives!'

Vitalievich asserts, "When all decision making has been concentrated in the hands of a single all-powerful ruler, such as Putin, only the physical removal of the tyrant can end the tyranny and lift the danger that he is posing. But perhaps we should be careful what we wish for. Lenin's death as a long-term consequence of Fanny Kaplan's gun shots paved the way for the rise to supreme power of Vladimir Putin's role model, Joseph Stalin."

I thought back through history and recalled an event that was possibly more pertinent to President Putin in his current circumstances. It was the successful killing of Tsar Paul I at St Michael's Castle in St Petersburg in March 1801.

The increasingly eccentric and isolated Tsar had alienated the circle of military and civil officials surrounding him by his erratic foreign policy decisions. Their discontent culminated in a conspiracy which ended with them entering Tsar Paul I's apartment and strangling him. He was succeeded by his son Alexander I who eventually ended the alliance with Napoleon and brought the rampaging French emperor down.

So, if the close circle of Generals and corrupt yes men on whom Putin's power rests are sufficiently alarmed by the increasingly chaotic course of the Ukraine war and the world's horrified reaction to it, is there the faintest chance that they will take a leaf from the playbook of Beria and Tsar Paul's officials and end the crisis by bumping off their boss?

That was precisely the question that Vitalievich was asking he oligarchs in the room.

Vitalievich asks, "As the options open to the those who Putin has protected and promoted begin to close down, and the rage of the Russian people themselves rise, the best hope for ending this crisis may well rest with bad men acting to bring down the still worse and possibly deranged dictator who is leading them and Russia into the abyss.

"To play Putin's own story of 'the rat fights back' against himself, he says he learned a valuable lesson about power in his youth when he cornered a rat in the apartment block where he grew up. Instead of submitting, the rat leapt at his face and attacked him. Now, it is time for the rats in the Kremlin to learn the same lesson and to attack Putin himself."

Vitalievich was trying to incite an assassination of Putin. Whether successful or not, I was stunned that I could not recall this from my Watcher's knowledge of history.

At precisely this moment, a text message arrives on several of the phones in the room. It is captioned from the Kremlin and reads:

"We, the Russian people will always be able to distinguish true patriots from scum and traitors and will simply spit them out like a gnat that accidentally flew into their mouths. I am convinced that such a natural and necessary self-purification of society will only strengthen our country."

The 'I', we decide, is pointedly referring to Vladimir Putin. It is the biggest bulk threat he can make to the people around the table. I wonder who else has received it.

Vitalievich looks around the table to his various oligarchs. It is obvious that they will fold. They can't play this superheated game with Putin, even if it is mostly a poker player's bluff.

One by one they stand, and their corresponding incoming security groups follow.

I sense that Mr Putin is weaker than he looks, but that makes him dangerous. He plays a grandiose game and is used to winning. His previous Ukrainian adventures came when the Russian economy was in trouble and his polls needed a boost. He has had to spend bandwidth on deterring this plot for a coup but seems to have crushed it with a few well-placed text messages.

I can see Christina scanning the room, looking for anyone who could break rank or attempt anything. I sense that she, Charlie, and I have all worked out that Putin knew about this meeting all along and planted Tsarsko Adam Borisovich to attempt to kill the ringleader. Word of his failure has already got back to the Kremlin which tells us that there must be at least one more Putin supporter in the room.

Vitalievich has worked it out as well. "Go," he says, "Go back to your dachas; we cannot speak of this again." He is broken and now a dead man walking. I wonder if he will see the dawn.

Today, Putin's personal polls continue to slide and barely a quarter of Russians support his party. The protests against the opposition's leader Mr Navalny and his arrest in January were the largest in a decade.

Even the situation in Ukraine's neighbour Belarus worry Mr Putin: President Lukashenko has been so weakened by protests that he now depends on Russian support to stay in power. If something similar were to happen to Mr Putin, he has no one to turn to. Facing protests at home, he may lash out abroad, in Ukraine, Belarus or elsewhere.

My headset crackles, Antanov speaks, "Visitors, Spetznaz, twelve, incoming."

Christina hears the message too, "Exit, Exit fast. To the pickup point, you too Antonov - bring the Bumerang," she says, picking up her weapon and leaving.

Antanov's voice is heard in the headset, "Copy that, engines are running."

Antanov has done well. The Spetznaz truck is still on the approach road to Vitalievich's property and Antanov's Bumerang APC is already in position.

We climb aboard, Christina and Charlie first, then me. Chantal is in the front cabin with Antanov. Christina swaps places with Chantal. I realise it is so Christina has access to the 30 mm 2A42 automatic guns and the adjacent controls for the Kornet anti-tank guided missiles. I hope we won't need them to clear the area.

Antanov sets a course, which ignores the road and ploughs directly through the fences separating this property from the next one. We can see the Spetznaz carrier arriving at the entrance just as we are making our escape. There are dull thuds and I realise that the incoming troops are using stun grenades, aimed toward the main building complex. Then I hear a receding rattle of sub-machine guns. I'm wondering whether there will be any survivors. Antanov has found a road by now and we bounce on to it. I'm surprised at how fast this monster truck can run and we are soon in open countryside.

"I don't suppose you left that helicopter anywhere useful?" asked Christina.

Antanov smiles, "Funny you should ask that, I dropped it off at Citicopter, which is just around 10 kilometres west, on the E22. We can be there is about 15 minutes."

I notice that Christina is giving Antanov instructions from her cellphone as we navigate through an equestrian centre, which is mysteriously decked out with a range of heavy artillery. Then we are at the Heli Club and Antanov gestures towards his helicopter, neatly stowed outside one of the hangers. We climb out of the back of the Bumerang and as I look back at our transport for the first time, I realise we have been travelling in a camouflaged armoured vehicle with a serious gun mounted forward. What was effectively a major tank.

Now, we walk peacefully across a daisy-strewn field toward the helicopter. Antanov and Christina have hurried ahead and are already starting it up.

The rest of us climb aboard. I wonder where we will be going next.

PART TWO

Never catch me now

On the lam from the law, on the steps of the capitol
You shot a plain clothes cop
On the ten o'clock and I saw momentarily
They flashed a photograph; it couldn't be you

You'd been abused so horribly
But you were there in some anonymous room
And I recall that fall, I was working for the government

And in a bathroom stall off the National Mall
How we kissed so sweetly
How could I refuse a favour or two?
And for a tryst in the greenery
I gave you documents and microfilm too

It was late one night, I was awoken by the telephone
I heard a strangled cry on the end of the line
Purloined in Petrograd, they were suspicious
Of where your loyalties lay,
So, I paid off a bureaucrat
To convince your captors, they're to secret you away

And at the gate of the embassy
Our hands met through the bars
As your whisper stilled my heart
No, they'll never catch me now
No, they'll never catch me
No, they cannot catch me now
We will escape somehow, somehow

It was ten years on
When you resurfaced in a motorcar
And with the wave of an arm
You were there and gone

Colin Meloy

Putin's (alleged) Palace

Now we are again in the air, Christina and Charlie are talking on the headset. They say we will need to visit Putin's Palace.

Antanov wants to fly right out of Russia and to land in Cyprus. He explains he and Cristina 'know some people there.'

Meanwhile, it becomes increasingly obvious to me that Charlie and Christina are planning an assault on Putin's Palace. They are asking Chantal if she is prepared to use some of Limantour's powers. Limantour, the Mistress of Chaos. I worry that we are getting into this situation very deep, but with only a sketchy plan.

To top it all, the Russian President is on record as denying owning the palace, although he is probably using more of his Saint Petersburg tactics of concealment. I wonder if we will get into somewhere that is the wrong place.

I check the internet from my phone. The current owner of the Palace is shown as Arkady Romanovich Rotenberg, a Russian billionaire businessman and oligarch. With his brother Boris Rotenberg, he was co-owner of the Stroygazmontazh (S.G.M. group), the largest construction company for gas pipelines and electrical power supply lines in Russia. He is a close confidant, business partner, and childhood friend of Putin. With an estimated fortune at $2.5 billion, Rotenberg became a billionaire through lucrative state-sponsored construction projects and oil pipelines. He has been implicated through the Pandora Papers leak in facilitating and maintaining elaborate networks of offshore wealth for Russian political and economic elites.

A video called 'Putin's Palace' was released by Mr Alexei Anatolievich Navalny, the leader of the Russian Opposition party. His investigation alleges the property cost £1bn ($1.37bn) and was paid for "with the largest bribe in history".

The palace, by the Black Sea, was allegedly financed by billionaires close to Mr Putin. It is said to have a casino, skating rink and vineyard.

"They built a palace for their boss with this money," Mr Navalny says in the video.

For years Navalny has castigated Mr Putin's administration on social media, accusing the Russian leader of feudal patronage and running a system riddled with thieves.

The video also alleges that Russia's Federal Security Service (FSB) owns some 27 square miles (70 square kilometres) of land around the palace, near the resort of Gelendzhik.

Navalny, Russia's most prominent opposition leader, is locked in jail by Putin. Navalny was the survivor of a near-fatal nerve agent attack, escaped Russia but was arrested upon his return to Moscow from Berlin.

Mr Putin called the palace video a 'compilation and montage' and said he found it 'boring'.

'Nothing that is listed there as my property belongs to me or my close relatives, and never did,' he said.

'No one should seek to advance their ambitious objectives and goals, particularly in politics, through protests,' President Putin said, in an apparent reference to Mr Navalny.

In 2017 Mr Navalny's Anti-Corruption Foundation (FBK) also accused ex-Prime Minister Dmitry Medvedev - one of Mr Putin's closest associates - of collecting luxury estates through a secret fortune. Mr Medvedev denied the allegations, dismissing them as 'nonsense'.

Antanov is landing the Kamov-62.

"Welcome to Kursk," he says, "We are around 550 kilometres from Leshkovo. I think we should trade up this helicopter now." He gestures to a white plane standing outside of a hanger. My first thought of its appearance is 'menacing'.

"We'll be using this plane for the next leg. I expect by now they have organised to look for the Kamov. But I don't think they'll be expecting us in a strategic bomber. Let's climb into the Tupolev Tu-160M2. Notice how it has the reflective paint to deflect nuclear blasts. We can fly this as fast as a fighter jet and at a good height. Sorry, but there's only four proper seats, someone will have to ride on the jump."

"How are we getting this into Cyprus?" asks Chantal. "I mean it is somewhat noticeable, and I don't want us to get shot down. A Russian jet in Cypriot airspace?"

"Taken care of," answers Christina, "We've called up Blackbird and he has arranged diplomatic clearance. It's obvious that he is rooting for us, although I fear his days in the Secret Service may be numbered."

Antanov checked around the plane. I realised he was looking at the on-board weaponry. "We've picked up this plane, which seems to be carrying 4 AS-15 missiles, but nothing else. The AS-15 or Kh-55 is an air-launched cruise missile developed by the Soviet Union. They can deliver a 250 Kilo-ton nuclear warhead 2,500 km. So, we'd better be careful how we fly this thing."

We climb aboard and I uprate my review of the plane from 'menacing' to 'lethal' as I look back into the weapons hold and see missiles hanging in their release apparatus.

"250 Kt is about 5 times the power of the bombs detonated in World War II, and this plane is carrying 4 of them. It has capacity to carry 16. That's enough to wipe out the entire eastern seaboard of the USA," observes Charlie, "We are flying World War III".

"Scrive, you can fly planes?" asks Antonov, "How would you like to be flight engineer on this one?" He points me to a seat and panel of dials and controls, and I realise that the 'glass cockpit' design still has some traditional controls further back for each of the engines. Christina quickly shows me around the controls and even fires up a couple of the engines. I realise she knows her stuff about big planes too.

Antanov is readying for take-off and has Christina seated

as co-pilot again. She appears to be concentrating in a way I've not seen before.

Then we taxi to the end of the military runaway and I notice several ground staff stop what they are doing and come over to watch the take-off.

I think I'm prepared, but this plane is built to a military specification and has four after-burning turbofan engines, to blast it along the runway. Christina and Antanov seen unperturbed as they pull this killing machine into the air. I feel the screech from the four jets as we climb to our cruising altitude. I just hope we won't get shot down.

In around an hour we are over Cyprus mercifully escorted in by two NATO F-35A fighter planes. I'm getting used to this drill now, but it is reassuring to have Antanov in the pilot's seat, who has been in similar situations. I find it fascinating that so much of the so-called air superiority is in planes that were designed 10 or 20 years ago. The F-35 is already 15 years old and the Tu-160 jet plane we are riding is around 35 years old.

We land and the same thing happens with ground crew here in RAF Akrotiri, where many people come out to look at this sleek, menacing white plane bearing Russian insignia. We are clearly a rare spectacle even among the hardened aircrews stationed at this base.

I look at Antanov and can see he is relieved to be back on the ground. He's had a tough day, with stakeouts, shootouts, driving a tank, a helicopter, and a jet bomber to bring us all to this island in the middle of the Mediterranean.

But I sense that Charlie and Christina are tightening the

screws even further.

Cyprus - Artem and Dakis

I'm acutely aware that we are still travelling on Dutch passports, with glued-in Russian visas.

Christina says something to the base security and we are all admitted without any further questions.

"I was here during one of the terrorism troubles and know the woman who is now the base Commander," explains Christina, "I've asked to meet her at some point while we are here."

"Well, we should hit the town tonight, to discuss our next moves," says Chantal, "I can already sense that Charlie has a plan, maybe with Christina too."

"Yes, and don't be surprised if you are roped into it too," says Charlie, "And you, Scrive!" She looks at me and I know this will be something wild.

Although a military base, RAF Akrotiri has the trappings of a small commercial airport, with a coffee shop, lounge area and for us, on this occasion, the all-important cab rank.

We take a large taxi, which is like minibus, and Christina asks for Limassol centre. We are on the hunt for somewhere to relax. The taxi driver, who introduces himself as Artem, suggests the Dionysus Mansion and we all agree to go there. He says that his brother Dakis is a waiter there, and that the food is very good.

He drives us out of the base and then turns east along the coast, before heading slightly inland to what I imagine is the centre of Limassol. At least, it is where the post office is situated. Then, he takes us to the restaurant, the last few metres on foot, and introduces us to his brother.

Call me a cynic, but I'm used to taxi drivers having that kind of line to get extra commission from the tavernas, but this seemed genuine. His brother shows us to a beautiful corner table, outside in a small sun bathed square and shaded by a large olive tree. Perfect. I can feel the tension drain away. I can see the same effect in the others too.

We order food in what is now the customary way, a selection that we can place in the middle of the table and then select our own, buffet style. Dakis brings us some Maratheftiko red wine and pours us each a glass.

"The first is on the house," he says, "You all look as if you need some good hospitality. After the first bottle you will all start to experience the magic of Cyprus."

Late afternoon turns into evening as we sit around this table. I notice, still that the sky has turned a green colour, before it darkens and I wonder if there has been some pollution of which I am unaware.

"I'm troubled that I can't remember this time," I say, "When I try to think about what happens, all I can remember is the Klima Wars, but I can't remember how they start, let alone anything about Putin."

"Me also," says Charlie, "It is so unusual for us, as Watchers to not have any sense of recall of this time."

Chantal speaks, "I've been wondering that too. I wonder if we have somehow been dropped into a Variation. Remember that stuff about when the world was supposed to be flat, or the scientists thought that the sun orbited the Earth.

"As Watchers we all say that the cosmic dealer smiles within his darkened room. He doesn't let on that there are other darkened rooms next door. You just must think of them."

She smiles cheekily, "It's a principle of my Chaos, as Limantour."

Then she continues, "Other darkened rooms can contain different frames of reference. In the Lepton Epoch he may have stumbled upon Light and his form of physics, but only in one of the frames of reference."

"At least one of these dark areas of deep shadow contains a key to explain what is happening.

"Think of parallel universes.

"We all know that the wrinkle in human thinking is that although multiple metaverses exist, it is not possible to travel between them. But maybe, just maybe, we have."

I understand Limantour's thought spoken by Chantal. It explains the small variances we are experiencing. The green sky, the larger-than-they-should-be cats and dogs.

The sudden, pervasive use of electric road vehicles. Chantel's theory could explain why we can't remember. If we are in a parallel metaverse, then we won't have experienced it before.

Charlie nods, "I think you could be right, Chantal, but this world is so detailed and torn with the same anguish as the one we left."

"I assume it is an effect from the Trigax, which fired us all back to this point. I can't understand how it would have the precision to bring us all to the same point unless it was somehow significant. Something controlled."

I remember something from a long time ago, before I became Scrive and when I was a Watcher. I'd been in a Chinese wet fish market with Lepton and Lekton. Lepton was a projection of Light and Lekton was a projection of Darkness, teaming up with Scheppach to create ever-increasing vestiges of warfare.

Scheppach the illegal arms dealer - a Bladerunner - provided her escalating devices of warfare, which began with knives, then guns and continued to increase in magnitude.

"Could it be that something has happened between Lepton and Lekton? " I ask.

Chantal looks concerned, "Yes, I can almost sense it. I wondered why my normal Limantour chaos moves were not available. Everything I must call on is now distressing instead of playful. It makes sense if we are in a different reality."

Christina and Antanov look on puzzled whilst we have this conversation.

"Look, have you had some magic mushrooms, or maybe the wine is too strong?" suggests Antanov.

Christina speaks, "No, Antanov, I think these people really have a gift. Remember, in our Russian culture: *nechistaya magiya* - unclean magic, and *produktivnaya magiya* - good, productive magic. Most productive magic was homeopathic, meaning that a symbolic action was performed with the hope of evoking a related response from reality.

Antanov agrees, "Yes, I remember that children's springtime ritual of carrying around branches with artificial birds or cookies. It was thought to help bring about the bird flight associated with the coming of spring."

This is all getting multi-layered.

I wonder if we need to do a symbolic act to evoke a response from our original reality.

That we have been flipped by the Trigax railgun into a historical instance that seems real but is one step removed from the Earth that we all know.

Maybe we have somehow slipped from Lepton's world filled with Light into an adjacent one of Darkness? Lekton and Scheppach's dark room.

Too much paperwork

We finish the meal, but the last discussion had thrown us all off kilter. We still don't have plan, other than to stay somewhere overnight.

Dakis calls his brother who arrives with the taxi-bus and takes as all to a spectacular apartment complex.

"It's my cousin," Artem explains, "She owns the whole complex here, I'm sure you will get a deal. Here, let me talk to them."

He dials a number and speaks quickly in a Greek dialect to someone. We'd all been speaking in English and I don't think he realised that three of us could also follow his conversation in Greek.

True to his word, he arranges for us to have a floor of apartments, and then he explains to us that it was considered 'off-season' by the owner. He had therefore obtained them for us at a special rate. We would need to go to the reception, but to say that Artem had sent some 'special guests'.

"You know something," said Christina, "I think we

should stay here for a couple of days. Recharge and get off the radar of Putin and Vitalievich.

"Won't they miss the long-range strategic bomber?" asks Charlie.

"Relax," says Antanov, "Blackbird and I have settled it with the right people. They think the stealth bomber is secretly in the air somewhere, and the Brits on the base won't be declaring they have it."

'Too much paperwork,' I surmise.

We check in and take the elevator to our floor. We have five modern and well-appointed bedrooms spread across three studio suites. The views are breath-taking and look out across the Mediterranean. We have balconies with sofas and can sit in the evening air.

I settle into my room and switch on the television. Unlike in Russia, we once again have news feeds which describe the war in the Ukraine. Now it has moved on to allegations of biological and chemical warfare.

A Russian foreign ministry spokesperson is accusing Ukraine of developing biological weapons with the assistance of the United States.

The US vehemently denies these claims, arguing that it is Russia who wants to use these weapons of mass destruction.

I muse that it is a classic use of misinformation to normalise a potentially aggressive move. The US Press Secretary is publishing tweets (diplomacy gone mad?) stating that Russia is using these allegations as propaganda in a bid to pave the way for its own use of biological and chemical arms.

In Chantel's words, "This Version of the world is even crazier that the one where we have been Watchers."

I think that it would not be the first time that Russia has accused another state of poisonous warfare as a means of justifying these weapons' use, either by itself or an ally.

I remember that when President Bashar al-Assad used chemical weapons in Syria, Vladimir Putin backed up the allegation that it was rebels who had used them in attacks.

US officials described this as a classic Soviet trick: accuse the enemy of what you want to do yourself to legitimise your later actions.

I have a hazy recollection of something, some kind of vaccine permanently triggered and connected to the arm by a small cartridge. Tropus? Something like that. Something to counteract the effects of nerve agents and lethal viruses. I realise it is something that must happen in the future and it dawns on me that it must be a consequence of the Klima Wars. I check with Charlie and she has a similar recollection.

But would Putin use these contentious and highly destructive weapons? If so, Scheppach the arms dealer would have her work cut out.

Putin is already said to have employed other extremely controversial weapons, such as thermobaric weapons (also known as a vacuum bombs) and cluster bombs.

These attacks have not been limited to military targets such as weapon stockpiles but have also been used against civilians. Attacks have been recorded against civilian targets such as hospitals and a maternity ward. There are also reports that Putin has now used a chemical weapon: white phosphorus.

Putin considers chemical weapons effective at breaking down enemy defences. Assad used chemical arms in Syria for precisely this reason. Chemicals were deployed against rebel defences when conventional bombings had not been sufficient to overcome them – for example, at Douma near Damascus.

If Putin has employed white phosphorous in Ukraine, it would be no surprise if he escalated to using other forms of chemical, such as nerve agents. Russia has, in the past experimented with every type of weapon. Choking agents like phosgene attack the lungs and respiratory system, causing the victim to drown in their lungs' secretions. There are blister agents, like mustard gas, which burns the skin and blinds people.

And then there is the most lethal category of all: nerve agents, which interfere with the brain's messages to the body's muscles. A tiny drop of these can be fatal. Less than 0.5mg of VX nerve agent, for example, is enough to kill an adult.

All these so-called chemical agents can be used in warfare in artillery shells, bombs, and missiles. They can even be sprayed from the sky. But all are strictly prohibited by the Chemical Weapons Convention of 1997, signed by most nations, including Russia.

Russia says it destroyed the last of its chemical weapons stocks in 2017 but since then there have been at least two chemical attacks blamed on Moscow.

The first was the Salisbury attack of March 2018 when a former KGB officer and defector, Sergei Skripal, was poisoned along with his daughter by the nerve agent Novichok. Russia denied responsibility and came up with over 20 different explanations for who could have done it.

But investigators concluded it was the work of two officers from Russia's GRU military intelligence and as a result 128 Russian spies and diplomats were expelled from several countries. Then, in August 2020, the prominent Russian opposition activist Alexei Navalny was also poisoned with Novichok and narrowly escaped death.

Later, during the Ukrainian war, a sanctioned Russian oligarch Roman Abramovich reportedly suffered from symptoms of poisoning after meeting with Russian and Ukrainian representatives and attempting to broker peace negotiations between the two countries. Abramovich allegedly suffered from a temporary loss of eyesight, problems eating, and peeling skin on his face and hands after a meeting with Ukrainian and Russian representatives in Kyiv. He said he'd only drunk water and eaten chocolate at the meetings.

These were similar symptoms to those suffered by Ukrainian President Viktor Yushchenko when an attempt to assassinate him failed during the Ukrainian elections. Yushchenko was confirmed to have ingested tetra-chloro-dibenzo-dioxin (TCDD), the most potent dioxin and contaminant in Agent Orange

If Russia were to use weapons like poison gas in its war, this would be seen as crossing a major red line, most probably prompting calls for the West to take decisive action.

The problem is there is a potentially a grey area between working on ways to protect your population from harmful pathogens, and secretly working on how they could be used as a weapon. Russia did not produce any immediate evidence of Ukrainian misdoings in this area. But it called for an emergency UN Security Council meeting on Friday to discuss its claims.

Russia, when it was part of the Soviet Union, controlled a truly massive biological weapons programme, run by an agency called Biopreparat. that employed about 70,000 people.

After the end of the Cold War, scientists went in to dismantle the Soviet Union's massive biological weapons programme, run by an agency called Biopreparat. They found the Soviets had mass produced and weaponised anthrax, smallpox, and other diseases after testing them on live monkeys on an island in southern Russia. They had even loaded anthrax spores into the warheads of long-range inter-continental missiles aimed at Western cities.

Finally, in this grim roll call of non-conventional weapons, there is the "dirty bomb" - a normal explosive that is surrounded by radioactive elements. It is known as an RDD - a radiological dispersal device. It could be a conventional explosive carrying a radioactive isotope such as Caesium 60 or Strontium 90.

It wouldn't necessarily kill any more people than a normal bomb, initially at least. But it could render a huge area uninhabitable for weeks, until it had been fully decontaminated.

A dirty bomb is almost like a psychological weapon,

designed to cause panic among a population and undermine the morale of a society.

The attacks on single people took place after the claim that Russia had disarmed – although Russia denies that it was involved in these assaults. However, on that basis, we can assume that Putin has access to chemical stockpiles which could be deployed in Ukraine.

I think back to what Dr Gvasalia briefed, "Putin accepts, like some psychopaths, that if they are going to lose, so must everyone else. Putin will want to preserve the way he believes Russia should be.

No wonder Vitalievich wanted to stop Putin. If Putin is playing a psychopathic endgame, it will hurt everyone.

Dr Gvasalia had speculated that Putin must realise there is going to come a time when he is going to have to hand over the reins.

But this creates contradictions. Putin wanting to bring back old Russia, even with many around him against this. Some that he trusts - even from his boyhood - are now in their seventies and may not be useful to his cause for much longer. He has created a citizenship living under his misinformation believing one thing which is becoming increasingly difficult to manipulate.

No wonder he is rummaging in the back of his weapons locker.

I wonder how I can recall this information with such clarity, yet don't know about the start of the Klima Wars.

I also remember that biological weapons are more difficult than chemical arms because biological agents are contagious. Contagion means that if an aggressor uses a biological weapon, there is every chance that their own troops will catch the disease they have released. This "boomerang effect" may deter Putin.

However, the Vector laboratory in Koltsovo (near Novosibirsk)—formerly part of the Biopreparat complex—intends to continue its research on pox viruses and is very dangerous because of possible changes in the Russian government.

When General Lebed was asked about the nuclear and biological capability of Russia, he replied that because the Russian Army is very weak, Russia still needs these weapons to protect itself.

Imagine the situation in a mountainous region like Chechnya or Afghanistan. It's very difficult to fight in the mountains using conventional weapons. But a single plane or cruise missile armed with biological weapons could kill absolutely everybody in any deep valley in the mountains. So, the Russians consider these weapons highly effective for certain types of conflict.

I think again of the cornered rat that Putin so frequently references.

Plans

We all go to meet in Christina's room. She has the best suite, we decide, with a long sofa set up on the balcony overlooking the sea. They have designed Limassol with a dual carriageway along the seafront and most of the buildings, including ours, set back maybe 50 metres. It's a similar setup to the South of France - maybe Cannes - but not as elegant.

Christina, Charlie and Antanov are engrossed in a conversation. I can see they are plotting our next moves.

Chantal walks in and when we are all seated, Charlie summarises.

"We've got several assets. A long-range Russian stealth bomber with four nuclear cruise missiles. We know where Putin's Palace is and we are within comfortable flying distance. Even from here, one of the cruise missiles could make its way to Krasnodar Krai without interruption. The problem is with the 250 Kiloton nuclear warheads, it's not as if we can dial them back a little.

"Set to stun?" says Chantal.

Charlie continues, "Antanov also reminds us that at the eastern end of Cyprus we have access to drones. They are Turkish Bayraktar UAVs, based at Geçitkale Air Base from where they are tested."

Antanov interrupts: "They are Bayraktar TB2 combat tactical unmanned aerial vehicle systems and are already being used by Ukraine, from Ukrainian soil, to target Russian tank convoys. They can carry Roketsan's MAM (Smart Micro Munition)'s and TUBITAK-SAGE BOZOK laser-guided bombs."

"But how would we get to the weapons?" asked Christina, "The stealth bomber we stole, so it can be re-used, but the drones are under Turkish control. The Turkish are on the fence about this war now, because of the direct threat to their own country."

Charlie continues, "So we have access to weaponry, but some of it is too powerful and the rest is inaccessible and probably too lightweight."

Antanov continues, still intent on developing his hybrid plan.

"Often during heightened conflicts, it is hard to distinguish between factual events and misinformation - like the way some videos of drone attacks have already been exposed as fakes. Given the chaotic events on the ground, it is almost impossible to assess how often and how successfully Ukraine has utilised its Turkish drones so far. We do know that Ukraine bought Bayraktar-TB2 over the past years and that Turkey and Ukraine signed an agreement for the production within Ukrainian borders of the TB2."

"Now, sitting right here in Cyprus, is a major flight

base for these Bayraktar drones. And ever since 2019, Kyiv has bought dozens of these drones from Ankara."

"We could create a haze of confusion by sending a whole squadron of these things across to Russia. Purely as a diversion

"Then, it will depend on Russian air defences. Drones like the TB2 are vulnerable to anti-air defence systems. To be effective, they need to be employed in a savvy way, in coordination with other electronic warfare systems that 'blind' enemy radars and through appropriate tactics. However, we know that Russian Air-superiority is suspect. They may have planes, but how many of them are functional?

"In Libya, Russian forces figured out effective ways to counter Turkish tactics and shoot down their drones. The same has been observed in Syria and Nagorno-Karabakh,

"Nonetheless, the fact that Ukraine could strike Russian ground forces with TB2 suggests either that Russian forces are advancing without air defence – which is very well possible, considering the logistical and organisational problems Russia has encountered so far.

Charlie shakes her head, "But we've no control over the drones. The Turks are hardly likely to do what we say."

"Looking like this, I agree," says Chantal, "But what about if we changed our appearance?" she walks out of the room with Charlie and a few moments later returns in the clothing of a Turkish Tümgeneral. Two star plus crossed swords epaulette insignia and a blue beret.

"Wow," says Christina, "That is some transformation.

You are suddenly a hot high-ranking Turkish NATO officer."

"The very model of a modern major general!" says Antanov.

We all laugh.

Charlie continues, "So, with Chantal's revised identity, we can probably influence the way that the drones get used. Also, either approach - the bomber or the drones - will look deliberate and can set off further chain reactions. We think that Vitalievich's plan to the oligarchs was to get closer to Putin and then to act. Putin seems to deploy poison against his enemies and that could be a better way to stop him."

Antanov grimaces, "But poisoning or similar is rather more 'hands on' than either of the other approaches, both of which can be run from a safe distance."

Charlie adds, "Yes, but that's where our other asset comes in. Three assets. Three of us are Watchers and would get bounced back on the time-line."

Chantal adds, "I've been thinking and it may also be a way to break us out of this alternative world-view, back onto our original one, even with all of its faults. Where skies are blue and cats are small enough to look cute on the internet."

"So how many of us can you shapeshift like this?" asks Antanov.

"I expect I can do you all, but only one at a time," says Charlie, "It uses considerable energy. And now I need to turn Chantal back into her own clothes."

Hey remember (mashup)

Hey, remember that time when I would only read
Shakespeare
Hey, remember that other time when I would only read
the backs of cereal boxes
Hey, remember that time when my favourite colours
was pink and green
Hey, remember that month when I only ate boxes of Tangerines,
so cheap and juicy

Tangerines

Hey, remember that time when I would only smoke Parliaments
Hey, remember that time when I would only smoke Marlboros
Hey, remember that time when I would only smoke Camels
Hey, remember that time when I was broke
I didn't care I just borrowed from my friends

Bum bum bum bum bum
Bum bum bum bum bum

Regina Spektor, adapted

Remember that time (Lepton reprise)

I looked towards Chantal.

"You remember that time?" I asked, "When we were Watchers and you summoned up Lepton and Lekton?" It was just before we ran our Intervention that saved Earth?"

"Oh yes, " said Chantal, "As Limantour, I was able to bring about all manner of chaotic events. Like that dingy Huanan apartment over the wet market. Although the unintended consequence was a global pandemic."

At that time, we, as Watchers, had been flipped to Southern China, when suddenly Lepton and a stranger - who I discovered was Lekton - had appeared.

Lepton had beamed when he saw me. "Hello, Farallon, it has been a long time! You asked to see me and to witness for yourself that I have continued to enjoy a good living beyond your metaverse."

Metaverse. That was the key word. He was referring to us being in one frame of reference, but that there could be others.

I could see what Lepton had done. Lepton was the originator of one of the earliest Interventions - the provision of Light for the universe. The Lepton Epoch.

Lekton was almost Lepton's opposite. Watchers knew that Lekton's intervention was to bring warfare to humanity. I had stood with the bringers of Light and Darkness.

Lekton had gone on to explain that we were in the workshop of Scheppach. She lived outside laws and created necessary evil in the form of ever-escalating methods of warfare.

When we were contemplating the first Intervention, Lekton had explained, "You should see now that whatever happens, you get placed in our metaverse after your Intervention. I'm sure you have seen that for Darnell, Bishop, Cardinal, Abbott, Lepton, and myself, it becomes a rewarding journey."

Then I remembered Scheppach's words: "We are seeing a shift now. Earth was able to keep itself balanced through increasingly vicious warfare. Now we are seeing a mixed effect. Part of it is through warfare, but this has been hybridised by the onslaught of other methods, such as the biological. The potential for 230 million cases of COVID and around five million worldwide deaths from the disease. Like the 25 million from Spanish flu."

Scheppach had gestured to some small canisters stood along a shelf. I realised they were various toxins.

Lekton spoke, "Earth rebalances, seeking a new equilibrium." I remembered Lekton was stuttering, like a poor-quality video, when I realised, I had run out of time with him.

Chantal spoke as if she was recounting Limantour's experience, "We left that place around when things started to stutter and become unstable. I think if we had been able to stay longer, we would have discovered more."

Christina and Antanov were looking towards us with their earlier suspension of disbelief.

Christina asked, "So what was in the canisters?"

And Antanov, "Did Scheppach start the COVID pandemic?"

Chantal answered, "Remember I, as Limantour am Mistress of Chaos. It means that sometimes things that don't happen seem very real, although I don't think this whole reality is a projection."

"Can you remember how to summon Lepton?" I ask.

"Oh yes but be careful what you wish for. In this metaverse some things are already different. I would not want Lepton to be totally changed."

"It's mainly our sense of the future that is missing," I said.

There was a shudder like a small earthquake. I noticed, from the balcony, a few larger than average waves on the sea. Waves that were moving against the tide and away from the shoreline.

I looked around the room, and an extra person was sitting amidst our group. It was Lepton.

"I am glad you called, Limantour!" he said, "Hello everyone, Farallon, Limantour, Tomales!" I could see that Lepton was already flickering slightly, like he was having trouble holding on to this reference frequency.

"You were right. This is an alternative metaverse. It is very close to the one you are accustomed to. After you left the fish market, a couple of further things happened. Lekton was already hostile and wanted revenge. He decided to try to reframe the worldview, using Putin as a human blunt instrument. Like a cheap movie with an arch villain, only played for real. Scheppach helped him, with the idea of increasing her arsenal."

"How do we get to Lekton?"

"He's gone. He finally met his end in the future, but within a few hundred years of now. He never visited this exact time and can't shed a direct influence upon it."

"What did Lekton do?

"He manipulated Putin, who was already a deeply troubled human."

"Then he called the four of you across to this world. He called for four of you and achieved three."

"And Scheppach?"

"She picked from the row of canisters. Selected COVID and let it run its course."

"But if we hadn't Intervened?"

"None of this would have happened. Remember unintended consequences."

Lepton continues, "I tried to put a marker in this metaverse, to keep people away and to make things difficult for Lekton. I modified my version of Lepton's atomic model and removed the tau and tau neutrino from the set of elementary particles. It was enough to mark this world as unstable. The green skies, the large pets, the early adoption of electric automobiles. Without the tau, the Chaos Frame will gently collapse."

"What? You condemned this reality?"

"Remember it is only a facsimile of the world view you have viewed for all of time. This copy is broken, thanks to Lekton and Scheppach.

"You can afford to do anything here; you will be bounced back into your own worldview at the end. You are fortunate, Drake is keeping your place in the other reality. Drake, now called Nathan, was outside the influence of the Trigax guns fired by Lekton. Nathan lives in in Bodø, Norway, with his partner Sheri. It is fortunate he lives within the Arctic Circle and the Trigax rail guns are orbiting the equator. He is out of their range."

We all look as Lepton's image sputtered and then disappeared.

Chantal speaks, "Don't worry, he's fine, it is just difficult to hold him in another reality view for more than a few minutes."

She walks across the balcony, and as she does so, her polka dot dress changes colour. The dots change from pink to blue.

"Sorry," she says, "Bringing Lepton here has drained me."

Charlie's Plan

Charlie speaks, "Seeing Lepton was incredibly helpful. Now we know that this variation, this metaverse, is a step away from the one we have watched."

Chantal nods, "The ancient Greek word 'chaos' means a chasm or void, and its opposite is 'cosmos', meaning the exquisite design of the world. We've been flipped from the cosmic into the chaotic. But hold that thought. We need some food. Or another way, I need some food after that draining Lepton experience."

Christina flipped her phone, "Here: Bolt. It's like Deliveroo or Doordash in many places. The Cypriot equivalent."

We all leaned over her phone, which displayed screen after screen of burger joints, with imaginative names like The B.B. King and The Big Cheese.

"Is that all they do?" asks Antanov.

"There's the Wings of Love, which is chicken wings, and Pizza Paradise," answered Christina.

In the end we decided we'd go for Most Popular Choices which was burgers and so we ordered five burgers, fries, Tasty Onion Rings, plus Cokes. It would be difficult for that to be a total failure. The burgers looked like they were well stacked with salads too. 20-25 minutes, it explained.

I remember that astrophysicists are struck by what they call the "fine-tuning" of the universe, a place which looks as though it saw us coming and framed its laws accordingly.

It's why Lepton has influenced this variant but held back some the components. The third order fermions. The tau leptons. It's an elegant form of sabotage. Maybe the pervasive burger joints are another sign. To bring a system to a gentle collapse.

I ask, "But what about Christina and Antanov? If all of this collapses we Watchers just get flipped to another position on the timeline. Christina and Antanov are humans."

Chantal nods, "But don't you see? One minute they are in the normal Earth metaverse. The one we watched for so long. Then, they manifest in this alternative. If it ends, they will find themselves back in their normal existence, at the point where they left. Except they will have some serious enlightenment! I've seen this before."

She continues, "Trust me, as Limantour, the Mistress of Chaos, I have seen other universes a lot more chaotic than our own. Chaos Theory deals with systems whose behaviour is random and unpredictable. One could even think of Vladimir Putin as such a system.

Antanov adds, "One curious feature of the Russian

president's attack on Ukraine is its recklessness. Former KGB agents may be callous, thick-headed or sadistic, but one would not expect them to be impetuous."

I can't help thinking that Putin was a ruthless KGB man a long time ago and has had plenty of time to go mad and be surrounded by sycophants. His motive in flattening Ukraine is partly ethnic. He thinks that the country is a fiction, a cardboard cut-out nation, and should be wound up as soon as possible. He sees Ukraine as a void, a non-entity, and Russia will impose some order on this chaos by incorporating it into itself, making him a mini-tsar.

Chantal adds, "I've walked on the rough ground of what Ludwig Wittgenstein called social existence, even if there are those who try to walk on the pure ice of some flawless vision of order.

"But I see the catch. The two are not simply opposites. Now I see in Ukraine how easily such flawless visions create the rough ground of a city brought to ruin."

The Czarlings

The entry phone rings. It's a guy with the food. We ask him to bring it up to our floor. Then there is due ceremony as it is handed out. The guy has brought some 'optional' cold beers too. He sells them to us in a separate deal. Genius.

Charlie speaks, "The oligarchs know that Russia is far less important than China, either to the world economy or to climate talks. But it still matters a great deal.

"These oligarchs, like the ones assembled by Vitalievich, also know that Russia, because of Putin, is the single most prolific stoker of instability on Europe's borders. The Kremlin has become the most energetic troublemaker in rich democracies, funding extremist parties, spreading disinformation, discord and eroding these oligarchs' wealth.

She continues, "Now, it becomes a question of how the West deals with Putin. It will set a precedent. China's leaders are certainly watching, even as they straddle both relationships. If the world lets Russia roll over Ukraine, the Chinese may assume that Taiwan is fair game, too."

I think it is one way that the Klima Wars could start. I wish I could remember. I wish any of us Watchers could remember.

Antanov speaks, "We can consider a hybrid attack on Putin. Blend together our assets and run some confusion. By the sound of it, Chantal will be good at assisting that?"

Chantal lets out a squee of delight and jumps up. I notice she has suddenly changed into new clothes. Earthy colours, a green, grey, and brown geometrically blocked top, and matching green shorts. Suddenly she has blonde shoulder length hair and I realise she is pulling a full chaos move on us all.

'Sorry," she says, "but that encounter with Lepton took it out of me. I've changed my look to generate some new energy."

Charlie looks at her enviously, "Those clothes..."

"I know, I look like a boy now. I'm calling this look 'boyish femininity!"

I realise that Chantal still has Limantour's full Watcher powers.

Christina continues, "Right, now if we can get to the north eastern end of Cyprus, we can try to gain access to many drone weapons."

Antanov adds, "And we still have the bomber. It's a problem here in Cyprus because the island is divided into so many nation states. The plane is landed in the United Kingdom, we are in Greek Cyprus and the drones are in Turkish Cyprus."

"I know, I could have designed this country," says Chantal, "It's so - multifaceted."

"Well, we need to think about moving that Tu-160 stealth bomber before someone misses it," says Antanov, "We put it on a 'round the globe' mission, but it would have run out of fuel by now. The paperwork to find it will have started in Russia."

Christina nodded, "And we don't want to lose it as an asset for this campaign."

Charlie asked, "The thing is, Putin could be in any of several places. You know he has gone to ground and is only making pre-recorded TV speeches now. He could be in Moscow, The Kremlin, at one of the dachas or at his Palace."

"Or in the Altai Mountains, in Siberia, in his underground bunker - the one where he has sent his family," says Chantal.

"Ah you know about that?" asks Antanov, "I had to go there once. It isn't a bunker, more like an underground city, equipped with the latest science and technology. It's supposed to be a sprawling mountain resort built by energy company Gazprom around a decade ago in the Ongudaysky district of the Altai Republic, a region of Siberia bordering Mongolia, China and Kazakhstan."

Antanov is looking toward the sea, "I was there, in Altai, with a secure team checking the exit paths. Blackbird requested it. As a simulation, we had to create an incident and attempt to stop the occupants from leaving. They had multiple escape routes; road, a double heliport, even the river, but most of the complex is hidden underground. It is like a hidden city. For us, it was like a version of cat and mouse. Think of a mountain hideout, with luxurious above ground facilities and then enough power to run a small city, most of which is underground. "

Antanov continues, "I'll say this is rumour, but Putin's 'other' wife was there when we were, Alina Kabaeva - some 30 years younger. She is alleged to be the Russian leader's secret spouse - although a Russian newspaper was closed down for reporting it. She's that Olympic Gold-winning rhythmic gymnast, who became a wealthy politician in the Duma and then a high ranking official in the National Media Group. Sometimes you just can't make this stuff up.

Putin's two daughters - The Czarlings - I think their names are Maria and Katerina, are from his previous marriage to former flight attendant Lyudmila Shkrebneva. They are almost the same age as Alina Kabaeva. Putin is alleged to have two sons and twin daughters with Kabaeva.

He looked around, "The Kabaeva family have Swiss passports and are probably in Putin's Swiss chalet.
"Then he has another daughter - Elizaveta from a previous relationship with cleaner-turned-multimillionaire Svetlana Krivonogikh, now a part-owner of a major Russian bank. See how the money rubs off."

Antanov shrugs, "I reckon if Putin has sent them to the chalet, then it is partly for their protection and partly so they can't see what he is doing.

"Defence Minister Shoigu is currently in personal control of Russia's military operation in Ukraine. It was him that arranged for Putin to attend a shamanic ritual in Siberia which involved the sacrifice of a black wolf in a rite to improve the president's health.

"Is that both physical health and mental?" asks Christina.

Antanov says, "According to my sources, which are largely men on the ground, they are saying that Putin was suffering from early-stage Parkinson's disease, and that he has a secret life-threatening illness - maybe throat cancer being treated with steroids, but they have no proof. It ties in with the medication rumours."

Christina answers, "We must see Putin for what he is. A psychopathic killer. But he also keeps communications open, although it is deceitful and only Putin's attempt to boost his prestige. He doesn't want to de-escalate military tensions nor to signal resolution. He just wants to pretend to negotiate to buy time for his hapless and bullied troops to regroup and continue their frightened onslaught.

She continues, "In the end it will not be outsiders who decide Russia's future. The long, hard task of creating an alternative to Mr Putin's misrule can be performed only by Russians themselves. Meanwhile, democracies should lend Russian democrats their moral support, just as they did in the Soviet era. The U.S. President should press hard for Mr Navalny to be released, immediately and unharmed.

Christina continues, "The world needs dissidents like Navalny to hold the Kremlin to account. Without such checks, Russia will remain a thuggish kleptocracy, and its neighbours will never be safe."

Russkiy Mir

Christina explains, "Look I know you won't have as much depth as Antanov and me on Russian matters. We have both been through the training in Russia. I can see it from the inside, as well as later from the outside when we both stepped away from the ideology

Antanov speaks, "Yes, 'Project Russia' is a Russian nonfiction book series written during the first decade of the 21st century. The books have placed highly on bestseller lists in Russia. These books discuss aspects of political, economic, and ideological life in Russia and the world, and try to predict the future.

"Project Russia touched on social questions, saying that Russia faced collapse and needed a unifying idea. State structural problems in Russia were also analysed, exploring the possibility that the Russian Federation might cease to exist.

"The authors of Project Russia predicted the financial crisis of 2007–08 in their 2005 book, describing the approaching unrest and analysing its causes and the fall of the world order."

Christina continues, "Putin's crusade against a liberal European future is being fought in the name of Russkiy mir—'the Russian world', a previously obscure historical term for a Slavic civilisation based on shared ethnicity, religion, and heritage.

The Putin regime has revived, promulgated and debased this idea into an obscurantist anti-Western mixture of Orthodox dogma, nationalism, conspiracy theory and security-state Stalinism."

Antanov adds, "When we used to read the books we were given back in Arkhangelsk, it was noticeable how many were old and skewed in their view of the west. In a hothouse environment this worked, but really it was at odds with reality.

Christina continues, "The war with Ukraine is the latest and most striking manifestation of Putin's revanchist ideological movement. By that I mean Putin's retaliation, to recover what he believes to be lost territory. This time it has brought to the fore a dark and mystical component within it, with Putin a bit in love with death."

It could be a link with Scheppach, I thought.

Eckhart adds, "Putin's first public appearance since the invasion was a rally at the Luzhniki stadium packed with 95,000 flag-waving people, mostly young, some bussed in, many, presumably, there of their own volition.

"And Putin is playing the 'God on their side' card too. An open octagonal structure set up in the middle of the stadium served as an altar. Standing at it Mr Putin praised Russia's army with words from St John's gospel: "Greater love hath no man than this, that a man lay down his life for his friends."

"Putin's oration made much of Fyodor Ushakov, a deeply religious admiral who, in the 18th century, helped win Crimea back from the Ottomans. In 2001 he was canonised by the Orthodox church; he later (and I know this will sound a little crazy) became the patron saint of nuclear-armed long-distance bombers.

"He said that the storms of war would glorify Russia. That is how it was in his time; that is how it is today and will always be!'

Christina adds, "Putin's words and actions were direct from Stalin's playbook. After the Soviet Union was attacked by Germany in 1941, Stalin rehabilitated and co-opted the previously persecuted Orthodox church as a way of rallying the people. He also created a medal for outstanding service by naval officers called the order of Ushakov."

Antanov nods, "Putin's rally is not a mere echo or emulation; there is a strand of history which leads quite directly from then to now. Links between the church and the security forces, first fostered under Stalin then stronger after the fall of Communism.

"The allegiance of its leaders, if not of all its clergy, has now been transferred to Mr Putin. Kirill, the outlandish patriarch of the Russian Orthodox church, has called his presidency 'a miracle of God'; he and others have become willing supporters of the cult of war. No wonder the Christian church is trying to disown Russian orthodoxy.

"Kirill has declared the current war a Godly affair and praised the role it will play in keeping Russia safe from the horrors of gay-pride marches. More zealous churchmen have gone further.

"Elizbar Orlov, a priest in Rostov, a city close to the border with Ukraine, said the Russian army 'was cleaning the world of a diabolic infection'.

Antanov continues, "Putin has painted a world where the main culprits in this aggression are Britain and America—no longer remembered as allies in the fight against Nazis but cast instead as backers of the imaginary Nazis from which Ukraine must be saved."

He adds, "More important to the cult even than the priests are the siloviki of the security services, from whose ranks Mr Putin himself emerged. Officers of the FSB, one of the successors to the KGB, have been at the heart of Russian politics for 20 years. Like many inhabitants of closed, tightly knit, and powerful organisations, they tend to see themselves as members of a secret order with access to revealed truths denied to lesser folk. Anti-Westernism and a siege mentality are central to their beliefs. Mr Putin relies on the briefs with which they supply him, always contained in distinctive red folders, for his information about the world."

Christina remembers, "Just like Project Russia. The FSB delivered the book by courier services to various ministries dealing Russia's relationship with the world, warning them that democracy was a threat and the West an enemy."

Of course, the rest of the world just sat on its hands.

How to find Putin?

"With so many places to hide, how can we even find Putin?" I ask.

Charlie lists them out: His dacha in Ozero, his Palace, his Swiss ski lodge, the underground bunker in the Altai mountains, The Kremlin, his yacht, a plane?

Antanov adds, "There's also his command centre - The National Defense Control Center (NDCC) in Moscow - It is like a brightly lit version of the control bunker used by the Americans in that movie Dr Strangelove - But I doubt whether it would suit Putin to be holed up there. It is more the kind of place where he'd place his Generals."

He continued, "Putin is one for grandiose sets, big spaces with his courtiers in distant attendance. The yacht or plane don't sound too useful for that,"

Christina adds, "The ski lodge in Switzerland may be useful to park his family, out of harm's way, but I don't think he'd go there and if he did, then the news would certainly get out."

Charlie says, "So we are left with Ozero, where we were all a few days ago, his Palace and the underground bunker at Altai. I think we can rule out Ozero. I can't imagine Vitalievich and his buddies wouldn't know if Putin was holed up there."

Antanov says, "We've reduced it to a short list of two: The Palace on the Black Sea coast near Gelendzhik, Krasnodar Krai, and the bunker complex at Altayskoye Podvorie, in the Altai Mountains, on the River Katun."

Christina adds, "And we can't just follow Putin. There are his main followers, his absolute inner circle. We need to find out where they are."

Christina raises a good point. Vladimir Putin's Russia is not a one-man show. To understand how governance works, we will need to consider the power and complexity of the bureaucracy. Russia and Putin have been clever to tempt everyone to believe that Vladimir Putin makes all important decisions in Russia on his own. That politicians and bureaucrats then execute Putin's commands without fail in a system known as the 'power vertical;' and those political institutions serve merely to implement Putin's wishes.

This simplistic myth relates to how Russian decision-making is understood, to the implementation of decisions in Russia, and to the nature of the country's political institutions.

Putin's 'Direct Line' – an annual televised question-and-answer session during which the president hears from, and responds to, the problems of Russians across the country – combines all elements of the myth.

Putin appears to make decisions alone and, on the spot,

to solve callers' woes. He instructs officials to carry out these orders. And he engages directly with citizens, without the need for mediating institutions such as political parties or parliament.

To the extent that it reinforces misperceptions of Russia, this 'all-powerful Putin' myth can be framed in two ways. The 'positive' version – Putin as the 'good tsar' – suggests strong and competent leadership.

This myth makes Putin appear a more potent and unconstrained political force than is the case.

The 'negative' version of the myth, no less detrimental to a realistic understanding of Russian politics, highlights the pathologies of personalised decision-making and thus supports cartoonish Putin-as-dictator characterisations in the West.

Antanov speaks, "Since the Ukraine invasion began, Russian Defence Minister Sergei Shoigu and armed forces Chief of Staff Valery Gerasimov have become the central figures in Putin's war. For example, when Putin is not alone on the screen, they are usually around. Since the beginning of the Ukraine invasion, Shoigu and Gerasimov have become the faces of the war.

Christina adds, "The two are extremely close to Putin. They were Putin's military cortège during his television announcement on February 28 about having put Russia's nuclear forces on heightened alert. It is not surprising that the Kremlin decided to put Shoigu and Gerasimov in the spotlight. In Putin's eyes, they are the architects of the successful campaign to annex Crimea in 2014, Russia's military strategy in Syria as well as the support for the pro-Russian rebels in the Donbas region. They are also perceived to be being among the most loyal of

Putin's followers."

Antanov adds, " Shoigu wasn't even a full-time soldier. Some say he is the heir apparent to Putin. He is one of the rare members of the first circle of power to have had as much influence under Boris Yeltsin at the end of the 1990s as under Putin. He's a few years younger than Putin and began his political career at the end of the Soviet era, becoming defence minister in 2012 despite lacking military experience.

Christina speaks, "It is a peculiarity not uncommon under Putin, who is keen to keep senior officers out of this position. However, Shoigu also does not have any experience of the secret services, which is much less common among those close to Putin.

Antanov adds, "They say his great quality is that he is "a servant to the tsar and a father to soldiers,"

..."Or a 'perfect chameleon' capable of transforming himself at will to suit the pleasure of his leaders." adds Christina.

Antanov adds, "Shoigu is described as responsible for the fiction of the vast modernisation of the Russian army. As defence minister, Shoigu also supervised the feared Russian military intelligence service or GRU, which is suspected of having stepped up assassination operations in Europe in the 2010s, including poison attempts."

"Why the fiction of modernisation?" I ask.

Antanov answers, "They may have created ever more fearsome weapons, but they still run badly on old technology which has seldom been maintained. See a field of tank armour but know that only half of it can move and that a few of those units will have jammed guns."

"It's the same with their logistics. It may be boring to have to think about how to get supplies to the front - fuel, food, bullets etc., but without the supply chain the war grinds to a halt, just like that 40 km long Russian armoured convoy which ran out of fuel, " says Christina.

"Yes, it explains the lack of air superiority as well. A military force ill-prepared for an actual war," agreed Antanov.

"What about Valery Gerasimov?" asks Charlie.

"He's a career soldier," answers Antanov, "He served in the armoured divisions of the Red Army throughout the former Soviet Union."

"Gerasimov was also one of the commanders of the North Caucasus army during the second Chechen war (1999-2009). Anna Politkovskaya, a journalist who was gunned down in 2006 in her own apartment's elevator, called out the brutal methods of the Russian Army, including the threats they made toward their own troops.

Christina adds, "Gerasimov's international fame is based on a misunderstanding. He is said by some to be the inventor of Russian 'hybrid warfare', which combines the use of conventional weapons with non-military methods – such as disinformation or cyberattacks – to prepare the ground for soldiers. There is even a 'Gerasimov doctrine' named for this military approach. I had to learn some of it when I was being trained in Arkhangelsk. Gerasimov didn't invent it but he doesn't dispel the illusion."

Charlie speaks, "Okay, so we now have two sites to

watch and an additional two Generals to track."

"At least two, " says Antanov, "You know that the original 'Dead Hand' Perimeter systems are still running in Russia?"

I remembered this, but not that it was ever used. Like something out of villain central casting, the Soviet Union developed a world-ending mechanism that would launch all its nuclear weapons without any command from an actual human.

So called MAD - Mutually Assured Destruction.

Russia currently has an estimated 1,600 deployed tactical nuclear weapons, with another 2,400 strategic nuclear weapons tied to intercontinental ballistic missiles. This makes Russia the largest nuclear power in the world. All of these weapons are tied into the Perimeter, an automatic nuclear weapons control system.

In a crisis that might mean a first strike from the United States, high-ranking government officials or military commanders could activate the Perimeter. Perimeter would guarantee that the Soviet Union (and now, Russia) could respond even if its entire armed forces were wiped out.

Once switched on, the Perimeter system can launch the entire Russian nuclear arsenal in response to a nuclear attack. It was part of the Cold War doctrine of mutually assured destruction, a means of deterring nuclear attacks by ensuring the side who initiated a first strike also would be annihilated.

Called "Dead Hand" in the West, the theory is that a command and control system measures communications on military frequencies, radiation levels, air pressure, heat, and short-term seismic disturbances. If the measurement points to a nuclear attack, the Perimeter begins a sequence that would end in the firing of all ICBMs in the Russian arsenal.

Perimeter would launch a command rocket, tipped with a radio warhead that transmits launch orders to Russian nuclear silos, even with the presence of radio jamming. The rocket would fly across the entire length of the country. After several test launches to prove the viability of such a command rocket, the Perimeter system went online in 1985.

The Soviet Union never confirmed that such a system ever existed, but Russian Strategic Missile Forces General Sergey Karakaev confirmed it to a Russian newspaper in 2011, saying the U.S. could be destroyed in 30 minutes. Russian state media outlets suggest the system was upgraded to include radar early warning systems and Russia's new hypersonic missiles.

In the United States, similar technologies were developed. Seismic and radiation sensors are used to monitor parts of the U.S. and the world for nuclear explosions and other activity, but the U.S. military never created an automatic trigger for its arsenal. Instead, it ensured that American humans with the ability and authority to launch a second strike would survive a first strike.

Since the Perimeter is still active, the danger of an automatic, computer-generated nuclear strike still exists. Now that Russian President Vladimir Putin has put Russia's nuclear weapons on high alert, he might have placed Russia's doomsday device on notice as well.

"This is where Antanov and I can assist," says Christina, "It is risky, but I can ask Blackbird for their location. They must have a command centre somewhere, and it is presumably close to Putin."

"Okay," says Antanov, "I guess we will need to reconvene in the morning, when Blackbird can give us some direction."

Blackbird

The next morning, we all assemble in Christina's suite again. She explains she had been able to contact Fyodor Kuznetsov, her handler, codenamed Blackbird.

Christina made the call with Antanov, who was also run by Blackbird. Christina explains that she has 'left' the FSB now, although Antanov continues to have a role.

"It is delicate, because Kuznetsov will want to remain cautious, although he trusts me and Antanov implicitly," she explains.

Antanov nods, "It would have been very difficult to ask him if he could help locate Vladimir Putin, Sergei Shoigu and Valery Gerasimov. Blackbird is smart and would know instantly why we were asking."

Christina explains, "Yes, but Blackbird has not been impressed with the current regime, since before Serbia, when Russia rolled in for a take-over. He could see Putin was beginning to lose perspective, even then."

Christina looks around the room and then continues, "I didn't ask him a direct question, because I could tell he was worried. Blackbird is caught between loyalty and distaste for the current regime.

Now he is less well-regarded by the Kremlin, it would have put him in danger if I'd asked him directly about Putin."

"Instead, I asked him for the whereabout of my old

friends, Irina Morozova and Eckhart Bloch, from Saint Petersburg. He didn't know but said he would text me. About half an hour later I received a number. It began with '+7 727' and I realised they must still be in Saint Petersburg. "

"I called the number and it was answered after three rings!"

"Irina thought it was a hoax to begin with - me calling them - but then she realised it was really me when I started to talk about the rock-concert trip we all did together. I checked their current links into the Russian establishment, but they have both left the Service. It was all part of a planned exit when I worked with them several years ago. But that is a whole other story."

"I remembered how well wired-up Saint Petersburg is and so I had to explain our purpose to them in a kind of conversational code. Well, the good news is that they will be in on our plan and decided of their own will to come to Cyprus and to meet us. True to form they both had 'Go-Bags' ready and decided to make the last plane yesterday to Moscow and then to catch the once-a-day service here to Cyprus. It lands this afternoon in Paphos!"

"They must be some really good friends," said Charlie.

"Oh - they are the best," said Christina.

Antanov nodded, "I'd better call Artem, to get us a ride to Paphos - I think it is around 70 kilometres west of here."

Irina and Eckhart

Later, we walk out to the familiar sight of Artem's taxi, and climbed in. We struck lucky with Artem and his wonderfully air-conditioned taxi, and he told us that the journey should take less than an hour, because the road we would travel was the Limassol – Paphos highway, a modern four-lane, high-speed route. It links Limassol, the largest port, and Paphos, the top tourist destination on the island.

Soon enough we're on the road, which reminded me of a brand-new English dual-carriageway complete with white writing on green signage, except it was bathed in permanent sunshine. Artem told us that the road had a maximum speed limit of 100 km per hour, but also a minimum limit of 65 km per hour.

"I'll pick up your friends and then take you all to the beach in Paphos. There are some lovely restaurants along the waterfront," declares Artem.

I wondered how many of them were run by members of his family.

We soon arrive at the small airport, which appears to be a single terminal, although Artem is telling us all that there is another older terminal as well. He seemed to know his way around and is waved to by several of the staff. Then he produces an iPad from the front of the van and types in 'Irina Morozova' - which appears in big letters on the front of the iPad. He adds 'Eckhart Bloch' and then walks out into the terminal, reassuringly saying, "It's okay here," leaving us all in the minivan parked in a loading zone.

We work out that Irina and Eckhart would have arrived around 20 minutes ago. Sure enough, he quickly returns, with two extra people in tow.

"My god!" says Christina and runs to greet first Irina and then Eckhart. There is much rapid chattering as they reacquaint. I can see they are all good friends, if somewhat distant. Antanov does the same at a more leisurely pace and all four re-board the minibus with big grins.

"Let me introduce you to the others!" announces Christina and in turn we all say 'hello' and greet one another.

"Right," says Artem, "Now I take you to my cousin's wonderful bar on the waterfront in Paphos. It will take about 20 minutes. I will leave you there and maybe Charlie can call me when you want to leave."

We all agree this is a good plan and before we know it, we are climbing from the minibus into the hot sunshine and making our way to a balcony bar which overlooks the Mediterranean. There are screens above our head making the whole restaurant seem cool compared to the ambient temperature.

"Artem, this is perfect," says Charlie, and the rest of us nod in agreement.

"I called ahead to ask them to give you a good table and view, " says Artem, as they show us to a corner area which looks out onto the sea. I still can't get over the green sky, which is making the sea look green as well.

"Remember," says Artem, "Just call me and I'll be here in about 10 minutes!"

"Christina Nott! I had forgotten just how much travelling with you is always an adventure!" says Eckhart.

"This time we are right in the middle of something - Oh and thank you for coming along - you must have been in transit for hours!"

"It was a couple of flights; to Moscow and then to here, although the flight to Paphos was on a holiday plane!" answers Eckhart.

"Maybe this will be a holiday for us!" answered Irina, "So what has been happening?"

Christina and sometimes Antanov regale Irina and Eckhart with the story of how we all came to Cyprus, and what we are doing. Eckhart is particularly interested when we mention the Tupolev plane. Christina doesn't mention the unusual qualities that Charlie, Chantal and I possess.

"So, what do you think you need us for?" asks Irina, "You know already that we can be more than backup singers!"

Irina explains

"It's crazy in Russia now," Irina explains, "Most of us don't like what Putin is doing, and none of us can see it ending well. It is destroying our economy, our reputation and many of the wealthy Russians are losing money so quickly they cannot believe it."

Irina continues with a jaded voice, "Russian television is filled with propaganda, the independent channels have closed down and foreign channels are blocked. By some technicality, the Al Jazeera service in English is still getting through, and that's where we and many Russians are getting news."

Eckhart explains, "To give a sense of Putin's reach, after he jailed the leader of Russia's opposition, Alexei Navalny, we hear that Navalny has just been sentenced to another nine years imprisonment. Putin is a loose cannon. He surprised his generals with a war, is telling the mass of Russians a different story and has silenced his opposition party.

"Navalny will probably be moved from Vladimir, where he has been kept for more than a year, to a yet harsher maximum-security jail elsewhere. The crime for which he was sentenced is a trumped up one of fraud.

Eckhart continues, "We have the parallel that the Ukrainians want to embrace many, if not all, the values held dear by other European nations. Mr Navalny wants the same for Russia. Vladimir Putin cannot countenance either ambition and has made up a phoney excuse to go on a disgusting and violent bombing campaign against civilians."

Irina says, "Yes, Putin sees himself like a modern-day crusader against the liberal future offered in Europe. He is fighting in the name of Russkiy mir — 'the Russian world', that obscure historical term for a Slavic civilisation based on shared ethnicity, religion, and heritage.

"Putin's regime has revived, promulgated and debased this idea into an obscurantist anti-Western mixture of Orthodox dogma, nationalism, conspiracy theory and security-state Stalinism.

Irina continues, "You must remember, Christina, when we talked about Putin's rise, back when we sat in Literaturnoye Kafe in Saint Petersburg? It's just more of the same gangsterism, but now he has the overt power to swagger through with his crazy plans. His inner circles fear him. If they don't do what he says, they'll end up in a prison camp or their families will disappear."

Eckhart speaks, "Though Putin's ascension to the presidency in 2000 was helped by his willingness to wage war in Chechnya, his mandate was to stabilise an economy still reeling from the debt crisis of 1998 and to consolidate the gains, mostly pocketed by oligarchs, of the first post-Soviet decade.

Eckhart adds, "Putin's contract with the Russian people

was based not on religion or ideology, but on improving incomes. His new ideology of isolationism appeared in some of the darker corners of the power structure. It took two years before his new way of thinking became obvious to the outside world. His Munich speech in 2007 was when Putin formally rejected the idea of Russia's integration into the West. He went so far as to tell a press conference in Moscow that nuclear weapons and Orthodox Christianity were the two pillars of Russian society, the one guaranteeing the country's external security, the other its moral health."

"This was around the time I wanted to get out," said Christina, "You know, from the life, the FSB, all of it."

"And it is when I met you in Saint Petersburg, and you had already secured financial freedom and were using that 'one last job' to guarantee it," said Irina.

"It was shortly before the 'Snow Revolution' that we did that Tour," said Eckhart, "I can remember thinking 'What would Christina do?' during that time - but of course, you had gone..."

"Ah yes, I watched the Snow Revolution unfold from afar," answered Christina, "I was worried about Russia at that time, as Putin's madness and faux-ideology crept over the entire Motherland."

Eckhart continues, "Both Irina and I were still attached to the FSB at that time. We were not able to fully engineer our escape for another couple of years. Fortunately, in Russia, money talks. "

He adds, "The Snow Revolution began as protests against the Russian legislative election results. You remember, the vote stuffing and so on. They were motivated by claims by Russian and foreign journalists, political activists, and members of the public that the election process was deeply flawed and being run by gangsters.

"After a week of small-scale demonstrations, Russia saw some of the biggest protests in Moscow since the 1990s. The focus of the protests was against the ruling party, United Russia, and its leader Vladimir Putin, the current president, previous prime minister, and previous two-term president, who announced his intention to run again for President in 2012.

Irina adds, "Then, another round of large protests took place. They were named "For Fair Elections" and their organizers set up the movement of the same name. Initial protest actions, organised by the leaders of the Russian opposition parties and non-systemic opposition sparked fear in some quarters of anti-regime protest movements and an accompanying change of government."

Eckart adds, "This is when it became complicated! On the first days following the election, Putin and United Russia were supported by rallies of two youth organisations, the government-organised Nashi, and United Russia's Young Guard. They set about confusing everyone by running counter-protests."

He continues, "On 24 December Sergey Kurginyan organised the first protest against what was viewed as "orange" protesters in Moscow, though the protest also went under the same slogan "For Fair Elections".

Irina speaks, "By the following February, more protests and pro-government rallies were held throughout Russia. The largest events were in Moscow: the 'anti-Orange protest' (alluding to the Orange Revolution in Ukraine), aimed against 'orangism', 'collapse of the country', 'perestroika' and 'revolution'. Putin kept using that motif right through to the start of the bombardment of Ukraine."

Eckhart adds, "By 6 May 2012, protests took place in Moscow against rigged elections. It was the day before Putin's inauguration as President for his third term. Some called for the inauguration to be scrapped. The protests were marred by violence between the protesters and the police. About 400 protesters were arrested, including Alexei Navalny, Boris Nemtsov, and Sergei Udaltsov and around 100 were injured. On the day of the inauguration, 7 May, over 100 protesters were arrested in Moscow so that in June 2012, laws were enacted which set strict boundaries on protests and imposed heavy penalties for unauthorised actions.

Irina adds, "I can remember, around that time, that Alexei Navalny spoke and was greeted with an ovation. He said there were enough people present at the protest to overrun the Kremlin, but that they were committed to remaining peaceful, at least for the moment. I copied his speech to my phone at the time, and here it is:

"I can see that there are enough people here to seize the Kremlin and the Moscow White House right now. We are a peaceful force and will not do it now. But if these crooks and thieves try to go on cheating us, if they continue telling lies and stealing from us, we will take what belongs to us with our own hands. ... These days, with the help of the zombie-box (TV), they are trying to prove to us that they are big and scary beasts. But we know who they are. Little sneaky jackals! Is that right? Is that true or not?' "

Now Antanov speaks, "There it was: the last act of defiance before Putin tightened the screws on everything. After a hearing a message of 'Russia without

Putin' the securocrats and clerics started to expand their dogma into daily life."

Irina adds, "Like in Saint Petersburg years before, it quickly became a regime which sustained networks of corruption, rent extraction and extortion required religion and an ideology of national greatness to restore the legitimacy lost during the looting. "

Eckhart says, "Navalny showed a video of Putin's palace near Sochi - which Putin has since denied owning."

Antanov adds, "Yes. Covering up things of such size requires a lot of ideology. At that point it was still possible to see the ideology as a smokescreen rather than a product of real belief. Perhaps that was a mistake; perhaps the underlying reality changed."

I thought that, either way, the onset of the Covid-19 pandemic brought a raising of the ideological stakes. Putin arranged constitutional changes which removed all limits on his term in office. His people also installed new ideological norms: gay marriage was banned, Russian was enshrined as the 'language of the state-forming people' and God given an official place in the nation's heritage.

Eckhart says, "Putin's long periods of isolation sees him become preoccupied with history, paying particular heed to figures like Konstantin Leontyev, an ultra-reactionary 19th-century visionary who admired hierarchy and monarchy, cringed at democratic uniformity, and believed in the freezing of time.

He adds, "Putin spends much time with Yuri Kovalchuk, a close friend who controls a vast media group. According to Russian journalists they discussed Mr

Putin's mission to restore unity between Russia and Ukraine. Hence a war against Ukraine which is also a war against Russia's future—or at least the future as it has been conceived of by the Russia's sometimes small but frequently dominant Westernising faction for the past 350 years."

I think about all of this. Putin's war is intended to wipe out the possibility of any future that looks towards Europe and a liberal modernity. In Ukraine there would be no coherent future left in its place. In Russia, the modernisers would leave as their already diminished world was replaced by something fiercely reactionary and inward-looking.

He is building a new iron curtain and the countries such as Ukraine and Belarus become the buffer zones. No wonder there is much discussion of which could become members of NATO, and of Latvia and Estonia already on the inside.

The Russian-backed republics in Donetsk and Luhansk become a subverted model. There, like in the old days in Dresden or Saint Petersburg, crooks and thugs are elevated to unaccustomed status, armed with new weapons, and fitted with allegedly glorious purpose: to fight against Ukraine's European dream. In Russia they would be tasked with keeping any such dream from returning, whether from abroad, or from within a cell.

Дедовщина

Дедовщина Dedovshchina

Dedovshchina: literally: reign of grandfathers is the informal practice of hazing and abuse of junior conscripts historically in the Soviet Armed Forces and today in the Russian armed forces, internal troops, and to a much lesser extent FSB, border guards, as well as the military forces of certain former Soviet Republics. It consists of brutalisation by more senior conscripts, NCOs, and officers.

Dedovshchina encompasses a variety of subordinating and humiliating activities undertaken by the junior ranks, from doing the chores of the senior ranks, to violent and sometimes deadly physical and psychological abuse, not unlike an extremely vicious form of bullying or torture, including sexual torture and rape. When not leaving the army seriously injured, conscripts can suffer serious mental trauma for their lifetime. It is often cited by former military personnel as a major source of poor morale.

Often with the justification of maintaining authority, physical violence or psychological abuse can be used to make the "youth" do certain fatiguing duties. In many situations, hazing is not the goal, and senior conscripts exploit their juniors in order to provide themselves with a more comfortable existence, akin to slavery, and the violent aspects arise when juniors refuse. There have been occasions where soldiers have been seriously injured or killed.

From Wikipedia, the free encyclopaedia

Big Story

Now it was Eckhart's turn to speak.

"This is a very big story. But it's not only about Ukraine. It's about the world, about the politicians of the world and I think we can speak about it after Ukraine wins. On one level it's not about who has more weapons or more money or gas or oil.

"That's the very first thing that I understood, when Russia made a move on Crimea in 2014. I was tasked with a security job over there as part of the FSB but wearing an unmarked Russian uniform. In fact, most of the Russian troops deployed were not wearing any form of identification. Within hours of the treaty's signing, a Ukrainian soldier was killed when masked gunmen stormed a Ukrainian military base outside Simferopol.

"I knew we'd all been through clandestine training, but this was wholesale deception, by not showing our true colours to the Ukrainians running Crimea. I thought we behaved like cowards."

Eckhart continues, "Russian troops moved to occupy bases throughout the peninsula, including the Ukrainian naval headquarters in Sevastopol, as Ukraine initiated the evacuation of some 25,000 military personnel and their families from Crimea. On 21 March, after the ratification of the annexation treaty by the Russian parliament, Putin signed a law formally integrating

Crimea into Russia.

"As international attention remained focused on Crimea, Yatsenyuk, the head of Ukrainian Economy and later Ukraine's Prime Minister negotiated with the IMF to craft a bailout package that would address Ukraine's $35 billion in unmet financial obligations.

"He also met with EU officials in Brussels, and on 21 March Yatsenyuk signed a portion of the association pact that had been rejected by the previous leader Yanukovych in November 2013.

"The IMF ultimately proposed an $18 billion loan package that was contingent on Ukraine's adoption of a range of austerity measures that included devaluation of the Hryvnya and curbs on state subsidies that reduced the price of natural gas to consumers.

Eckhart sighs, "I honestly thought Yatsenyuk was fighting Ukraine's corner, even after the overrunning of Serbia. He spoke out against Russia, tried to get Ukraine funding and was trying to position for EU-acceptance of Ukraine.

Antanov speaks, "I remember this. Yatsenyuk was smeared in a political campaign run by Oleksandr Onyshchenko, former Ukrainian MP. Onyshchenko admitted to the UK's Independent newspaper that he organised and funded a US$30 million smear campaign against Yatsenyuk and his government, playing on a corruption line. Russia's National Bureau of Interpol requested that Yatsenyuk be put on the international wanted list alleging his violation of articles of the Criminal Code of Russia.

"In other words, Russia wanted to get Yatsenyuk back

under their control in a cell. I remember because we were all sent the bulletin."

Irina speaks, "Volodymyr Zelensky is the current leader of Ukraine and has strong spirit. He also says that Ukraine can't be part of somebody else. He says that Ukrainians are the same as people in the USA and Europe and Russia. He says it's not about who has more weapons or more money or gas or oil.

She adds, "Zelensky sees that some politicians live in an information vacuum. What we see is that this is a closed atmosphere with Putin now. It means Putin can't understand or he couldn't know what's going on outside. Now we see a Putin obsessed with his dream of being a mini-tsar.

Eckhart starts again, "Yes, but pre-emptive sanctions would have given more time to Ukraine's military to prepare for Russia's further invasion. They would have shown Belarus what could happen if pre-emptive sanctions involving Russian businesses, oil, and gas exports, and more were taken, and this considering that Belarusians do not support Russia's war against Ukraine.

Irina adds, "Zelensky raised the Nord Stream 2 pipeline with Biden and Merkel, when she was still in office, and later with Scholz. He said the first step will be to launch it, then Russia will block gas supplies to Ukraine, and next they will apply pressure, including on Moldova, and then Russia will block supplies to split countries within the EU. Like a gangster drug dealer, they know how to apply pressure."

Eckhart again, "Even now, weeks into the conflict, western partners have still not completed the sanctions on disconnecting the banking system from SWIFT, many more banks have not been disconnected. Much fuss is made but less happens. They have taken very important steps to support Ukraine, but the central bank of Russia has not been disconnected. They say they will impose an embargo on Russian oil and gas exports. All these sanctions are incomplete. They have been threatened, but not yet implemented.

Eckhart sounds angry, "They also kick the triggers for action along the road. Now we are hearing those certain decisions depend on whether Russia launches a chemical attack on Ukraine. It's disgusting. As if they think that Ukrainians are like so many guinea pigs to be experimented on."

Christina speaks, "Many countries view Russia now through a military-strategic lens and are using Ukraine as a shield. It is good that they are on the side of Ukraine, but they must stop being defensive in their dialogue with Russia. They can act offensively. SWIFT is still operating in Russia for the leaders of Russia. Don't forget that ordinary Russians are now isolated, deprived of information. They don't know what's going on.

Eckhart's voice is still angry, "Ukrainian people are dying. Russian people don't know what's going on. They don't understand. Social media have been shut down and a lot of people are watching state-run television. It's a big problem because the Kremlin controls all levels of power and all this information. And it is painting the Russian people as co-conspirators."

Irina adds, "Meanwhile, the Russians block supplies to Mariupol, Melitopol, Berdyansk, Kherson, Kharkiv, but they're not in the bigger cities. And what do they do? In Melitopol and Berdyansk they are switching to roubles.

"They are kidnapping the mayors of the cities. They killed some of them. Some of them can't be found. Some of them are dead.

"Most of the Russian military are scared, they are operating under Dedovshchina rules, where they are bullied, frightened and harassed. It is no way to run the military."

Eckhart adds, "And some of them were replaced. They are doing the same thing that they did in Donbas in 2014. The same people are carrying out these operations. It's the same methodology. The West can't say, 'We'll help you in the weeks to come.' It doesn't allow Ukraine to unblock Russian-occupied cities, to bring food to residents there. People are simply not able to get out. There is no food, medicine or drinking water there. Some small cities have been destroyed. There are no people and no houses. All that's left is the name.

Christina says, "Putin knows that many western politicians are afraid of him and therefore afraid of Russia."

Eckhart nods, "That is right. Everyone has varied interests. There are those in the West who don't mind a long war because it would mean exhausting Russia, even if this means the demise of Ukraine and comes at the cost of Ukrainian lives.

Christina adds, "This is definitely in the interests of some countries. For other countries, it would be better if the war ended quickly, because Russia's market is a big one that their economies are suffering because of the war.

Antanov speaks, "From the west, Ukraine needs aeroplanes, tanks and armoured personnel vehicles.

They don't have as many as they need. The Russians have thousands of military vehicles, and they are coming and coming and coming.

Christina summarises, "Putin has a 20th-century view of a 21st-century country."

"If that!" says Eckhart, "I'd say it was almost a 'Middle-Ages' mentality. The invaders do not even mourn their own casualties. This is something I do not understand. Some 15,000 Russian soldiers have been killed in one month. Ukrainians talk about a war that has lasted for eight years. Eight years!"

Eckhart continues, "And in eight years, Ukraine also lost 15,000 lives. And Russia loses 15,000 of its soldiers in a month! Putin is throwing inexperienced Russian soldiers like logs into a train's furnace. And they are not even burying them. Their corpses are left in the streets. In several small cities soldiers say it's impossible to breathe because of the smell. It is the stench of rotting flesh."

Antanov speaks, "Ukrainian soldiers defended Mariupol. They could have left if they wanted. But there were still others alive in the city along with their wounded. And then there were the dead, the fallen comrades. Ukraine's defenders say they must stay and bury those killed in action and save the lives of those wounded in action. If people are still alive, the Ukrainian army must continue to protect them. And this is the fundamental difference between the way the opposing sides in this war see the world."

I'm aware that most of this conversation has been between four Russians, all ashamed of their country's actions driven through fear by an increasingly isolated Putin.

Chantal describes the situation

Chantal summarises:

"We can't be certain this whole situation is being driven by Putin. Certainly, The Kremlin is making it look that way, but it could be to throw us off balance. Putin also has many bases for operation. We are trying to rule some of them out. There's his dacha in Ozero, his Palace at Krasnodar Krai, his Swiss ski lodge, the underground bunker in the Altai mountains, The Kremlin, his yacht, a plane, and the National Defense Control Center in Moscow."

Irina smiles, "Well, that is easy. It must be the place that the least people know about, ideally protected from air strikes. He will be at the bunker complex at Altayskoye Podvorie, in the Altai Mountains, on the River Katun."

Eckhart nods in agreement, "Yes, crazy as it may seem, Putin will have picked the bunker complex. It is incredibly well provisioned and most of it is underground. They brought in special German tunnelling equipment when they built it, and they used Dostoyevskaya on the Lyublinsko-Dmitrovskaya Metro Line as a motif during construction. Don't think of rough chiselled walls, instead think of a shiny metro station. Putin tries so hard to copy Stalin. His bunker is like the metro stations Stalin had built all over Moscow. "

"How do you know all of this?" asks Charlie.

"I was one of his protection officers back in the day," explains Eckhart.

"Christina. You know some interesting people," says Charlie.

"The scary thing about Putin was his developing craziness," adds Eckhart.

"We were told about this by Dr Maria Gvasalia, when we were at a meeting with Svalov Rollan Vitalievich," answers Christina.

"Vitalievich?" asks Irina, "Is he still alive?"

"Very much so when we saw him, but I can't be sure what happened when a truckload of gun carrying heavies arrived at his Dacha in Ozero," answers Christina, "in fact, Vitalievich was planning a hit on Putin."

"Well Putin is extremely paranoid, also," says Eckhart, "He will probably be wearing the latest personal armour. If you see him now, outside in public, like at the Luzhniki stadium, he will be wearing big bulky clothes. Not because it is cold, so much as because of the protection it offers. He distracts the message by wearing a 1.5 million rouble coat, and a 350 thousand rouble sweater. The press are more interested to find out it comes from Loro Piana than whether or not he has Kevlar underneath."

"Well, we think that the Russian elite are plotting to

poison Putin," explains Irina, "There would be a wry sense of irony if they did, after it is what he ordered for so many of his enemies."

Eckhart says, "It is why spy chief Oleksandr Bortnikov, head of the FSB agency, was removed. He was once thought to have been lined up as Putin's replacement. He used to be a top sidekick of the president, but 70-year-old Bortnikov is said to have fallen out of favour over 'fatal miscalculations' in the conflict with Kyiv.

"Putin's Chief Directorate of Intelligence said: 'It is known that Bortnikov and some other influential representatives of the Russian elite are considering various options to remove Putin from power. In particular, poisoning, sudden disease or any other coincidence is not excluded.'"

Christina comments, "But that is so much more difficult to do, than, say a close-range bullet, or an explosion."

"Not necessarily," answered Eckhart, "You remember that I was a protection officer? When they excavated Altayskoye Podvorie they also had to install hundreds of kilometres of HVAC. Heating, Ventilation and Air Conditioning. The difference with the HVAC used in Altayskoye Podvorie is that they built in security too. The Central Air Handling Unit could also pump a blend of other mixture around the system, or a selected part. It was designed as a ring defence for the whole complex. The outer rings, if breached, could inject a toxin into the HVAC, which would destroy incoming marauders in a similar manner to the boiling oil used in bygone times."

Irina adds, "It would be a little more subtle than the head-on attack planned by Vitalievich and could even be made to look like an accident."

Eckhart agrees, "There's a battle inside the Kremlin even in advance of Vladimir Putin's departure from office, with claims that the president presides over a secret multibillion-dollar fortune.

"Rival clans inside the Kremlin are embroiled in a struggle for the control of assets as Putin prepares to transfer power to his hand-picked successor, Dmitry Anatolyevich Medvedev.

Eckhart continues, "Western observers widely believed Medvedev was too liberal and too pro-Western for Putin to endorse as a candidate. Instead, they expected the candidate to arise from the ranks of the so-called siloviki, security and military officials many of whom were appointed to high positions during Putin's presidency. The silovik Sergei Ivanov and the administrator-specialist Viktor Zubkov were seen as the strongest candidates. We still don't know exactly why Putin proposed Medvedev although I guess at stake are billions of dollars in assets belonging to Russian state-run corporations. Additionally, details of Putin's own personal fortune, reportedly hidden in Switzerland and Liechtenstein, are being discussed for the first time. It's the siloviki vs the liberal clan, of which Medvedev is a member.

Irina says, "After eight years in power Putin has secretly accumulated more than $40bn (£20bn). The sum would make him Russia's - and Europe's - richest man.

"Putin owns vast holdings in three Russian oil and gas companies, concealed behind a "non-transparent network of offshore trusts".

Eckhart says, "Irina is right: Putin's name doesn't appear

on any shareholders' register, of course. The scheme is of successive ownership of offshore companies and funds with a final point buried in Zug, Switzerland, and Liechtenstein.

Irina says, "Discussion of Putin's wealth has previously been taboo. But the claims have leaked out against the backdrop of a fight inside the Kremlin between a group led by Igor Sechin, Putin's influential deputy chief of staff, and a "liberal" clan that includes Medvedev.

"The Sechin group is made up of siloviki - Kremlin officials with security/military backgrounds. It is said to include Nikolai Patrushev, the head of the Federal Security Service (FSB), his deputy Alexander Bortnikov, and Putin's aide Viktor Ivanov. I find it difficult to believe that Putin would favour the liberals over the siloviki, although that seems to be the outward appearance."

Christina queries, "Well maybe that is the idea. We know Putin is a master of deception and lies. Perhaps this is to keep everyone guessing?"

Eckhart adds, "Those associated with the liberal camp include Roman Abramovich, the Russian oligarch once the direct owner of Chelsea football club who is close to Putin and the Yeltsin family. Other members are Viktor Cherkesov, the head of the federal drug control service, and Alisher Usmanov, an Uzbek-born billionaire."

"And Abramovich has already been targeted in a most silovik manner," says Irina.

Christina adds, "I doubt the struggle has much to do with ideology. It is a war between business competitors. Putin's decision to endorse Medvedev as president - who has no links with the secret services - dealt a severe blow to the hardline Sechin clan - it will soon be full-on bratva wars all over again."

Irina adds, "Some analysts have said Putin would like to retire but has been forced to carry on in order to shield Medvedev from siloviki plotting. Others say Putin wants to stay in power. There is no secret any longer about his cancer nor about the steroids he is receiving as treatment.

I surmise Putin could be running interference to make things seem very uncertain for everyone.

"The siloviki are not nice," said Christina, "They play like tough and ruthless gangsters - exactly the way Putin himself used to operate"

"Used to?" queries Charlie, "His current actions seem no different."

Christina adds, "The wave of re-nationalisations under Putin transformed Putin's associates into multimillionaires. The dilemma now facing the Kremlin's elite is how to hang on to its wealth if Putin leaves power. Most of its money is located in the west. The pressing problem is how to protect these funds from any future administration that may seek to reclaim them?"

Eckhart adds, "Yes, the first hints of the intra-clan warfare gripping the Kremlin emerged when the FSB arrested General Alexander Bulbov, the deputy head of the federal drug agency, and part of the liberal group. His arrest saw a surreal standoff, with his bodyguards and FSB agents pointing machine guns at each other. Sergei Storchak - another liberal - was also arrested and charged with embezzling $43.4m. He is currently in prison."

Irina adds, "But the liberal group - one of several competing factions inside the Kremlin - has struck back. Oleg Shvartsman, a previously obscure fund manager, gave an interview to Kommersant newspaper claiming he secretly managed the finances of a group of FSB officers. These officers were involved in 'velvet reprivatisations', - in effect forcibly acquiring private companies at below-market value and transforming them into state-owned firms. The assets were redistributed via offshore companies."

Christina looks agitated, "But these are the same moves that Putin did in Sankt Petersburg and during his early days in Moscow!"

"Correct," said Eckhart, "The randomised corruption of the 1990s has given way to the systemic and institutionalised corruption of the Putin era. Members of Putin's cabinet personally control the most important sectors of the economy - oil, gas, and defence. Medvedev is chairman of Gazprom; Sechin runs Rosneft; other ministers are chairmen of Russian railways, Aeroflot, a nuclear fuel giant and an energy transport enterprise. Putin has created a new, more streamlined oligarchy where the crown jewels of the country's wealth have ended up in the hands of Putin's inner circle,"

"You know something?" says Eckhart, "Gazprom Neft built Putin's bunker in the Altai Mountains, The above surface version of it looks like a luxury resort. Above ground is maybe one tenth the size of the underground complex. And you know who is the owner of Gazprom? Alexander Ivanovich Medvedev! Wheels within wheels."

Irina adds, "And Roman Abramovich and Boris

Berezovsky acquired the Gazprom Neft for a knock down US$100 million, after bidding through several front companies that had been set up for this specific purpose. It is worth $35.2 billion now. That is some return."

"Crooked return?" queries Christina, "Even Blackbird thinks that the west has misunderstood Putin. Everyone has been distracted by his 'neo-Soviet' image. Putin is ultimately a 'classic' businessman who believes money can solve any problem, and whose psychology was shaped by his experiences working in the St Petersburg mayor's office in Russia's crime-ridden early 1990s."

Antanov says, "We shouldn't forget the degree to which the west was taken in by Putin. In 2007, he was Time Magazine's Person of the year and even in 2014 he was short-listed as a runner-up."

Eckhart adds, "An unprecedented silent battle is taking place inside the Kremlin in advance of Vladimir Putin's departure from office with claims that the president presides over a secret multibillion-dollar fortune."

Charlie tries to summarise, "So, I get it: Rival clans inside the Kremlin are embroiled in a struggle for the control of assets as Putin prepares to transfer power to a hand-picked successor. And now we see a new Iron Curtain is grinding into place. As Ukraine bears the brunt of heavy bombing, we can also see an economic war deepening as the military conflict escalates and civilian casualties rise."

Drones

"So how can we formulate a multi-point attack?" asks Charlie, "We'll need to deploy the drones as well."

I suggest, "Maybe we need our 'mistress of chaos', our shapeshifter and a couple of military people to go along to Geçitkale Airbase?"

"Good plan, and we'll need Artem to ferry us to the base, " answered Christina.

"At this stage, all we need to do is convince them that they have weapons which they can validly deploy," says Eckhart.

"Maybe I'll need that NATO uniform again?" asks Chantal.

"I think I might need one, actually," asserts Christina, "I'll know all the ranks and the terminology. I won't get caught out," she explains, "Besides, I know how to wear the beret correctly."

Antanov chuckles, both he and Christina had noticed a few uniform infractions when Chantel had been clothed by Charlie.

I notice Chantal look relieved at this news, but also that Charlie is ready for action. It looks as if Christina will be the high-ranking leader, with Antanov, Charlie and me as backup. I'm beginning to learn about being in Christina's band.

Artem is to drive us to the base, which is right at the other end of Cyprus in Turkish territory.

First Artem takes Irina, Eckhart, and Chantal back to our apartment block, and then drives the rest of us to Geçitkale, in Turkish Cyprus. I wonder about the border crossing, but Artem says everything will be simple.

Artem estimated 3 hours to cover the 200 kilometres to the airbase. But we have good traffic and the last section from Limassol to the airbase takes about an hour and a half.

Artem was also correct about the border crossing. We had to put a white paper into our passport to have it stamped, with our corresponding passport number. Artem had to show minibus insurance and then we were on our way and soon at the airbase.

This is where Christina takes control and with her Turkish NATO uniform directs us all to the entrance to the facility. Charlie pulls a couple of shapeshifts to get us through the barriers and we are then inside, driving along in Artem's minibus.

He has been to the base many times and knows the way to the main building. We disembark and Christina requests to see the Airbase Commander. She is speaking English now, but her NATO rank and beret are helpful and we are soon whisked into a holding lobby.

She explains that Charlie and I are civilians and Antanov, is a Colonel, but in plain clothes. She explains that I am a lawyer, sent to ensure that the entire process is legally executed. Charlie is a non-specific NATO representative. We have been tasked with gaining agreement that the Turkish NATO Bayraktar TB2s can be deployed, but that there is no immediate call for action.

Christina's Tümgeneral rank is a NATO OF-7, which is higher than the base commander, who is the Turkish rank of Albay or Colonel, which has more stars but is a lower rank of NATO OF-5. Christina asserts her demands, in English, to the surprised Commander, who, with a graceful dignity agrees to everything. It is clear t base has never seen such a powerful woman officer in their military, nor are they used to conducting business in English.

There are a few administrative details to clear up, but Christina's assertions with her clearly superior rank is all that is needed for us to have mobilised a significant air threat, which is poised but won't do anything without further orders.

Christina gestures for us to leave and we back out of the meeting room and into Artem's waiting minibus. He sedately pulls away as we all look at one another.

"We did it," says Christina, "I did lay it on pretty heavy and explained that failure to accept the order would be a court-martial offence. That's why I was travelling with a lawyer. What is happening is unprecedented and so people don't really know how to act. We now have the whole airbase ready to activate their Bayraktar drones on our instructions."

بث الأخبار

bathu al'akhbar
Doha, Qatar
English version

KYIV, Ukraine — President Recep Tayyip Erdogan of Turkey agreed to expand supplies of one of the Ukrainian Army's most sophisticated weapons, a long-range, Turkish-made armed drone whose use in combat for the first time in Ukraine last fall infuriated Russian officials.

Mr. Erdogan's decision to provide weapons and diplomatically back Ukraine was a public rebuke to Moscow and another complicating factor in the mix of cooperation and conflict between Turkey and Russia, historical rivals for supremacy in the region around the Black Sea.

The promise of more arms for Ukraine, especially an offensive weapon like the Bayraktar TB-2 Turkish drone, is an extremely sensitive issue for Moscow, which claims that its security is threatened.

An American airlift of anti-tank missiles and small-arms ammunition continued with the arrival of a seventh cargo jet of weaponry to Kyiv.

At the same time, Russia denounced the Biden administration's announcement that it would send additional troops to NATO countries, with the Kremlin spokesman, Dmitri S. Peskov, accusing the United States of "igniting tensions on the European continent."

Russia's defense minister, Sergei K. Shoigu, said that the Russian military would send additional troops and equipment for military exercises in Belarus, which borders Ukraine to the north, adding to tens of thousands of soldiers already deployed there.

Turkey is a member of NATO but also maintains economic and military industry ties with Russia. And the two countries are also on opposing sides in two Middle Eastern wars, in Syria and Libya, and in the conflict between Azerbaijan and Armenia in the South Caucasus region.

Turkey has sold Ukraine armed Bayraktar TB2 drones that the Ukrainian military used for the first time in combat in the war with Russian-backed separatists last October.

Earlier, in a bid to reassure Moscow, the Turkish defense minister, Hulusi Akar, stressed Ankara's commitment to a treaty that restricts NATO forces' access to the Black Sea through the Bosporus and Dardanelles Straits, which Turkey controls. The accord, 'the Montreux Convention', prohibits aircraft carriers from crossing the straits and limits other warships to short voyages in the Black Sea.

Seven American cargo jets have carried a total of about 600 tons of military assistance so far including anti-tank weapons and small-arms ammunition.

The shipments included additional Javelin anti-tank missiles from the United States. Britain has airlifted about 2,000 light anti-tank missiles, known as NLAWs, to Ukraine in the past two weeks.

With approval from the United States, the Baltic countries of Estonia, Latvia and Lithuania said they would transfer more Javelin anti-tank missiles and Stinger antiaircraft missiles, plugging some holes in Ukraine's air defenses. Poland has also said it will send antiaircraft missiles.

With the additional British and American supplies, Ukraine now has more anti-tank missiles than Russia has operational main battle tanks in its military, though Russia's total including tanks in reserve is still larger.

Ukraine does not have to reach a so-called 'capability parity' with the Russian Army — an impossibility in any case — to deter a military intervention.

Towards World War III

We drive back from the base at Geçitkale to our apartments in Limassol. It already feels as if we have been here a month, although I realise it is only a matter of days. Human time can be dense.

Christina has phoned ahead and explained to the others that everything had gone to plan. I notice she speaks in a semi-coded way to the others and realise she is still thinking about us being bugged.

Back in our rooms, Charlie calls a meeting for us to plan the next steps. Antanov is concerned about the Tupolev being discovered and we are all worried about what we'd unleash if we activated the drones.

It will be our third night in Limassol and we all think will need to move the next day. We rebalance the occupants of the apartments so that Christina and Antanov, and Irina and Eckhart can share rooms.

That evening we sit on Christina's balcony discussing options.

"We've little choice but to start something if we genuinely want to end this war. However, if Putin has really activated the Perimeter system, then there could be nothing left at the end of it," says Charlie.

"I think there's often a coup to finish this kind of tyrant," says Christina.

Antanov adds, "Think about it: In most cases, scenarios of future war have rarely come to pass as originally envisioned. At least two inter-related reasons can account for this.

"First, due to the incredibly large number of variables to consider – geopolitical, technical, human, etc. – it is simply impossible to calculate how they will interact with each other"

"Yes," agrees Christina, "The second reason has to do with distinguishing between 'future war' and the 'future battlefield.' Regrettably, too many scenarios and models, whether developed by military organisations, political scientists, or fiction writers, tend to focus their attention on the battlefield and the clash of armies, navies, air forces, and especially their weapons systems."

Antanov agrees, "You are right. The broader context of the war needs to be examined – the reasons why hostilities erupted, the political and military objectives, the limits placed on military action, and so on. They are given much less serious attention, often because they are viewed as a distraction from the main activity that occurs on the battlefield."

Eckhart joins in, "Putin's decision for war initiation is almost superficial. Most wars have a run-rate of costs, and the Americans usually factor that into their economic calculations.

"Yes, but those calculations are also off by billions of dollars," said Antanov, "And then the contractors brought in to tidy up are onto a huge money-making scheme. Halliburton, Kellogg, Brown and Root and Bechtel along with Brant Industry and others, have made huge contracts in countries that had been demolished."

Eckhart continues, "Putin's decision to attack Ukraine is not one based on a Kremlin desire for world conquest, but rather as a defensive move motivated by fears of the elite. The fears that the future 'correlation of forces' won't favour the Kremlin and that internal weakness will lead to state collapse. Putin wants to re-establish boundaries but with one less variable. He can't stand the thought of Ukraine becoming friendly with NATO."

Antanov adds, "Then there's the timing of the war initiation. It hardly came as a surprise, with all the protracted build-up of troops close to the Ukrainian and Belorussian borders. Of course, the whole build-up was covered in lies, which the west was only too grateful to believe. Putin on the other hand would say anything he could get away with.

Christina adds, "He messed up the war logistics though. Frightened soldiers under the reign of Dedovshchina and kept in the dark about their mission. Faulty main battle tanks with insufficient fuel and armaments that can't withstand modern weapons like the American Javelin and Anglo-Swedish Next-generation Light Anti-Tank Weapons (NLAWs). Planes that could hardly fly and weapons systems that were unreliable. Shoigu must have been mad to agree to start hostilities in that condition."

Eckhart adds, "Then there's the boundary conditions of the onslaught. Ukraine now, but Moldova and Latvia next? I'm sure the war planners have looked into both scenarios."

Christina adds, "More fundamental is whether Putin expects to cross the nuclear threshold. On the NATO side, Supreme Allied Commander Europe and Supreme Allied Commander Atlantic could be pressed to authorise the use of tactical nuclear weapons against Soviet ground and naval forces. In war games they always refuse, fearing a Kremlin escalation. On the Kremlin side, there's little in the path of a determined Putin and even less in the way should Perimeter be triggered."

Antanov continues, "If Putin really doesn't care, and his military losses suggest he doesn't, then I'd expect to see chemical weapons deployed next. We'll see"

Antanov adds, "Writing an ending to a third world war is as difficult as writing the beginning. In the scenarios discussed here, unlike in much of the nuclear fiction genre, the war does not end in global Armageddon. I predict that the war will end with a coup in the Kremlin, or wherever the main controllers of the war sit. One feature that is pretty much a constant in all these scenarios is that as the war is taking place, so too are diplomatic negotiations. The problem is that the man on one side of the equation is a proven liar and may just be stalling for time to reposition or refresh their troops."

Charlie interrupts, "That's just it. We need to hasten the coup. Make Putin's supporters very afraid of the consequences."

I once again wonder why - as Watchers - none of the three of us can remember anything about the Klima Wars, nor how they start.

PART THREE

Eve of Destruction

The Eastern world, it is explodin'
  Violence flarin', bullets loadin'
  You're old enough to kill but not for votin'
  You don't believe in war, but what's that gun you're totin'?
  And even the Jordan river has bodies floatin'

But you tell me
  Over and over and over again, my friend
  How you don't believe
  We're on the eve of destruction

Don't you understand what I'm trying to say
  Can't you feel the fears I'm feeling today?
  If the button is pushed, there's no runnin' away
  There'll be no one to save with the world in a grave
  Take a look around you boy, it's bound to scare you, boy

And you tell me
  Over and over and over again, my friend
  How you don't believe
  We're on the eve of destruction

Yeah, my blood's so mad, feels like coagulatin'
  I'm sittin' here just contemplatin'
  I can't twist the truth, it knows no regulation
  Handful of senators don't pass legislation
  And marches alone can't bring integration
  When human respect is disintegratin'
  This whole crazy world is just too frustratin'

And you tell me
  Over and over and over and over again, my friend
  You don't believe we're on the eve of destruction
  No no, you don't believe we're on the eve of destruction

Phil Sloan / Steve Barri

How to create Rage

"We'll need to create a rage in the Kremlin. Instead of pointing outwards, it will need to be internal," says Christina.

"Financial first?" suggests Charlie, "If all those Kremlin elite think they are going to lose their wealth..."

"Maybe we should add some property damage to the equation?" suggests Chantal, "Imagine if a stack of drones deployed themselves over Krasnodar Krai?"

"And perhaps that Tupolev can be rediscovered on a course for Moscow?" adds Antonov.

"We need to get ahead of the Perimeter system, somehow," Eckhart adds, "If *Mertvaya Ruka* - Dead Hand - gets actioned, then an ageing Soviet command missile will fly right across Russia triggering all of the nuclear silos."

"That sounds like enough," says Christina.

"Enough for what?" I ask.

Christina spoke, "Enough to bring forward whatever plan the Kremlin elite have created to depose Putin. Vitalievich can't have been the only one with plans. Their rage is heightened since we now know that both the siloviki and the liberals are already arguing in the Kremlin, but that everyone is living in an information famine. It is destroying the Motherland."

Antonov speaks, "That plane we borrowed, it is a Tu-160M1 with the new navigation system and ABSU-200 autopilot. We can set it on its way, parachute out and let it continue crewless. I suggest we set its course for either Putin's Palace or the underground bunker."

"The threat is only good if they know the plane contains 4 nuclear warheads," says Christina.

Antonov answers, "That's easy to arrange. I'll light it up for the piloted part and then go stealth for the second part of the journey, but with a clear vector."

"But won't you have to fly slow for the parachute exit?" I ask.

"Correct, I'll pre-program the flight to pick up speed after we are clear. Think of it as fire and forget," answers Antanov, "Hey, Christina and I have done this over the White Sea about a dozen times."

"Never in a supersonic bomber though," says Christina.

Charlie and I exchange a look.

"You know something," I say, "I think I should fly the Tupolev. Maybe with Charlie?"

Antanov looks surprised, "Why? Do you have experience

with this kind of plane?"

"Yes, I've flown thousands of air miles in dozens of different planes. I had to pilot a stolen Sukhoi-27 to get out of Ukraine on the way here. It also had a glass cockpit, so I'm used to them. In my head I keep thinking of a X-Blade plane too. Not something from the movies, but something futuristic that I think Scrive has flown before, with modifications applied to it by Charlie."

It's very strange at this moment, answering as Farallon via Scrive's Presence - and even weirder having 'jamais vu' / 'flash-forward' moments, yet not being able to recall Big History.

"My main thought is that even if I do it, it is impossible for me to be in direct jeopardy, and neither can Charlie. If something happens to us, we will be placed back on the timeline in our own metaverse, I guess we will be close to our friend Drake. I know, it sounds like we've been eating the magic mushrooms, but it is true."

Charlie adds, "Scrive is right. If you, Antanov, or you, Christina, get into trouble, you will find yourselves back on the almost parallel path in your own metaverse. The same goes for Irina and Eckhart."

"Leaving just me!" says Chantal, "But you know something? I can create quite a lot of chaos now I know this version of reality is expendable."

She continues, "As an example, I can create some new news. Imagine if the western press started to publish cover stories in Russian. We could set a cat amongst the pigeons!

Chantal looks gleeful, "I've already suggested it to a

leading western left-leaning tabloid in London. If the Daily Mirror goes large with something aimed at the Russian people, then it can only be a matter of time before a few copycat articles appear. They won't be able to hold back every news item unless they entirely shut Russia off from the world."

"And most of the Russian elite also speak English as well," says Charlie, "It will be quite unsettling."

"Not as unsettling as having your hometown bombed to extinction by Russian militia," says Chantal.

Russian Laundry

Chantal continues, "I think we can also up the economic stakes further. I know that Scrive and Charlie spent a lot of time in London, England. Well, so did I. I even had to work in one of those investment houses. I know that over the past two decades, London became one of the preferred investment locations for Russian oligarchs, as well a key financial centre for Russian companies, all encouraged by British governments.

"London was known as 'the laundromat': an apparatus that allowed billions of pounds — some of it obtained through illegal or questionable means — to be siphoned out of the Russian economy and into trophy assets in the UK.

"Then, to cement things, people started to call London 'Londongrad'."

I knew this, of course, but wondered what Chantal was planning.

"We can use the strength to become a weakness," she explains, "The first wave of Russians to invest in UK real estate was new money wanting very substantial properties. The properties on the wish list were the biggest and the best, including some very close to the monarch's property in Buckingham Palace."

Chantal continues, "This is how we - I - can exploit the situation. The lack of UK rules requiring full disclosure of foreign owners of British property was an attraction for the Russian super-rich, who appreciated the privacy. If you came along with a photocopy of your passport, that was enough to be deemed suitable.

"Then, Transparency International, the anti-corruption campaign group, identified £-billions of UK property — nearly 150 land titles — bought by Russians who have been accused of links to the Kremlin or corruption. That is where we can strike."

Chantal asks, "Do you see? Under -regulated acquisitions held anonymously, but with a list available from Transparency International? We can tilt the table. Make all of the balls run our way."

Chantal adds, "Transparency International UK has been collating information on questionable funds from around the world being invested in UK property since 2016. This figure now stands at £6.7 billion. Of this total, £1.5 billion worth of property was bought by Russians accused of corruption or links to the Kremlin. Nearly £430 million worth is in the City of Westminster, while £283 million is in Kensington and Chelsea. Their analysis of how this £1.5 billion is owned reveals over half is held by companies in Britain's Overseas Territories and Crown Dependencies. The secrecy provided by these offshore financial centres is often used by those seeking to hide their ownership of assets.

"Then, they identified over two thousand companies

registered in the UK and its Overseas Territories and Crown Dependencies used in around 50 Russian money laundering and corruption cases. These cases involved more than £80 billion worth of funds diverted by rigged procurement, bribery, embezzlement and the unlawful acquisition of state assets."

Chantal pauses, then adds, "If I start looking at the Transparency lists, I can arrange for most of the assets to progressively drift toward a single, newly minted entity."

"The oligarchs will be furious!" says Christina, "But to do it you'd have to do several large-scale illegal things."

"Welcome to my world of chaos," says Chantal, "It's why we can act but the world can't."

"I knew that some of Putin's inner circle used to favour those Caribbean regions - sometimes they would be arriving to meet Putin or leaving to go on a visit to the islands," says Antanov.

Chantal adds, "Yes, that is why so many banks have Caribbean desks, often specialising in the British Virgin Islands - usually called BVI. In fact, the most vocal resistance to beneficial ownership transparency has come from the British Virgin Islands. The Transparency International research highlighted how criminals favoured BVI companies as money laundering vehicles, with over 90 per cent - that's over 1000 - of the legal entities used in over 200 major corruption and money laundering cases incorporated in this jurisdiction. Their use by people smuggling illicit loot across borders is so prolific that 'BVI' risks becoming synonymous with questionable financial dealings."

She adds, "To give an idea of the depth of criminality, the BVI's special envoy may seek to renegotiate its constitutional relationship with the UK to avoid having to implement public beneficial ownership registers. Part of the BVI's reticence to introduce transparency is the perception that this would drive business from its shores, which would have severe consequences for its public finances. According to its latest budget estimates, receipts from its corporate register – the biggest by far in the UK's offshore financial centres – account for almost 60 per cent of its state revenue."

"This looks like an interesting place to drop some chaos," said Christina.

"Well, it looks as if it has already started," said Chantal, "UK Prime Minister Boris Johnson declared he wants to impose punitive sanctions on individuals and companies with links to the Kremlin as well as crack down on money laundering."

"But it's probably not true?" asks Antanov, "I mean, that man has too many links to Russia and aren't his cabinet so many puppets of the Kremlin?"

Charlie speaks, "Yes, despite the bold rhetoric, it is not clear the laggardly Johnson government has either the political will or the tools to completely strip illicit money from the UK financial system. And there would be too many senior fat cats embarrassed if their sources of funding dried up. Once more I sense hollow words."

Chantal says, "We can also get the 2020 report by the UK parliament's intelligence and security committee. It noted that several British politicians had business interests linked to Russia or worked directly for major companies with ties to the Russian state. The report also highlighted a supporting cast of advisers, including bankers, lawyers, and estate agents, who were on hand to provide their services to oligarchs and Russian companies. It said that Russian money had been used in extending patronage and building influence across a wide sphere of the British establishment — public relations firms, charities, political interests, academia, and cultural institutions were all willing beneficiaries of Russian money. The report found that certain members of the Russian elite who were closely linked to Putin had donated to UK political parties and had a public profile which positions them to assist Russian influence operations.

Charlie says, "Go Chantal! go Limantour! It looks as if you have identified a whole house of cards which we could assist to tumble."

"But if we can do it, then why can't the prevailing legislation?" asks Eckhart.

Chantal says, "It's the art of managing the long grass. Any scoundrel will kick anything controversial into the long grass. They can just say that successive governments have faced a consistent problem: how to distinguish between the Russians whose money has been made in a legitimate manner, and the crooked others. Take 2018, when the May government introduced a measure called unexplained wealth orders. When certain criteria are met, they require the owner of an asset worth more than £50,000 to explain how they could afford it. But, so far, only a handful of orders have been issued by the courts.

"It's all very well now to say UK is no place for dirty money, no more golden visas awarded to overseas millionaires, increased placing of sanctions, but everything moves so slowly."

"Like partygate?" asks Irina, "You know, where some illegal parties were held in 10 Downing Street?"

Christina nods, "Just like partygate. They took too long and issued too few warrants; the fines were often for a derisory amount. The whole issue of repeated lies to parliament was deftly side-stepped. These people know how to dodge just about everything."

"Well, we'll see about that," smiles Chantal.

Charlie's plan.

Charlie stands and walks around the balcony. We still had beautiful views of the green sea and sky. I was even getting used to the unusual colours.

"Let's recap," Charlie says,

"We can fly a stealth bomber towards Putin's bunker.

"We can set off some drones, from Turkish Cyprus, targeted on the Ozero commune.

"We can run confusion through their information channels

"We can disrupt the financial system, which will massively inconvenience the oligarchs.

"But we need someone else to be triggered to action too," says Christina.

"Don't you think this is enough?" asks Chantal, "Will the Kremlin elite, who must be plotting some kind of coup, be triggered into action when they see how bad things can turn for Russia?"

Irina speaks, "We must tell Blackbird about the HVAC at the Altai bunker. That the air-con is part of the bunker defences."

"Putin's enemies will be planning to use some form of OTC," answers Eckhart. "It is the most reliably available nerve agent in Russia."

"Organophosphate Toxic Chemicals?" queries Charlie, "But I thought they were made illegal many years ago?"

"They were," answers Eckart, "but it is no secret that disposal is expensive. Nor is it a secret that the Russian Biopreparat program identified more lethal airborne toxins than even the American VX toxin."

"Eckhart is correct," says Irina, "So far, it is not clear how to get rid of residues of toxic chemicals and toxic products because of their impact in the environment.

Irina adds, "The army does not know, because it did not deal with questions of decontamination of civilian facilities that were not involved in military operations, and army requirements for results of decontamination are many orders of magnitude less stringent. Nor has civilian 'applied science' worked out these methods and standards, because in the early eighties it was only planning to organize such research at Novocheboksarsk."

She continues, "We saw these plans die along with the elimination of the laboratory. With the fall of the Soviet Union, Soviet properties were scrapped or deserted, along with their missions. It wasn't just Putin burning records. There has been practically no organisation of special ecological monitoring around today's chemical weapons storage bases, even though chemical weapons will have to stay there for a very long time."

Eckart continues, "Irina is right, The Convention on Chemical Disarmament has so far not only been unfulfilled by Russia but has actually been violated through the theatrical open destruction of chemical weapons. The Russian answer to ceasing production of the weapons was to build stores for them instead. The original organophosphorus toxins produced paralytic nerve action: tabun (GA), sarin (GB), soman (GD) and V-gases. The first three toxins were developed in Germany at the turn of the nineteen-forties and are usually referred to as second generation nerve agents."

He adds, "Third generation agents were developed during the Foliant program and yielded five toxins of a new type. One of these was refined into a fourth-generation nerve agent (A-232, "novichok-5"). This turned out to be convenient for combat use and could be stored in a in binary form.

I realised that the binary storage meant it is not breaking any conventions. Only when they get combined do the separate agents become lethal.

"Okay, my revised recap," says Charlie.

"We can fly a stealth bomber towards Putin's bunker.

"We can set off some drones, from Turkish Cyprus, targeted on the Ozero commune.

"We can disrupt the financial system, to inconvenience the oligarchs by stealing their money.

"We can let Putin's elite enemies know about the weakness of the Altai complex, via Blackbird.

"And the information chaos is already running."

"Right, let's get moving," says Antanov.

Ride a white swan

Ride it on out like a bird in the sky ways
Ride it on out like you were a bird
Fly it all out like an eagle in a sunbeam
Ride it on out like you were a bird

Wear a tall hat like a druid in the old days
Wear a tall hat and a tattooed gown
Ride a white swan like the people of the Beltane
Wear your hair long, babe you can't go wrong

Catch a bright star and a place it on your forehead
Say a few spells and baby, there you go
Take a black cat and sit it on your shoulder
And in the morning, you'll know all you know, oh

Marc Bolan

If the money's no good

Chantal speaks up, "I've been looking at how we can move some of that money around. We should chaotically disrupt a few things. Their way to move funds was a repeating pattern. All the oligarch core-group companies appeared to be owned by proxies standing in for hidden owners. Even directors and shareholders of the companies were fake.

"Some payments did go to genuine companies for real goods – but the transactions were made not by their clients, but by 21 core companies using bogus copy-pasted paperwork which specified goods the company didn't sell.

"Some payments went to another layer of shell companies, like the core group making the payments from Trasta Komercbanka, in Riga, Latvia.

"But isn't this how all Russian clients do business?" asked Christina.

"Not exactly," answers Chantal, "Heavy users of the scheme were rich and powerful Russians who had made their fortunes. After all the middle moves, payments of laundered money slid easily into the world's biggest international banks."

I consider how this Laundromat illustrates that the world's banking system has been impotent, unable to staunch massive flows of illicit money. Bank officials offer several reasons as to why this is so – including that their Russian counterparts have not been helpful.

I recollect that HSBC, Deutsche Bank, Bank of China, Bank of America, Danske Bank, and Emirates NBD Bank all ended up with tainted money.

FSB representatives served on the board of at least one of banks that wired billions out of Russia as did Igor Putin, Putin's cousin. He was a manager and executive board member in the Russian Land Bank. This bank wired more than $9.7 billion to Moldindconbank in Moldova, most of which went on to Trasta Komercbanka and from there on to the world.

"Now that is an interesting pinch point," said Christina.

Chantal nodded, "Yes, if we can connect a fake BVI corporation to Russian Land Bank's outbound correspondent banks, we will have a way to siphon off large amounts of Russian money. We can place it in the new account and there it will be almost impossible to trace."

She smiled, "Of course, I'd have to do something illegal to make such a scheme viable,"

"What?" asked Antanov, intrigued.

"Just add a little something; something chaotic!" answered Chantal," I can just slightly tilt the way that Russia perceives itself.

She smiled, "We can also export chaos through information management. Like the earlier Project Russia papers, we can produce something new, and distribute it by the same method. It will look as if the FSB have engineered something. In this case, we can ensure that Russia and the west get access."

She continues, "The storyline needs to be about foreign companies leaving Russia. We can signal in our document that the departures are a torrent and of the unintended consequences for average citizens.

"When you think about a business, you think about the owners of the capital and the owners of the labour. A lot of these companies have employees in Russia, and these departures also directly affect the Russian people."

"But won't the Russian people be indoctrinated with the Russian line that everyone in the west is writing propaganda to fool them?" asks Charlie.

Chantal answers, "I agree, that can happen, but we are not so much targeting the average Russian, as we are targeting the oligarchs and elite. The people who are already inside the circle and can see what is really happening. This will reinforce their views. I think we may need a news source to uncharacteristically publish something in Russian to help spread the word."

Chantal continues to speak, "We don't even need to fabricate much. Think about it. On February 24th, Russia launched a full-scale military attack on Ukraine, and faced a wall of global outrage. Three days later, BP announced that it was abandoning its stake in the Russian oil giant Rosneft, at a cost to the company of up to twenty-five billion dollars. The next day, Shell announced that it was leaving, too, withdrawing from a partnership with Gazprom and the Nord Stream 2 natural-gas pipeline. Then Germany revoked its interest in Nord Stream 2. Alone those few moves will have crippled the Russian equity holdings in those companies.

"The following day, Exxon announced that it would leave as well. Shell's C.E.O. seemed to speak for more than just himself when he said he was shocked by the deplorable loss of life in Ukraine resulting from a senseless act of military aggression which threatens European security.

Christina agrees, "We need to lay it out for the elite of Russia. The fossil-fuel giants lobbed some of the earliest salvos in what has since amounted to a private-sector declaration of war on Russia, which has now seen an astonishing four hundred and fifty companies announce a withdrawal, suspension, or scaling back of business in Russia.

She gestured towards her laptop, "Look, I've found this list of business departures from Russia. After the oil giants, the next big surprise was the speed which major consulting firms such as Bain & Company, Boston Consulting Group, and McKinsey & Company, decided to pull out.

"They would usually rather jump off a cliff than get involved in political conflict or geopolitics. They were followed by the major accounting firms, and a long list of global law firms.

"And then you had big tech companies. Dell, I.B.M., Apple, HP, Google, Meta, and Twitter suspended some or all operations in Russia."

Christina interrupts, "Yes, I can see where this is going. A report could crystallise such a situation for the Kremlin elite. We will need to make a few basic statements. The kind which would get a source arrested. Except we will be using the same tricks that the FSB did to launch Project Russia. That includes anonymity. Our report can say that the global business exodus from Russia serves as a powerful condemnation of Russian President Vladimir Putin.

"It will also underscore that his military invasion of Ukraine is not only going to devastate that country but will also cripple the Russian Motherland."

Yarost

"We'll need to set the Tupolev back on a course, maybe to Putin's Palace. We could remodel the Black Sea with its payload," said Charlie, "Although we don't think we'll have to use it, will we?"

"Correct, " I say, "We need them to think we could use it. That should be enough."

"But will they try to shoot us down?" asks Charlie.

"Unlikely," answers Antanov, "They will want to persuade you to surrender and then bring the plane down without any explosions. But they will be worried that you could activate one of those warheads. And, think about it, you'll be flying a Russian plane in Russia."

Christina adds, "It will be at the same time as a squadron of drones are flying to the dachas of Putin's elite. They will not like it."

"Well, I wondered if those co-ordinates would come in useful," says Antanov, "Remember, when we were at the Ozero. I saved the GPS for each of the properties. Now we have some useful drones to point at them. This is going to take a convincing phone call from Tümgeneral Nott to the base commander at Geçitkale. She even left them with a secret trigger code word. Ярост -Yarost'"

"It means 'Rage'."

"I will speak to them in English - except for the code word," says Christina, "And we had better make it urgent, so that they don't have time to query things. I'm sure they will love to send a whole flight of their Bayraktars skyward,"

"But will they have enough range?" asks Charlie.

"Sure," says Antanov, "Those things have a 6000 km range at an altitude of up to 8 kilometres.

Christina adds, "But they are quite slow, only 200 km per hour."

Antanov speaks, "It's about 3,100 kilometres to Leshkovo from Cyprus, so that would take around 16 hours at the drone's cruising speed. We can certainly introduce some chaos."

Chantal had been studying the financial systems and scribbling onto a small black notepad.

"It's been slightly difficult, but I've found several transactions which will help us plug into their money transfer system," she explained, "The transactions have been listed for other reasons, usually because they are exhibits in a court case but didn't get redacted."

"Because I know I'm looking for a certain Russian Bank, it means I can filter down the accounts to a few suspects. Then I can rewire routing to come into our account instead of the originally intended one. It is total fraud and theft but is the fastest way to shut down the money laundering. I'm also attacking the bulk endpoint of these transactions. Where it will really hurt."

Irina asks, "But wait a minute, if you can do this, then why can't anyone else? Maybe a government, or even a bank wanting to stay legal?"

"It's because I'm prepared to break the law spectacularly to achieve this," explained Chantal. "Any government would have to go through so much legislation to get to this point, and any bank would have lawyers crawling all over such a process. Don't forget, the people doing these things want time to react. I'm just closing that door."

"Is that why it took so long for initial sanctions to be imposed?" asked Charlie, "It seemed to take weeks?"

"Exactly," answered Chantal, "It's also heavily political."

Christina nodded her agreement, "It's a KGB Rulebook 101 - and I do mean KGB -it has been around for so long. None of these kleptocrats want anything to be too speedy. They want time to move their money, their yachts, mistresses, families, and their planes. A large part of the financial and legal infrastructure in certain countries revolves around assisting these criminals."

"It was interesting when that UK report came out in 2019," said Antanov, "The so-called Russia Report. That buffoon of a UK Prime Minister tried to suppress it whilst the Kremlin was laughing at the name Londongrad given to London."

Eckhart speaks, "If the kleptocrats think they can sit out these latest sanctions, then they will. Their model is to steal money leveraged by the Russian state and then to offshore it, paying a fealty to the Kremlin. But fall behind with payments and you'll be likely to receive a plutonium sandwich."

"It's why I'm moving the money to a hedge fund," said Chantal. "Партнеры треугольника

Partnery treugol'nika" - Triangle Associates. It will be impossible to find anything when I've finished, just a huge ball of interlinked nominees and directors. Only we will have access to the available funds, but no-one else. Look, I can set up multiple companies in minutes for payments of £12 per company. And I'll never have to pay any tax, because I'll have them all showing a loss and closed inside of a year."

"I like your choice of name, Christina, by the way, it's very much in the mould of other funds like: Blackrock Advisors, Bridgewater Associates, Renaissance Technologies, Two Sigma Investments, Millennium Management, Citadel Advisors. You'd hardly know that, for example, Citadel started out of a 19-year-old's dorm room in 1987 trading convertible bonds.

"But what about us? Asked Christina.

"Oh no, we have a thoroughly respectable back story and a wonderful web site. I've just scraped a competitor and chaotically renamed everything. I can't fake the comparison ratings though, but there are plenty of other funds that show as unrated."

Chantal continues, "Hedge Funds are still an instrument of choice when involved with this class of fraud. Christina picks a Glock, and I pick a Hedge Fund. They can both do serious damage. Hedge Funds can suppress pro-social behaviour, which is ideally suited to the lifestyle of a kleptocrat. It is a case of the strength becoming a weakness. They already have dead social consciences, now there is a worrying personal impact as well."

Christina laughs, "SIG Sauer, actually,"

Chantel continues and lists: "Example 1: Authority Doesn't Care About Ethics. Since the days of Stanley Milgram's notorious electric shock experiments, behavioural science has shown that people do what they are instructed to do. Hedge fund traders are routinely instructed by their managers and investors to focus on maximising portfolio returns. Thus, it should come as no surprise that not all hedge fund traders put obeying federal securities laws at the top of their to-do lists.

"Example 2: Other Traders Aren't Acting Ethically. Behavioural experiments also routinely find that people are most likely to 'follow their conscience' when they think others are also acting pro-socially. Yet in the hedge fund environment, traders are more likely to brag about their superior results than their willingness to sacrifice those results to preserve their ethics.

"Example 3: Unethical Behaviour Isn't Harmful. Finally, experiments show that people act less selfishly when they understand how their selfishness harms others. This poses special problems for enforcing laws against insider trading, which is often perceived as a 'victimless' crime that may even contribute to social welfare by producing more accurate market prices. Of course, insider trading isn't victimless: for every trader who reaps a gain using insider information, some investor on the other side of the trade must lose. But because the losing investor is distant and anonymous, it's easy to mistakenly feel that insider trading isn't really doing harm.

Chantal continues, "I just need to mimic the behaviour of other bad apple traders and we should soon have turned off all of the funding taps to those oligarchs. They will not be pleased but won't know where their money has gone."

The mistress of chaos smiles at me. Limantour was enjoying this.

Eckhart speaks, "I just heard from Sorokoput - Irina's and my handler. It is very dangerous, because although I think Christina and Eckhart's handler Blackbird is damaged goods, Sorokoput maybe playing both ends against the middle."

Chantal asks, "Do you think you can trust him?"

"Irina speaks, "Her, actually. Sorokoput - butcher bird - is a woman. I never have really trusted her, but now I guess she can see the writing on the wall."

Eckhart speaks again, "Well, Sorokoput says that Altai have just received a consignment of nerve agent. It was ordered by the President."

Chantal asks, "But where would they get it from?"

Antanov speaks, "Indeed, but you have to remember Russia once held 40,000 metric tons of chemical weapons and then very slowly started to dispose of them.

"My guess is the last nerve agent store house or maybe the last couple have remained in use. It would also explain where The Kremlin could get novichok and other nerve agents used in various assassinations and assassination attempts. Think about it: Yuschenko, Navalny, Litvinenko, Skripal, Abramovich...the list goes on."

Eckhart speaks again, "This sounds like a whole truck load of dangerous chemicals has been shipped to the Altai bunker, from Shikany-2."

Antanov speaks, "Shikany-2 is a military town. I was continuously scared when I was posted there for a couple of weeks. I had to run security for a couple of young scientists named Panyaev and Kudryavtsev who worked for FSB's Criminalistics Institute – also known as its poisons factory. They had a special project, but I was pulled out before it ran to conclusion. I think they were deliberately moving us around so that no-one knew the big picture. The town itself was a closed town. No-one allowed in unless they had a strong reason."

Antanov adds, "Shikany-2 is close to the Volga and about 100 kilometres north-east of Saratov. It used to have a code name Tomko. The Shikany-2 facility includes a chemical weapons burial site with approximately 4,000 metric tons of adamsite - That's the riot control agent that causes vomiting - and the adamsite was collected from all over the former Soviet Union.

Antanov continues, "Shikany-2 was where Col Stanislav Makshakov, a top-notch military scientist was based. He worked at the State Organic Synthesis Institute. Makshakov used to report to Gen Kirill Vasilyev, director of the FSB Criminalistics Institute. Vasilyev then reported to Maj Gen Vladimir Bogdanov, former chief of the Criminalistics Institute and deputy director of the FSB's Scientific-Technical Service. Vasilyev's direct superior is the FSB's director Alexander Bortnikov. He still, in turn, reports to Putin."

Christina adds, "Okay, so now we are seeing a truckload of one of their products being shipped to Putin's bunker? This can only end badly."

"Assuming that Sorokoput is telling the truth."

"You know something, her knowledge of that is probably the reason why she is suddenly open.

"Do we know whether Putin ordered it? Or someone else?"

"No, we don't, but either way, it helps our case."

A hard rain is gonna fall.

Charlie summarised "Okay, so now we have the pieces in place.

"We can send the stealth bomber to Putin's Palace
"We can target the drones on the Ozero Community
"We can stop the entire money flows of the kleptocrats
"We can disrupt the security of the Altai base.

If some of these conditions happen, I think we will see a change from the oligarchs. The elites will realise that this too much. That the situation is out of control.

And it is not obvious where the impacts have come from either. It won't look as if America or the United Nations has intervened.

"We will have chaos," says Chantal gleefully, "That's when things could change suddenly."

"We three, Scrive, Charlie and me; we have a way to get back into the correct timeline and metaverse. I should hitch a ride with Scrive and Charlie. They can show me how to be the flight engineer.

"The rest of you, Christina, Antanov, Eckhart and Irina,

will be switched back into your original times and places. You will be safe, but much of what has happened here may need to play out again. I doubt you will remember much, but it will feel instinctive to each of you to take certain actions."

"I wish we could tell you how this ends," said Charlie, "But none of us have recall of these events. But when we are back on our correct lines we will know. Like we will all remember about the Klima Wars and how they started."

We all stood and looked at one another. I was aware that the sky was an even deeper green than normal, and the sea reflected the colour.

Chantal put her hand into the middle of us group in a circle. We all did the same.

"Slava Ukraini", she said. We all repeated it and then hugged one another. And again a second hug.

"At least, in this metaverse," Chantal added.

"Okay," said Christina, "Let's hit it, then."

Chantal and Christina make busy with the phones. Chantal shows us the capital in the new Triangle Associates Hedge Fund. It is already over half a billion dollars. That's in a few hours. And I think that it is half a billion dollars that is not with the Kremlin elite. I sense a few fuses blowing.

Christina has been onto the Bayraktar base and has now sent a dozen drones toward their GPS co-ordinates.

She speaks, "I added Putin's Palace as well, and the

Kremlin for good luck. Those last two will be shot down, but it shows our intent. The hard rain is gonna fall,' "

I was thinking, 'Yes, rain from 8 kilometres high'.

Antanov called Artem, and in ten minutes he was outside to take me, Chantal, and Charlie back to RAF Akrotiri. We say good-bye to him at the base, and he says he'll look forward to seeing us the next time. I wonder what he means.

Now, we are both back in green flight suits looking at the Tupolev, which has been moved into a hangar. I can understand why its nickname is The White Swan.

We climb in for what we realise will be the last time. Almost the last time in this version of Earth.

We are about to start up when Charlie says something.

"Remember our original intervention? When we used your gravity wave to trigger another event? - We talked then about unexpected consequences?"

I remembered, even although the last part of what had happened seemed to have been erased.

"Well, you, Farallon, haven't been using your gravity power much since we arrived here."

I wasn't sure where this was leading. Did she want me to levitate the plane or something?"

"I've some co-ordinates here,' She pulls open her zipper pocket and pulled out a folded sheet of A4. On it were some GPS numbers.

50.78129882, 86.48117964

"These are Putin's bunker co-ordinates. Remember it is full of tunnels and may now be storing some unstable chemical from a poison lab?"

I realised what she was asking.

"I guess you could shake things up a little around there. Maybe a new hole in the ground. The river Katun could find a new path and it could all be mixed up with some of that chemical?"

"But it would be another Intervention," I say.

"Yes, but look where we are. This is like the last reel of a movie, or the last few pages of a novel. We know that there must be another way. We are just trying to force it."

I look at Charlie.

"Oh, okay, give me the co-ords."

I knew I'd need to concentrate, and that I could either fly the plane or do some kind of gravity shift.

The weak force won. We watched as the sky rippled.

"Thank you," said Charlie,

The headphones crackled: 'Line up and wait'; 'Cleared for runway two seven '; 'Fly heading two three zero,' 'Runway two seven, Cleared for take-off,' then he added 'Godspeed'.

Our Tupolev sped forward. We'd be climbing to an obscene altitude and flying at a crazy speed pulling a sonic boom in our wake.

It didn't matter if this variant of Earth was ready to implode.

I looked first at Charlie and then behind me to Chantal. They were both smiling. We'd set so many hares running now. The Bayraktars. The diverted oligarch money. Misinformation. A natural catastrophe at Putin's bunker, with toxins blended with river water. Maybe a crazy man's Perimeter system, triggering the Dead Hand. We would never know.

It was no longer 'If', just 'When'.

"Look!" said Charlie. I almost hadn't noticed. We were at 40,000 feet, still climbing. The sky had changed. No longer green, it was restored to blue.

A judder.

"See you both on the other side!" called Charlie.

.

Farallon and Limantour

Another judder.

He heard the apartment judder from the impact. A mournful sigh. This one had been close, but not that close. He knew the building was meant to take it.

He looked towards the window. Grey night skies, something resembling clouds, thin trails, raked towards the horizon.

It had happened, he was someone new. He thought he'd have been restarted close to the spot where everything and become chaotic, courtesy of Chantal. Now he looked at the clock. Ten minutes to midnight.

He realised he had moved off-world. This was no longer Earth. Wherever he was, the place was being battered by extreme weather conditions. His new Presence knew how to handle it and was reassuring him that he was better indoors. Going out just added to the tension. If he could stay inside, he could watch some transmissions to take his mind off the situation.

He moved from his bedroom into the main living area. He flipped a switch and could suddenly hear the weather. A gentle rain and a rustling of leaves. The occasional spatter of water dripping from branches.

The main screen started. Not the full screen but the one designed to show just entertainment transmissions and data. It opened on a standard news transmission and he surprised himself as he gestured for it to move across to his messages.

The main room had noise cancellation and so he was now no longer aware of the crashes from outside. Just a slight feeling underfoot as the building absorbed more impacts.

"Peter give me status," he asked.

A small pop-up window appeared on the top right of the screen. Every condition was green. At this rate, he did not need to do anything at all.

He walked across to the kitchen area, flipped a tap, and drank some water. The tap illuminated the water as it poured. The blue colour signified that the source was both pure and cold. He remembered that they had built his block in the 40s and it was still good at the management and monitoring functions. He knew it had originally been built for the military as an offshoot of the nearby base.

Farallon realised that his full range of faculties seemed to be functional. He'd been projected here as an entity but had a full backstory as Roelof too.

When he arrived in the city, they had given him a choice of either staying on the base or moving out if the commute was less than 30 minutes. He had opted for off-base because it was already like living in a bubble and on the base was like living in a bubble inside another bubble.

A little information light on the screen briefly flickered to amber. The moment later it had returned to green. He realised another advantage of being away from the base was that smaller incidents were handled autonomously by the base management systems.

"Hi Peter," he said, "please provide an update on base status."

"Full base status is green. There was a short incident with a meteorite, but they cleared it with a grid gun. Incident duration 1.2 seconds. There are zero requests for your attendance at the base."

He walked to the kitchen cupboard and flipped open a compartment.

"Peter dispense modafinil. Two units."

Two small capsules appeared in the compartment. He placed them in his mouth and took a small drink from the water glass. He could feel the rush at once. His senses heightened as if he had been over-clocked like a computer.

The modafinil was for mission use. He remembered he had someone fix Peter's system so that there was always a modest threat level running such that Peter would always dispense the drugs. The same fix meant that Peter also lost track of how many drugs were dispensed.

He just needed to remember not to get the automatic updates for the health-care system in the apartment. That was another advantage of being off base. Living quarters on the base would always run with the latest and greatest versions of everything.

A chime sounded from the streamcom. "Peter accept," he said.

A small repeater screen in the kitchen showed the face of one of his colleagues.

"Hi Roelof, it's Jasmijn. There's something very unusual happening here."

He knew instantly that it was Limantour and felt a part of his cortex spark the recognition. They'd made it out of that crazy place with the bomber.

She continued to speak, "The incoming meteor shower seems to be concentrated on our control centre. We've already lost the above-ground units and now the incoming is creating a crater where the underground centre is located. At this rate we'll have lost everything within another 15 minutes."

"What about the HSDA?" asked Roelof.

"I know. This is one of the times where our fast reflex friends should be able to solve this without us even noticing. I've seen the high-speed defence array running today almost non-stop. There's no question it's been working but it just doesn't seem to be enough to stop this. It's almost as if the meteors have their own avoidance telemetry."

"Do I need to come in?" asked Roelof.

"I don't think you would be in time to make any difference," said Jasmijn, "We are all being backed into a corner here. They've already given the order to flip command to another centre."

Peter interrupted the transmission, "I am stabilising the display, it exceeds my tolerance levels."

"Hi Peter, remove video stabilisation," requested Roelof.

Roelof watched Jasmijn on the display as the stabilisation was removed. He had never seen such a level of erratic framing. Most of the base was designed to withstand just about anything that could be thrown at it. Quakes, powerful winds, floods, fire. The original designers had borrowed the triple X symbol from the Earthside town of Amsterdam. Fire, flood, and pestilence. Three Xs. Three times "No".

Triple X Protection.

Jasmijn looked back towards the camera. "I'm gonna bail," she said. "I'm guessing this place is only going to be around for a few more minutes."

He heard the noise of a siren. Then a bleep and the screen terminated.

"Transmission terminated," said Peter.

"Peter please give me externals," requested Roelof, "Put it on the main wall."

He stepped back in the living space. Across the wall was a scene showing distant clouds, a red sky, and white streaks of light focused on a smoking central area.

Roelof walked towards a console in the living space. He sat in a swivel chair and grabbed the controls. He looked around the sky and locked on to two monitor drones.

Requesting access to their video channels, he zoomed

the drones towards the distant control centre. The external centre disappeared and that an ominous hole in the ground suggested the Secondary control centre was also compromised.

"Jasmijn, Jasmijn, do you copy?"

He repeated the request a couple more times.

Then a voice. "Copy that, Jasmijn here - I can hear you."

" What is your status?"

" The pod is secure, and I am outside the main ring of damage. Another 20 seconds and it would be very different. It looks as if some of the others have made it too."

"Okay, follow the protocol and join me here," said Roelof.

"Copy that"

Roelof knew that the protocol had been designed to protect as many people as possible on the base. Everyone had been paired, and he had been selected to pair with Jasmijn. He was officially English, and she was officially Belgian, although neither of them had spent much time in their designated home countries.

Now he realised that Jasmijn was his accomplice from before his last exit. He'd been paired with Limantour and Jasmijn must be Limantour. He wondered where the other two would be. Tomales and Drake. He suddenly realised that the reason they had come to this point was because of Drake. He'd thought Drake was on earth in

Bodø Norway. It was the wrong information, but it would still be good to get the four of them back together.

Roelof flicked through some of the observation systems to check the wider impacts of what had been happening. This was one of the worst storms he had seen since he had been active on Ganymede. There was also something very unusual its focus. Usually anything that appeared in the weather systems was predictable in the way that it travelled across the winds of the surface. Although violent, the normal storms dissipated across large geographical tracts. This protected the mines and other constructions from acute damage.

A paradox was that the very substances wanted from Ganymede and the adjacent Europa for use on Earth were also capable of being harnessed within Ganymede's own biosphere.

For around two hundred years the magnetosphere of Jupiter's largest moon had been observable from Earth. It had only been for the last 40 years that dependable space transit had been possible. The discovery of two complimentary passive minerals that when combined created a magnetic field like that within an electricity generator had been a breakthrough discovery.

Small amounts of the minerals could be used to make powerful generators which could be used for domestic and commercial purposes back on Earth. The same technology could be used in-situ on Ganymede to create the required defence shields to protect the mining and other operations from danger. For planet Earth this had been a life-saving discovery such that as fossil fuels declined, the new availability of magnetite had become a complete game changer.

The original predictions of a six-year flight from Earth had been dramatically reduced to three years in each direction augmented with the creation of SkyTrains to provide a near continuous round-trip service. For a two year stay on Ganymede base there was the prospect of considerable wealth for those that pioneered the creation and exploration of the bases.

The sovereign structure of Ganymede had been incorporated into Earth's United Nations although a series of different and sometimes very unconventional procedures had been allowed. The Earth Council had superseded the United Nations although the exact sequence of events and their timing was hazy.

The jurisdiction was not so much 'out of sight, out of mind' as a series of procedures to support the necessities of developing a base to support the future of humankind so far from Earth.

Pioneers to Ganymede had taken the longer and slower six-year outbound trip, then 2+ years working and then the faster three-year return cycle using newer technology driven by Ganymede's own propulsion devices. In practical terms this was an 11-year absence and during that time the first settlers used a range of techniques to create the necessary labour capabilities for the mining to be successful. The roundtrip with work time was now reduced to eight years. Three outbound, two moonside and then three to return.

Most people on earth were unaware of the change taking place on Ganymede. It was much further than a distant small country and as long as the requisite technologies arrived in time to be useful than the main debates were about the rise in fortunes of those that had made the return trip.

Roelof and Jasmijn did not know much about the situation on earth. Their memories of it were very dim, as were the memories of many of the people they worked with. There were some individuals, sometimes referred to as the Sharps, who seemed to have a much better knowledge of life on Earth. Curiously, the Sharps were perceived by people like Roelof and Jasmijn as dim-witted and slow thinking.

The buzzer to Roelof's landing deck signalled the arrival of Jasmijn.

"Peter, please guide her in."

"Acknowledged," responded Peter.

A few minutes later, Jasmijn buzzed again, and Peter opened the main door to the apartment.

"Are you okay?" asked Roelof.

"Everything is fine," said Jasmijn, "That was a close thing, but I think most of us had evacuated each area before it was destroyed."

"It's still a very worrying change of situation," said Roelof, "It's the worst I remember, after nearly two years and despite the hostile environment, there has been nothing like this."

Roelof could sense a nagging question in the back of his mind, but each time he was distracted away from it, the question seemed to be less well formed.

At that moment Peter interrupted, "I have an incoming transmission for both of you."

"Okay Peter, put it on the wall."

A newsflash appeared on the whole of the living space wall. It was accompanied by newscaster soundtrack music. There was a flash and both Roelof and Jasmijn momentarily tipped their heads sideways. Four seconds later, the news broadcast resumed with a good news story from Perth about a pet dog that had been found after it had run away from home.

"Okay then," said Roelof to Jasmijn. "I'll meet you at the alternate control centre tomorrow."

"That's fine," said Jasmijn, as she left the apartment.

Tomales and Drake

Earth.

"These system updates are taking longer and longer," said Sam Walker, "This time we had to wait for nearly four hours to get the new command centre online."

"I know," replied Cindy Shaw, "They told us this time it was the new extraction modules that were being introduced."

Cindy sensed an electro stab in her brain. She realised what it was. Someone was trying to communicate. It must be a Watcher making contact. Maybe it was Farallon, seeking Tomales.

"Anyway," said Sam, "We seem to have everything back now. Just about every system is already green and a couple of the minor ones are still restarting."

"There are still some discrepancies, though," said Cindy, "If I add together the time for a reload plus the transmission times, even with those new modules, we should see the return to ready state within maybe a couple of hours. There is no hint that the systems were ready - it looks like a complete restore. "

They both studied the console for moment. Sure, the transmission time for the command up to Ganymede were about 34 minutes. That made a round trip of just over an hour. All the new software had already been transmitted so it should have just been a case of firing it up.

Then the stab again, this time with a shower of polka dots inside of her head. Playful. She realised this time. "Hey Tomales! Limantour here, are you both okay?"

"You bet. I've been with Farallon this evening. When we were all on that Tupolev at 40,000 feet. Then there was a bang. Do you know what happened?"

"No idea. Everything just ended. We'd sent off the drones, had over a billion in the Hedge Fund and then...Wipeout. We don't know what happened."

"It's probably for the best," answered Limantour.

"Sorry reception is bad," said Tomales, "Probably a space storm."

Tomales beamed another message to Limantour, "I've found Drake. We are together. I guess it is because of him that we are at this intersection. Shame you are such a long way away."

"Let's take a look at the log," said Sam. He could see that Cindy was engrossed in her thoughts.

Cindy was aware that she had known Sam for a long time, but right now it seemed like a whole new experience. She also realised that she could communicate directly with his thoughts, as he could to her. She told Drake she had just received a communication from Tomales.

Limantour was concluding that she had just been propelled into this situation, and that her Presence here was yet to stabilise.

"Yes," said Cindy to Sam, "I see this was an update that created a new release level. We are on release seven now. It still seems strange that when we go through minor release levels, they take about two hours, but the major levels are adding increasing amounts each time.

"See here," said Sam "There's this whole extra section for transfer..."

He looked at what was an extra section which had inserted itself into the update.

"Yes, that only seems to happen when we do one of these big levels," said Cindy.

Sam reached across to a mug which contained a kind of vegetable soup. As he lifted it from the work surface, it made a resonant chink sound which cut across the sounds from the faintly whirring technology.

"What is it?" Asked Cindy. She peered towards the brownish liquid with little white, green, and orange pieces floating in it.

"It's Italian," said Sam, "They call it minestrone. It's not bad for a sub."

Cindy grinned, "Happy Nutrition."

Cindy peered towards the observation windows.

"One day these subs will have proper vegetables in them again."

Outside she could see the land. An orange-brown colour. It was only just daybreak. She could still make out the outline for the moon and across the sky from it the second much smaller moon which had been created by man. Small pinpricks of light twinkled between the two moons indicative of transiting space hardware.

She looked across to the Meteo display. 40C degrees already.

"It is going to be a hot one today."

Sam nodded.

Their base was in New Delaware on the east coast of the United States. The whole island area of what had once been called Delaware and what had been the eastern half of Maryland had been re-designated as New Delaware when the efforts to bolster the space program had redoubled.

Global warming had affected the original sites further south in the deserts and across on the eastern seaboard of Florida. The move further north still had the advantages of nearby sea as well as a convenience for any military reinforcement that may have been required.

New Delaware had then aggressively become a TEZ - total exclusion zone - permitting the wholesale development of first lunar and then interplanetary transport vehicles.

Secondary developments had sprung up around the bases providing supplies and other technologies for the agency. In the early 22nd Century it had been a race to find power sources to keep those functional and to avoid major global instabilities.

The very necessary race to space had itself created huge new industrial footprints across many parts of the globe.

Cindy and Sam had met at IPX school. Interplanetary Exploration was a career choice for the very brightest. They were selected early and then encouraged to form friendship groups and ultimately to pair off. The process was part of the selection for further duties, where couples were always selected together for space mission work.

Earlier attempts with longer flights and separated spouses had failed for all manner of reason and there was usually salacious reporting of the unfortunate outcomes. It had culminated when an early high-profile mission to the intermediate planet of Mars had been destroyed by an unhappy astronaut who had realised his wife was cheating on him back on earth.

Sam and Cindy had been deselected from space travel part way through the programme. The official story was that they were too precious to be gambled in space travel and that there were others more suited to the roles required.

It was a blow to them both after what had been training since their childhood. They'd been through a full process that was not disclosed to many earth dwellers and they now, as Watchers, knew the concluding fate of planet Earth.

Awaken reality

Christina awoke. She wasn't entirely sure how she came to be in this Paddington hotel. Oh yes, there had been a bad storm and the trains had been delayed. She had booked into the hotel until transport resumed.

She flipped on the television.

A reporter was speaking, "As the Russian invasion of Ukraine continues, it is becoming abundantly clear that the Kremlin's maximalist geopolitical aims of regime change and a 'greater Russia' which includes Ukraine and Belarus are no longer achievable. The question now is how much damage Russian forces will inflict on Ukrainian cities and their brave defenders before Putin and his advisors lower their ceasefire conditions to terms that Ukraine's leaders and population can accept. Ukraine is in a strategically stronger position than many in the West appreciate, but the war on the ground is still stacked in Moscow's favour in the short term."

The view cut to a street scene littered with burnt out tanks and the remnants of a local population trying to recover from devastating fighting.

A voiceover began, "The plan to decapitate the Ukrainian state at the national and local levels with infiltrated special forces and operatives, while seizing key points with airborne assaults and surrounding the major cities with ground forces, failed spectacularly during the first week. Having been given next to no warning or time to plan, the Russian army advanced down major roads in poorly coordinated columns and the lead elements were largely obliterated by stiff Ukrainian defences."

More scenes of roadblocks and destroyed armaments.

"Airborne assaults, most notably at Hostomel airport west of Kyiv, were almost unsupported and were rapidly destroyed or scattered by Ukrainian rapid reaction forces. As a result, many of Russia's best trained and motivated VDV (paratrooper) and special forces units suffered huge casualties in the first week of the invasion without achieving significant results."

A picture of a helicopter, sliced in two by ground fire.

"In the north of Ukraine, regular Russian army formations found themselves stuck on congested roads due to the extremely muddy off-road conditions. This has allowed Ukrainian forces to conduct ambushes with artillery, UAVs (drones) and the numerous anti-tank guided missile (ATGM) launchers provided by western countries."

Pictures of Ukrainian military with 'tank busting rocket launchers.

"Due to the lack of planning, most Russian frontline units were sent into Ukraine with very limited food, fuel and ammunition. In the north and northeast, Ukrainian forces have successfully exploited and aggravated this initial weakness by ambushing and destroying resupply convoys travelling along supply routes. These tactics have been aided by the failure of the Russian Aerospace Forces (VKS) to provide effective cover for the ground troops or for resupply convoys."

Rows of boy soldiers being rounded up and put onto Ukrainian trucks.

"Some of them barely look like they have left school," said Christina, nudging her slumbering partner.

"As a result, the second week of the invasion saw Russian forces in the north and northeast of Ukraine largely pause to regroup, try and sort out their logistics nightmare, and complete the encirclement of Kharkiv and Sumy. Ukrainian forces have even conducted successful counterattacks to take back towns to the north and west of Kyiv at Chernihiv and Irpin, which are now the scene of renewed heavy fighting."

Christina nudged Antanov again, "This could drag on, the Russian people were completely unprepared for Putin's barbarism."

Antanov nods, "I agree, what could the cold-blooded little madman expect to achieve apart from horrific bloodshed?"

"I suspect he has sealed his fate," said Christina.